LONDON MATHEMATICAL SOCIETY LECTURE NOTE SERIES

Managing Editor: Professor N.J. Hitchin, Mathematical Institute,
University of Oxford, 24–29 St Giles, Oxford OX1 3LB, United Kingdom

The titles below are available from booksellers, or from Cambridge University Press at www.cambridge.org

London Mathematical Society Lecture Note Series. 310

Topics in Dynamics and Ergodic Theory

Sergey Bezuglyi
Institute for Low Temperature Physics, Kharkov

Sergiy Kolyada
National Academy of Sciences, Kiev

CAMBRIDGE
UNIVERSITY PRESS

CAMBRIDGE
UNIVERSITY PRESS

University Printing House, Cambridge CB2 8BS, United Kingdom

One Liberty Plaza, 20th Floor, New York, NY 10006, USA

477 Williamstown Road, Port Melbourne, VIC 3207, Australia

314-321, 3rd Floor, Plot 3, Splendor Forum, Jasola District Centre, New Delhi - 110025, India

103 Penang Road, #05-06/07, Visioncrest Commercial, Singapore 238467

Cambridge University Press is part of the University of Cambridge.

It furthers the University's mission by disseminating knowledge in the pursuit of
education, learning and research at the highest international levels of excellence.

www.cambridge.org
Information on this title: www.cambridge.org/9780521533652

First published 2003

A catalogue record for this publication is available from the British Library

ISBN 978-0-521-53365-2 Paperback

Contents

Preface

In the 60s and 70s, Katsiveli was the place where regular summer schools and various conferences took place under the aegis of the Academy of Sciences of Ukraine. An International Conference and US-Ukrainian Workshop "Dynamical Systems and Ergodic Theory" were held in Katsiveli (Crimea, Ukraine) during the period 21-30 August, 2000. The meeting which was dedicated to the memory of one of the best lecturers of the Katsiveli's schools - Vladimir Mikhaïlovich Alexeyev, is an attempt to revive this tradition.

The conference and workshop has had no permanent funding. Their existence and function rely on financial aid from international and national organizations supporting the development of fundamental sciences.

There were 86 participants, coming from Australia, Austria, Azerbaijan, Canada, Czech Republic, Denmark, France, Germany, Israel, Korea, The Netherlands, Poland, P.R. China, Russia, Slovakia, Spain, Ukraine, United Kingdom, and the United States and from about 50 institutions working in many fields of Dynamical Systems and Ergodic Theory.

The meeting was consisted of 5 mini-courses (17 50-minute lectures), 30 50-minute lectures, 40 30-minute talks and the "Open Problems" section (see http://iml.univ-mrs.fr/~kolyada/opds/).

The proceedings of the conference will be published in two parts: 1) research papers & 2) mini-courses and survey papers. Research papers will be published in one issue of the journal *Qualitative Theory of Dynamical Systems*, mini-courses and survey papers in this book.

This meeting was organized by the Institute of Mathematics and the Institute for Low Temperature Physics and Engineering of the National Academy of Sciences of Ukraine together with Institut de Mathématiques de Luminy and Université de Tours, France.

The organizers of the International Conference and Ukrainian-US Workshop on Dynamical Systems and Ergodic Theory gratefully acknowledge the following organizations for their financial support: U.S. Civilian Research & Development Foundation (CRDF), INTAS OPEN-97 grant 1843 and INTAS Monitoring Conference Grants, Centre National de la Recherche Scientifique (CNRS, France), Ministry of Science and Education of Ukraine, European Science Foundation (Scientific Program PRODYN), Center for Dynamical Systems at the Penn State University (US), European Mathematical Society, ProMathematica (France).

Organizing Committee: P. Arnoux (Institut de Mathématiques de Luminy, France), V. Bergelson (Ohio University, US), S. Bezuglyï (co-chairman, Insti-

viii

tute for Low Temperature Physics and Engineering, Ukraine), F. Blanchard (Institut de Mathématiques de Luminy, France), S. Ferenczi (Université Francois Rabelais, France), B. Hasselblatt (Tufts University, US), S. Kolyada (co-chairman, Institute of Mathematics, Ukraine), E. Lesigne (Université Francois Rabelais, France), M. Malkin (Pedagogical University of Nizhny Novgorod, Russia), C. Mauduit (Institut de Mathématiques de Luminy, France), M. Misiurewicz (IUPUI, US), Z. Nitecki (Tufts University, US), A. Sharkovsky (Institute of Mathematics, Ukraine)

Sergeĭ Bezuglyĭ and Sergiĭ Kolyada

INTRODUCTORY TALK AT THE OPENING OF THE CONFERENCE

ANATOLE M. STEPIN

Dear colleagues and friends! This conference is devoted to the memory of Vladimir Mikhaylovich Alexeyev, professor of the Moscow State University, who untimely passed away in 1980.

Vladimir Mikhaylovich was one of the lecturers at Katsiveli Mathematical School in 1971. Such schools (and conferences) were regularly conducted by the Institute of Mathematics of the Academy of Sciences of Ukraine since 1963. This regularity was broken because of certain political changes in the former Republics of the USSR. I would like to express my hope that our meeting in Crimea is a step towards restoring Crimean mathematical schools and conferences.

The organizers of the present conference suggested that I give a talk today, on the first day of our work here, and tell you what I recall about Vladimir Mikhaylovich Alexeyev. I am grateful to the Organizational Committee for the invitation to participate in the conference and for the honor to present my recollections of V. M. Alexeyev, a brilliant mathematician and personality.

V. M. Alexeyev was born on June 17, 1932. His father comes from a well known family of Russian merchants, the Alexeyev family, who gave the world K. S. Stanislavsky, a famous reformer of theatrical art. While in the ninth grade, Volodya Alexeyev started attending Math Club meetings at the Moscow State University, and in the next year he was honored with the first prize at Moscow Mathematical Olympiad for high school students. He then entered the Moscow State University to major in Mechanics and Mathematics. A. N. Kolmogorov was his advisor. His Master's Diploma (1955) and Ph.D. Thesis (1959) were devoted to the rigorous proof of the possibility of satellite exchange in the three body problem – the phenomenon discovered for the first time with the help of numerical methods (L. Bekker, 1920).

My first meeting with Alexeyev took place in 1964 when I was a student. The point is that the students who were majoring at Mechanics and Mathematics were supposed to have pedagogical practice, semester-long one, and my advisor, Felix Alexandrovich Berezin, suggested that I practice at the Kolmogorov Boarding School in Physics and Mathematics, founded a year before. V. M. Alexeyev was one of the lecturers for ninth graders at this school. He was lecturing on Mathematical Analysis while we, his assistants, were conducting practical sessions with students on this and other subjects as well.

As a lecturer, V. M. Alexeyev was of course interested in his students thoroughly practicing in class all the theoretical material. He recommended problems appropriate for solving in class, yet at the same time never fixed the manner in which the practical sessions were actually conducted. Let me put it straight here – the teachers and students were using this freedom quite a lot. However sometimes, during the so-called "pedagogical meetings" and personal discussions, Vladimir Mikhaylovich kept us from excesses of the system. In general though, the main ingredient of the working atmosphere of the Kolmogorov School was the feeling of common creativity, "co-creativity", between teachers and students, and V. M. Alexeyev was among the most appropriate persons to help us to form such an atmosphere.

After finishing my practice as an assistant at the Kolmogorov School I received an offer from A. N. Kolmogorov who suggested that I continue teaching in the School. From that time on our pedagogical contacts with V. M. Alexeyev continued. Some of students of this school later became active participants in seminars on dynamical systems at the Moscow State University; let me name, e.g., A. Krygin, Yu. Osipov, Ya. Pesin, E. Sataev (by the way, Eugene Sataev participates in the present conference).

In 1964-1965 academic year V. M. Alexeyev and Ya. G. Sinai invited me to give a talk at their seminar in dynamical systems and ergodic theory (they were running this seminar after V. A. Rokhlin moved to Leningrad). The thing was that by that time, working at F. A. Berezin's seminar, I constructed a periodic Abelian group of mesure-preserving transformations whose maximal spectral type did not subordinate its convolution square.

The problem about group property of spectra of ergodic dynamical systems dates back to A. N. Kolmogorov; V. M. Alexeyev and Ya. G. Sinai at that time were actively interested in spectral theory of dynamical systems. The discussion at the seminar concerned my construction, in which both Alexeyev and Sinai participated, and certainly assisted in a much deeper understanding of the phenomenon which I discovered.

It turned out that the following fact was responsible for the spectral group property breaking (and, as it was found out later, in some other examples of unexpected or even unusual behavior of dynamical systems): the group of all automorphisms of the standard probability space, endowed with weak topology, is not complete with respect to one-sided uniformity (this circumstance had been noticed already by Halmos but was never used before). The points of the corresponding completion (or even compactification) are closely related to the notion of joining introduced later on. Especially important role in this relation is played by the joinings which are the limit points of Koopman operators corresponding to the dynamical systems.

In the fall of 1965 V. M. Alexeyev together with a large group of Moscow mathematicians participated in the famous Humsan conference. It took place in the village Humsan, close to the Tyan'-Shan' mountains and the city of

Tashkent. Three of us, Vladimir Mikhaylovich, A. Katok and myself, lived there in a big bright room of a comfortable mansion.

Our common accommodation in Humsan is still fresh in my memory, and I have a good reason for that. The point is that the experience in dealing with periodic transformation groups and the advice of F. A. Berezin not to stop research in this direction prompted me to study general dynamical systems as perturbations of a sequence of periodic transformations, and at that time I actively thought about the possibility of such an approach. A. Katok joined me in this at that time.

We used every opportunity to talk to V. M. Alexeyev about our (not yet embossed distinctly) ideas and preliminary arguments. He spared a lot of attention to us, in general approved the idea of approximation, made critical remarks and in some cases insistently requested formal proofs (though sometimes finding heuristical geometrical arguments satisfactory). Once V. M. made an interesting comparison of the newly born method of periodic approximations with the theory of approximation of functions. It so happened that Vladimir Mikhaylovich was one of the first who got acquainted with the initial outline of the method of periodic approximations; his friendly criticism was very helpful, and our paper, joint with A. Katok, published in Proceedings of the Academy of Sciences (1966) received a lot of attention from mathematicians.

Alexeyev himself at that time continued thinking over various questions about the asymptotic behaviour of motion in the three body problem. Here it would be appropriate to mention two stages in the research of final types of motions in this problem. The first stage is characterized by the usage of methods from perturbation theory developed precisely for these purposes. Alexeyev summarized the results of this stage in his paper published in the collection "The problems of movement of artificial celestial bodies" (1967).

In Humsan Vladimir Mikhaylovich was thinking about the existence of movements with "temporary capture", when a comet following a trajectory, co-asymptotic to a certain parabola, turns around the Sun prescribed number of times, despite passing another celestial body close by. He constructed examples of such movements with the help of the so-called discontinuous solutions of the ideal Kepler problem. Vladimir Mikhaylovich gave a talk about these results at the International Congress of Mathematicians in Moscow (1966). At the time he also started thinking about applications of the methods of symbolic dynamics and the theory of hyperbolic dynamical systems to the problem of classification of two-sided final movements in the three body problem.

Let me go back to the time of the Humsan conference. It should be said that the participants of this meeting managed to also have a good time during the conference. For example, Vladimir Mikhaylovich turned out to be a champion in swimming across the nearby mountain river with a very strong and fast current. Nadya Brushlinskaya, the wife of V.I.Arnol'd at the time, was very excited about these races praising Volodya Alexeyev for his success.

One more recollection about V. M. Alexeyev is also related to the same river in which he splashed quite often. Once, an acquaintance of mine, one of the participants, managed to dive in the river with his glasses on. And sure thing, the glasses fell off, and even though we both kept diving in the river and the water was crystal clear we could not find the glasses. Next morning I modeled the loss of the glasses: I made a model using aluminium wire, then dove and dropped "the glasses" exactly where they fell yesterday and observed their trajectory. Next to the spot where my model landed I found the true glasses in a perfect condition. When Vladimir Mikhaylovich learnt about this he told me rather seriously: "Tolya, you took into your heart the trouble of another man".

After the Humsan School Vladimir Mikhaylovich concentrated upon a complete solution of the two-sided version of the classification problem as regards asymptotic types of movements in the three body problem. The point is that the author of the classification of one-sided final types of movements, the French astronomer Chazy, formulated the statement (1929) on the coincidence of the final types as $t \to \pm\infty$. The existence of asymptotically symmetric movements of various types (covering all the possibilities) was established by Lagrange, Euler, Poincare, Birkhoff, Chazy, K. A. Sitnikov and V. I. Arnol'd.

The first rigorous result showing the possibility of asymptotic asymmetry of movements in three body problem was obtained by K. A. Sitnikov (1953). He implemented a partial capture (and, therefore, complete break-up), i.e. the possibility of combination of types: hyperbolic as $t \to -\infty$ and hyperbola-elliptic as $t \to +\infty$; with the help of numerical methods this was discovered earlier (1947) by O. Yu. Schmidt, well-known algebraist, polar explorer and the author of certain cosmogonical hypothesis. This achievement together with the aforementioned result of V. M. Alexeyev, concerning the possibility of satellite exchange, completely solved the problem of two-sided classification of final types in the case when the energy of the system is positive.

The main problem on the agenda then became the question about the possibility of complete capture (partial decay), i.e. the existence of movements of hyperbola-elliptical type as $t \to -\infty$ and bounded as $t \to +\infty$. This question called for the application to the three body problem the techniques of constructing and investigating (with the help of symbolic dynamics) of hyperbolic sets, developed in well-known works by D. V. Anosov, Ya. G. Sinai, S. Smale.

In 1968 Vladimir Mikhaylovich Alexeyev constructed an example of complete capture as well as examples of hyperbola-elliptical (or bounded) as $t \to -\infty$ and oscillating as $t \to +\infty$ (the latter means that for some pair of bodies the distance between them is unbounded but does not tend to infinity). Thus, it was established that all combinations of final types of movements can be realized (of course, taking into account the sign of the energy constant). This result as well as the classification of types of movements of one-dimensional oscillator in the force field, periodically depending on time, were presented in a series of papers published by Vladimir Mikhaylovich in Sbornik.Mathematics (1968-1969) and soon became famous.

Let me notice here that the existence of solutions to differential equations of second order having, in a sense, a random distribution of zeros, was observed earlier by Cartright, Littlewood and Levinson (1957). They constructed the solutions that admit coding by arbitrary sequences of zeros and ones in such a way that for some T_0, $T_1 \in \mathbb{R}^+$ zero (one) is associated with the interval between consecutive zeros of the corresponding solution, approximately equal in length to T_0 (respectively T_1).

Vladimir Mikhaylovich was invited to give a talk on his results at the International Congress of Mathematicians in Nice (1970). Shortly prior to the time of the Congress Jean Leray phoned to Alexeyev. Leray praised very highly the mathematical achievements of V. M. Alexeyev and asked him to be merciful to Chazy. It must be said that by the time the text of Alexeyev's talk have already been prepared and after mentioning Chazy's contribution the following was written: "C'est pour rendre hommage à cet éminent mathématicien et astronome français, dont les travaux ont stimulé en grande partie ce qui est expose ci-dessous, et aussi pour soulinger la continuité de l'effort des diverses generations de mathematiciens, que j'ai donné à cette conférence le titre même de deux de ses Mémoires". It so happend that V. M. Alexeyev was not included in the group of Soviet participants to the Congress, and he passed the text of the talk to me before my departure for Nice. At the Congress I reported about Alexeyev's results concerning final motions.

Let me point out a fact showing how focused V. M. Alexeyev was when he worked on the three body problem. It can be seen from his papers that he was a master of rigorous style, and at the same time, was capable to capture the attention of the reader. This made the work of the editor of translations of mathematical literature into Russian very important to him, and he devoted a lot of his time to this work. However, there was a clear seven-year (1963-1970) gap in this activity important especially for students; it was during this period the idea to attack the three body problem first ripened in V. M. Alexeyev's mind, and then was realized with the help of the new methods of dynamical systems theory.

V. M. Alexeyev actively participated in the life of Moscow Mathematical Society; he supervised the "Communications to MMS" section of the journal "Russian Mathematical Surveys" and gave talks at the meetings of MMS. The topic of one of his talks was the discussion of the recent result by Schweitzer who solved the Seifert problem on the nonexistence of closed trajectories for smooth vector fields on the 3-sphere. For several years V. M. Alexeyev was elected the Secretary of MMS. Once, in 1971 as far as I remember, Vladimir Mikhaylovich said to me: "You should also work in this capacity". I answered directly to him that for me at that time the main problem was to get visiting position abroad in order to earn money for buying an apartment for my family. Vladimir Mikhaylovich was sympathetic and took no offence at my de-facto refusal to his offer.

After coming back from visiting abroad I was in touch with V. M. Alexeyev regarding problems upon which our students were working (Yu. Osipov, S. Pidkujko, A. Tagi-Zade). The last time I visited Vladimir Mikhaylovich was in September of 1980 when I told him about the Mathematical School on Differential Equations which had been organized by the Institute of Mathematics of Ukrainian Academy of Sciences and Uzhgorod University and had taken place that summer in Carpathian mountains.

On December the 1st, 1980, Vladimir Mikhaylovich passed away. By the proposal of Ya. G. Sinai the annual lecture in memory of Alexeyev was established at the Mathematics Department of the Moscow State University. One of these lectures was delivered a few years ago by S. Smirnov, a colleague of Vladimir Mikhailovich and one of the members of the Organizing Committee of Moscow Mathematical Olympiads for high school students. I would like to end my talk quoting from that lecture by S. Smirnov.

"Alexeyev died when he was only 48, the age when Tolkien's hobbits reach their maturity. We do not know whether Alexeyev had found the time to read the Great Book of the Ring – it was not yet translated into Russian, but he read English easily and liked science fiction. If VMA read the biography of Frodo Baggins, then surely he must have felt his spiritual kinship to all the gentle-hobbits and to their creator..."

DEPARTMENT OF THEORY OF FUNCTIONS AND FUNCTIONAL ANALYSIS, FACULTY OF MECHANICS AND MATHEMATICS, LOMONOSOV MOSCOW STATE UNIVERSITY, VOROB'EVY GORY, MAIN BUILDING MSU, MOSCOW 119899, RUSSIA

E-mail address: stepin@mech.math.msu.su

MINIMAL IDEMPOTENTS AND ERGODIC RAMSEY THEORY

VITALY BERGELSON

1. INTRODUCTION

What is common between the invertibility of distal maps, partition regularity of diophantine equation $x - y = z^2$, and the notion of mild mixing? The answer is: idempotent ultrafilters, and the goal of this survey is to convince the reader of the unifying role and usefulness of idempotent ultrafilters (and, especially, the minimal ones) in ergodic theory, topological dynamics and Ramsey theory.

We start with reviewing some basic facts about ultrafilters. The reader will find the missing details and more information in the self-contained Section 3 of [B2]. (See also [HiS] for a comprehensive presentation of the material related to topological algebra in the Stone-Čech compactification).

An ultrafilter p on $\mathbb{N} = \{1, 2, ...\}$ is, by definition, a *maximal filter*, namely, a nonempty family of subsets of $\mathbb{N}$ satisfying the following conditions (the first three of which constitute the definition of a *filter*):

(i) $\emptyset \notin p$;

(ii) $A \in p$ and $A \subset B$ imply $B \in p$;

(iii) $A \in p$ and $B \in p$ imply $A \cap B \in p$;

(iv) (maximality) if $r \in \mathbb{N}$ and $\mathbb{N} = A_1 \cup A_2 \cup ... \cup A_r$, then for some i, $1 \leq i \leq r$, $A_i \in p$.

The space of ultrafilters, denoted by $\beta\mathbb{N}$, and equipped with appropriately defined topology, is nothing but Stone-Čech compactification of $\mathbb{N}$ and plays

This work was partially supported by NSF under the grants DMS-9706057 and DMS-0070566.

an important role in various areas of mathematics including topology, analysis and ergodic Ramsey theory.

In what follows we will find it useful to view ultrafilters as finitely-additive, $\{0,1\}$-valued probability measures on the power set $\mathcal{P}(\mathbb{N})$.

Given an ultrafilter $p \in \beta\mathbb{N}$, define a mapping $\mu_p : \mathcal{P}(\mathbb{N}) \to \{0,1\}$ by $\mu_p(A) = 1 \Leftrightarrow A \in p$. It is easy to see that $\mu_p(\emptyset) = 0$, $\mu_p(\mathbb{N}) = 1$ (follows from (i), (iv) and (ii)), and that for any finite collection of disjoint sets $A_1, A_2, ..., A_r$, one has $\mu_p(\bigcup_{i=1}^{r} A_i) = \sum_{i=1}^{r} \mu_p(A_i)$. Indeed, note that if none of A_i belongs to p, then both sides equal zero. Also, it follows from (i) that at most one among the (disjoint!) sets A_i may satisfy $A \in p$, in which case both sides of the above equation equal one.

One of the major advantages of viewing the ultrafilters as measures is that one can naturally define the convolution operation which makes $\beta\mathbb{N}$ a compact semigroup. Given two σ-additive measures μ and ν on a topological group G, the convolution is usually defined as $\mu * \nu(A) = \int_G \mu(Ay^{-1})d\nu(y)$. In particular, $\mu * \nu(A) > 0$ iff for ν-many y one has $\mu(Ay^{-1}) > 0$. Taking into account that a value of ultrafilter measure on a set $A \subseteq \mathbb{N}$ is positive iff it equals one, we make the following definition in which for a reason to be explained in the remark below, we denote the convolution by $+$.

Definition 1.1. *Given $p, q \in \beta\mathbb{N}$, the convolution $p + q$ is defined by*

$$p + q = \{A \subseteq \mathbb{N} : \{n : (A - n) \in p\} \in q\}.$$

In other words, A is $(p + q)$-large iff the set $A - n = \{n \in \mathbb{N} : m + n \in A\}$ is p-large for q-many n.

It is not too hard to check that $p + q$ is an ultrafilter and that the operation defined above is associative (see, for example, [B2], p.27).

Now we shall explain the reason for denoting this operation by $+$. For any $n \in \mathbb{N}$ define an ultrafilter μ_n as a "delta measure" concentrated at point n:

$$\mu_n(A) = \begin{cases} 1, & n \in A \\ 0, & n \notin A. \end{cases}$$

The ultrafilters $\mu_n, n \in \mathbb{N}$, are called *principal* and it is clear that for any $n, k \in \mathbb{N}$ the convolution of μ_n and μ_k equals μ_{n+k}. In other words, the principal ultrafilters μ_n, $n \in \mathbb{N}$, form a semigroup which is isomorphic to $(\mathbb{N}, +)$ and the convolution defined above extends the operation $+$ to the space $\beta\mathbb{N}$, the closure of $\mathbb{N}$. At this point it will be instructive to say a few words about the topology on $\beta\mathbb{N}$. Given $A \subset \mathbb{N}$, let $\overline{A} = \{p \in \beta\mathbb{N} : A \in p\}$. It is immediate that for any $A, B \subset \mathbb{N}$ one has $\overline{A \cap B} = \overline{A} \cap \overline{B}$, $\overline{A \cup B} = \overline{A} \cup \overline{B}$. Also, since $\overline{\mathbb{N}} = \beta\mathbb{N}$, one has $\bigcup_{A \in \mathcal{A}} \overline{A} = \beta\mathbb{N}$, where $\mathcal{A} = \{\overline{A} : A \subset \mathbb{N}\}$. It follows that the set $\mathcal{A}$ forms the basis for the open sets of $\beta\mathbb{N}$ (and the basis for closed sets too!). One can show that with this topology $\beta\mathbb{N}$ is a compact Hausdorff space and that for any fixed $p \in \beta\mathbb{N}$ the function $\lambda_p(q) = p + q$ is a continuous self map of $\beta\mathbb{N}$ (see for example Theorems 3.1 and 3.2 in [B2]). In view of these facts, $(\beta\mathbb{N}, +)$ becomes a compact *left topological* semigroup. We remark in passing that the operation $\rho_p(q) = q + p$ is, unlike $\lambda_p(q)$, continuous only when p is a principal ultrafilter, and that the convolution defined above on $\beta\mathbb{N}$ is the unique extension of the operation $+$ on $\mathbb{N}$ such that $\lambda_p(q)$ and $\rho_p(q)$ have the properties described above.

Before going on to explore additional features of the semigroup $(\beta\mathbb{N}, +)$ that are important for us we want to caution the reader that while having various nice and convenient properties, the semigroup $(\beta\mathbb{N}, +)$ is in many respects an odd and counterintuitive object. First, the compact Hausdorff space $\beta\mathbb{N}$ is too large to be metrizable: its cardinality is that of $\mathcal{P}(\mathcal{P}(\mathbb{N}))$. Yet, in view of the fact that $\overline{\mathbb{N}} = \beta\mathbb{N}$, it is a closure of a countable set $\mathbb{N}$. Second, the operation $+$ on $\beta\mathbb{N}$ is highly non-commutative: the center of the semigroup $(\beta\mathbb{N}, +)$ contains only the principal ultrafilters. (Here the analogy with the convolution of σ-additive measures on locally compact abelian groups fails. The reason: the ultrafilters, being only *finitely* additive measures, do not obey the Fubini theorem which is crucial for the commutativity of the convolution of σ-additive measures).

By a theorem due to Ellis [E1], any compact semigroup with a left-continuous operation has an idempotent. Actually, $(\beta\mathbb{N}, +)$ has plenty of them, since any compact subsemigroup in $(\beta\mathbb{N}, +)$ should have one and there are 2^c disjoint compact subsemigroups in $\beta\mathbb{N}$. As we shall see below, of special importance for combinatorial and ergodic-theoretical applications are *minimal* idempotents, which we will define and apply later in this section. In a way, idempotent ultrafilters in $\beta\mathbb{N}$ are, in a way, just generalized shift-invariant measures. Indeed, if $p + p = p$, it means that any $A \in p = p + p$ has the property that $\{n : (A - n) \in p\} \in p$, or, in other words, for p-almost all n, the set $A - n$ is p-large.

It is easy to see that principal ultrafilters are never idempotent and hence, if p is an idempotent ultrafilter, any p-large set A is infinite, as is the p-large set $\{n : (A - n) \in p\}$. As we shall presently see, the members of idempotent ultrafilters always contain highly structured subsets which can be viewed as generalized subsemigroups of $\mathbb{N}$.

Let $A \in p$, where $p + p = p$. Since

$$A \cap \{n : (A - n) \in p\} \in p,$$

we can choose $n_1 \in A$ such that $A_1 = A \cap (A - n_1) \in p$. (Note that this is nothing but a version of Poincaré recurrence theorem; the important bonus is that $n_1 \in A$. By iterating this procedure one can chose $n_2 \in A \cap (A - n_1)$, $n_2 > n_1$, such that

$$A_1 \cap (A_1 - n_2) = A \cap (A - n_1) \cap (A - n_2) \cap (A - n_1 - n_2) \in p.$$

Note that $n_1, n_2, n_1 + n_2 \in A$. Continuing in this fashion, one obtains an increasing sequence $(n_i)_{i=1}^{\infty}$ and inductively defined sets $A = A_0, A_1, A_2, ...,$ such that $n_1 \in A$, $n_{i+2} \in A_{i+1} := A_i \cap (A_i - n_{i+1})$, $i = 0, 1, 2, ...$ One readily checks that this construction implies that A contains the set of *finite sums* of $(n_i)_{i=1}^\infty$:

$$FS(n_i)_{i=1}^\infty = \{n_{i_1} + n_{i_2} + ... + n_{i_k}, \; k \in \mathbb{N}, \; i_1 < i_2 < ... < i_k\}.$$

Such sets of finite sums are customarily called IP sets (IP stands for IdemPotent) and are featured in the following important theorem due to N.Hindman [Hi1].

Theorem 1.2. (N.Hindman). *For any finite partition $\mathbb{N} = \bigcup_{i=1}^r C_i$ one of the cells of partition contains an IP set.*

Proof. Fix any idempotent ultrafilter $p \in \beta\mathbb{N}$ and observe that one (and only one!) of C_i belongs to it. Now use the fact proved above that any member of p contains an IP set. □

Let $\mathcal{F}$ denote the family of non-empty finite subsets of N. Noticing that the mapping $\mathcal{F} \to \mathbb{N}$ defined by $\{i_1, i_2, ..., i_k\} \to 2^{i_1} + 2^{i_2} + ... + 2^{i_k}$ is 1-1 and that elements of IP sets are naturally indexed by elements of $\mathcal{F}$, we have that each of the following two theorems implies Hindman's theorem, each revealing yet another facet of it.

Theorem 1.3. (Finite unions theorem). *For any finite partition $\mathcal{F} = \bigcup_{i=1}^r C_i$, one of C_i contains an infinite sequence of non-empty disjoint sets $(U_i)_{i \in \mathbb{N}}$ together with all the unions $U_{i_1} \cup U_{i_2} \cup ... \cup U_{i_k}$, $i_1 < i_2 < ... < i_k, k \in \mathbb{N}$. In addition, one can assume without the loss of generality that for all $i \in \mathbb{N}$ one has $\max U_i < \min U_{i+1}$.*

Theorem 1.4. *For any finite partition of an IP set in $\mathbb{N}$ one of the cells of the partition contains an IP set.*

Exercise 1. Prove that Theorems 1.2, 1.3, 1.4 are equivalent.

In the proof of Hindman's theorem above IP sets emerge as subsets of members of idempotent ultrafilters. One may wonder whether given an idempotent p and a set $A \in p$, it is possible to find in A an IP set which is itself p-large. It turns out that this is not always the case. For example, the minimal idempotents which we will define below, can not have this property. The following theorem shows that, nevertheless, any IP set is a support of an idempotent.

Theorem 1.5. *For any sequence $(x_i)_{i \in \mathbb{N}}$ in $\mathbb{N}$ there is an idempotent $p \in \beta\mathbb{N}$ such that $FS((x_i)_{i \in \mathbb{N}}) \in p$.*

Sketch of the proof. Let $\Gamma = \bigcap_{n=1}^\infty \overline{FS((x_i)_{i=n}^\infty)}$. (The closures are taken in the natural topology of $\beta\mathbb{N}$). Clearly, Γ is compact and non-empty. It is not hard to show that Γ is a subsemigroup of $(\beta\mathbb{N}, +)$. Being a compact left-topological semigroup, Γ has an idempotent. If $p \in \Gamma$ is an idempotent, then $\overline{\Gamma} = \Gamma \ni p$

which, in particular, implies $FS((x_i)_{i=1}^{\infty}) \in p$.   □

The above definitions and theorems readily extend to general semigroups. Given a semigroup $(G, \cdot)$, one defines βG as the set of ultrafilters on G. The semigroup operation naturally extends to βG by the rule

$$A \in p \cdot q \Leftrightarrow \{x : Ax^{-1} \in p\} \in q$$

(where $Ax^{-1} := \{y \in G : yx \in A\}$).

Exercise 2. Verify that $(\beta G, \cdot)$ is a left-topological compact semigroup.

The IP sets, which, in the case of multiplicative notation, become *finite product sets*, are defined as follows. Given any sequence $(x_n)_{n \in \mathbb{N}} \subset G$ and $F \in \mathcal{F}$ denote by $\prod_{n \in F} x_n$ the product of $x_n, n \in F$ in the *decreasing* order of indices. Then the IP set generated by $(x_n)_{n \in \mathbb{N}}$ is defined as $FP((x_n)_{n=1}^{\infty}) = \{\prod_{n \in F} x_n, F \in \mathcal{F}\}$. As in the case of $(\beta \mathbb{N}, +)$, the IP sets in general semigroups are closely related to idempotent ultrafilters (whose existence follows from Ellis' theorem, alluded to above). In particular, one can check that the proof of Hindman's theorem given above transfers verbatim to give a proof of the following theorem and its corollary. Note that the Corollary 1.7 below can also be obtained from the finite unions theorem (Theorem 1.3 above).

Theorem 1.6. *Let $(G, \cdot)$ be a discrete semigroup and let p be an idempotent in $(\beta G, \cdot)$. Then for any $A \in p$ there exists a sequence $(x_n)_{n \in \mathbb{N}}$ in βS such that $FP((x_n)_{n=1}^{\infty}) \subseteq A$.*

Corollary 1.7. *For any finite partition $G = \bigcup_{i=1}^{r} C_i$ one of the C_i contains an IP set.*

Exercise 3. Give a detailed proof of Theorem 1.6 and Corollary 1.7.

Since $\mathbb{N}$ (and, hence, $\beta \mathbb{N}$) has two natural structures, namely, those of addition and multiplication, it follows that for any finite partition $\mathbb{N} = \bigcup_{i=1}^{r} C_i$ there are $i, j \in \{1, 2, ..., r\}$ such that C_i contains an additive IP set and C_j contains a multiplicative IP set. The following theorem due to Hindman shows that one can always have $i = j$.

Theorem 1.8. ([Hi2]) *For any finite partition $\mathbb{N} = \bigcup_{i=1}^{r} C_i$ there exists $i \in \{1, 2, ..., r\}$ and sequences $(x_n)_{n=1}^{\infty}$ and $(y_n)_{n=1}^{\infty}$ in $\mathbb{N}$ such that $FS((x_n)_{n=1}^{\infty}) \cup FP((y_n)_{n=1}^{\infty}) \subseteq C_i$.*

Proof. Let Γ be the closure in $\beta \mathbb{N}$ of the set of additive idempotents. We claim that $p \in \Gamma$ if and only if every p-large set A contains an additive IP set. Indeed, if $A \in p \in \Gamma$, then $\overline{A}$ is a (clopen) neighborhood of p. It follows that there exists $q \in \overline{A}$ with $q + q = q$. Then $A \in q$ and by Theorem 1.2 A contains an IP set. Conversely, if $\overline{A}$ is a basic neighborhood of p and for some $(x_n)_{n=1}^{\infty}$, $FS((x_n)_{n=1}^{\infty}) \subseteq A$, then by Theorem 1.5 above there exists an idempotent q with $FS((x_n)_{n=1}^{\infty}) \in q$, which implies $A \in q$, and hence $p \in \Gamma$.

We will show now that Γ is a right ideal in $(\beta\mathbb{N}, \cdot)$. Let $p \in \Gamma$, $q \in \beta\mathbb{N}$, and let $A \in p \cdot q$. Then $\{x \in \mathbb{N} : Ax^{-1} \in p\} \in q$ and, in particular, $\{x \in \mathbb{N} : Ax^{-1} \in p\}$ is non-empty. Let x be such that $Ax^{-1} \in p$. Since $p \in \Gamma$, there exists a sequence $(y_n)_{n=1}^{\infty}$ with $FS((y_n)_{n=1}^{\infty}) \subseteq Ax^{-1}$, which implies $FS((xy_n)_{n=1}^{\infty}) \subseteq A$ and so $p \cdot q \in \Gamma$. We see that Γ is a compact subsemigroup in $(\beta\mathbb{N}, \cdot)$ and hence contains a multiplicative idempotent. To finish the proof, let $\cup_{i=1}^{r} C_i = \mathbb{N}$ and let $p \in \Gamma$ satisfy $p \cdot p = p$. Let $i \in \{1, 2, ..., r\}$ be such that $C_i \in p$. Then, since $p \in \Gamma$, C_i contains an additive IP set. Also, since p is a multiplicative idempotent, C_i contains (by Theorem 1.6) a multiplicative IP set. We are done. $\square$

Remarks (i) For an elementary proof of Theorem 1.8 see [BH2].

(ii) Theorem 2.12 below shows that for any finite partition $\cup_{i=1}^{r} C_i = \mathbb{N}$ one C_i has interesting additional properties. In particular, one C_i can be shown to contain in addition to an additive and a multiplicative IP sets, also arbitrarily long arithmetic and arbitrarily long geometric progressions.

2. MINIMAL IDEMPOTENTS, CENTRAL SETS, AND COMBINATORIAL APPLICATIONS

Let $\sigma : \mathbb{N} \to \mathbb{N}$ denote the shift operation: $\sigma(x) = x + 1$, $x \in \mathbb{N}$. As we saw above, all that it takes to prove Hindman's theorem is to apply a version of Poincaré recurrence theorem to the "measure preserving system" $(\mathbb{N}, p, \sigma)$, where p is an arbitrary idempotent in $(\beta\mathbb{N}, +)$. Indeed, the idempotence of p implies that any p-large set A has the property that for p-many $n \in \mathbb{N}$ the set $A - n = \sigma^{-n}(A)$ is also p-large and hence for some n (which in our situation can be chosen from A) one has $(A \cap \sigma^{-n}(A)) \in p$. The rest of the proof is just a routine iteration.

We shall introduce now an important subclass of idempotents which will allow us to make a connection with another basic dynamical notion, namely, that of a minimal dynamical system.

A topological dynamical system (with "time" $\mathbb{N}$) is a pair (X, T), where X is a compact space and $T : X \to X$ is a continuous map. The system (X, T) is called *minimal*, if for any $x \in X$ one has $\overline{(T^n x)_{n \in \mathbb{N}}} = X$. One can show, by a simple application of Zorn's lemma, that any system (X, T) contains a minimal compact non-empty T-invariant subset Y which, consequently, gives rise to minimal system (Y, T) (by slight abuse of notation we denote the restriction of T to Y by the same symbol). Extending the shift operation σ from $\mathbb{N}$ to $\beta\mathbb{N}$ by the rule $q \to q + 1$ (where 1 denotes the principal ultrafilter of sets containing the integer 1), we obtain a topological dynamical system

$(\beta\mathbb{N}, \sigma)$. The following theorem establishes the connection between minimal subsystems of $(\beta\mathbb{N}, \sigma)$ and minimal right ideals in $(\beta\mathbb{N}, +)$.

Theorem 2.1. *The minimal closed invariant subsets of the dynamical system $(\beta\mathbb{N}, \sigma)$ are precisely the minimal right ideals of $(\beta\mathbb{N}, +)$.*

Proof. We first observe that closed σ-invariant sets in $\beta\mathbb{N}$ coincide with right ideals. Indeed if I is a right ideal, i.e. satisfies $I + \beta\mathbb{N} \subseteq I$, then for any $p \in I$ one has $p+1 \in I+\beta\mathbb{N} \subseteq I$, so that I is σ-invariant. On the other hand, if S is a closed σ-invariant set in $\beta\mathbb{N}$ and $p \in S$, then $p+\beta\mathbb{N} = p+\overline{\mathbb{N}} = \overline{p + \mathbb{N}} \subseteq \overline{S} = S$, which implies $S + \beta\mathbb{N} \subseteq S$.

Now the theorem follows from a simple general fact that any minimal right ideal in a compact left-topological semigroup $(G, \cdot)$ is closed. Indeed, if R is a right ideal in $(G, \cdot)$ and $x \in R$, then xG is compact as the continuous image of G and is an ideal. Hence the minimal ideal containing x is compact as well. (The fact that R contains a minimal ideal follows by an application of Zorn's lemma to the non-empty family $\{I : I$ is a closed right ideal of G and $I \subseteq R\}$). □

Our next step is to observe that any minimal right ideal in $(\beta\mathbb{N}, +)$, being a compact left-topological semigroup, contains, by Ellis' theorem, an idempotent.

Definition 2.2. *An idempotent p in $(\beta\mathbb{N}, +)$ is called minimal if p belongs to a minimal ideal.*

Theorem 2.3. *Any minimal subsystem of $(\beta\mathbb{N}, \sigma)$ is of the form $(p + \beta\mathbb{N}, \sigma)$ where p is a minimal idempotent in $(\beta\mathbb{N}, +)$.*

Proof. It is obvious that, for any $p \in (\beta\mathbb{N}, +)$, $p + \beta\mathbb{N}$ is a right ideal. To see that any minimal right ideal is of this form, take any $q \in R$ and observe that $q + \beta\mathbb{N} \subseteq R + \beta\mathbb{N} \subseteq R$. Since R is minimal, we get $q + \beta\mathbb{N} = R$. In particular, one can take q to be an idempotent.

Before moving to some immediate corollaries of Theorem 2.3 we want to remind the reader that a set $A \subseteq \mathbb{N}$ is called *syndetic* if it has bounded gaps, or equivalently, if for some finite $F \subset \mathbb{N}$ one has $\bigcup_{t \in F}(A - t) = \mathbb{N}$. A set $A \subseteq \mathbb{N}$ is called *piecewise syndetic* if it can be represented as an intersection of a syndetic set with an infinite union of intervals $[a_n, b_n]$, where $b_n - a_n \to \infty$.

An equivalent definition of piecewise syndeticity (and the one which makes sense in any semigroup) is given in the following exercise.

Exercise 4. Prove that a set $A \subseteq \mathbb{N}$ is piecewise syndetic if and only if there exists a finite set $F \subset \mathbb{N}$, such that the family $\{\bigcup_{t \in F}(A - t) - n : n \in \mathbb{N}\}$ has the finite intersection property.

Exercise 5. (i) Prove that if (X, T) is a minimal system then every point $x \in X$ has a dense orbit.

(ii) Prove that if (X, T) is a minimal system then for any $x \in X$ and any neighborhood V of x the set $\{n : T^n x \in V\}$ is syndetic.

Theorem 2.4. *Let p be a minimal idempotent in $(\beta \mathbb{N}, +)$.*

(i) *For any $A \in p$ the set $B = \{n : (A - n) \in p\}$ is syndetic.*

(i) *Any $A \in p$ is piecewise syndetic.*

(iii) *For any $A \in p$ the set $A - A = \{n_1 - n_2 : n_1, n_2 \in A\}$ is syndetic.*

Proof. Statement (i) follows immediately from the fact that $(p + \beta \mathbb{N}, \sigma)$ is a minimal system. Indeed, note that the assumption $A \in p$ just means that $p \in \overline{A}$, i.e. $\overline{A}$ is a (clopen) neighborhood of p. Now, by Exercise 5, in a minimal dynamical system every point x is *uniformly recurrent*, i.e. visits any of its neighborhoods V along a syndetic set. This implies that the set $\{n : p + n \in \overline{A}\} = \{n : A \in p + n\} = \{n : A - n \in p\}$ is syndetic.

(ii) Since the set $B = \{n : A - n \in p\}$ is syndetic, the union of finitely many shifts of B covers $\mathbb{N}$, i.e. for some finite set $F \subset \mathbb{N}$ one has $\bigcup_{t \in F}(B - t) = \mathbb{N}$. So, for any $n \in \mathbb{N}$ there exists $t \in F$ such that $n \in B - t$, or $n + t \in B$. By the definition of B this implies $(A - (n + t)) \in p$. It follows that for any n the set $\bigcup_{t \in F}(A - t) - n$ belongs to p, and consequently, the family $\{\bigcup_{t \in F}(A - t) - n : n \in \mathbb{N}\}$ has the finite intersection property. By Exercise 4 this is equivalent to piecewise syndeticity of A.

(iii) Observe that $n \in A - A$ if and only if $A \cap (A - n) \neq \emptyset$. But then it follows from (i) that the set $\{n : A \cap (A - n) \in p\}$ is syndetic. We are done. $\square$

Exercise 6. Let $(\Gamma, \cdot)$ be a compact left-topological semigroup and let $p \in \Gamma$. Verify that $p \cdot \Gamma$ is a compact right ideal and that any minimal right ideal is representable in this form.

For minimal idempotents in $(\beta G, \cdot)$ (i.e. idempotents belonging to minimal right ideals) one has an analogue of Theorem 2.4. In order to properly formulate it, one has first to define in a general context the notions of syndeticity and piecewise syndeticity since in a non-commutative situation one has left and right versions and needs to make "right" choices (pun not intended!).

Since we work with the left-topological semigroup $(\beta G, \cdot)$ the correct notions, which allow one to painlessly transfer the theorems above to the general set-up, turn out to be the left ones. Recall that given a set A and an element x in a semigroup $(G, \cdot)$ one has, by definition, $Ax^{-1} = \{y \in G : yx \in A\}$.

Definition 2.5. *Let G be a semigroup. A set $A \subseteq G$ is called syndetic if for some finite set $F \subset G$ one has $\bigcup_{t \in F} At^{-1} = G$. A set $A \subseteq G$ is piecewise syndetic if for some finite set $F \subset G$ the family $\{(\bigcup_{t \in F} At^{-1})a^{-1} : a \in G\}$ has the finite intersection property.*

The following exercise establishes a general form of Theorem 2.4.

Exercise 7. Let G be a discrete semigroup and $p \in (\beta G, \cdot)$ a minimal idempotent. Prove:

(i) For any $A \in p$ the set $B = \{g : Ag^{-1} \in p\}$ is syndetic.

(ii) Any $A \in p$ is piecewise syndetic.

(iii) For any $A \in p$, the set $A^{-1}A = \{x \in G : yx \in A$ for some $y \in A\}$ is syndetic. (Note that if G is a group, then $A^{-1}A = \{g_1^{-1}g_2 : g_1, g_2 \in A\}$.)

The notion of minimality for idempotents can also be expressed in terms of a natural partial order which we will presently introduce.

Definition 2.6. *Let p, q be idempotents in a semigroup $(G, \cdot)$. We shall say that p is dominated by q and write $p \leq q$ if $pq = qp = p$.*

Exercise 8. Check that $\leq$ is transitive, reflexive and antisymmetric relation on the set of idempotents in G.

Theorem 2.7. *Let $(G, \cdot)$ be a compact left topological semigroup. Then an idempotent p is minimal with respect to the order $\leq$ if and only if it belongs to a minimal right ideal.*

Proof. (i) Assume that $p \leq q$ is minimal with respect to the order $\leq$. By Exercise 6, pG is a right ideal which contains a closed minimal right ideal R. Let $q_0 \in R$ be an idempotent. Since $q_0 \in pG$, $q_0 = pg$ for some $g \in G$. We have $pq_0 = ppg = pg = q_0$. Therefore $q_0pq_0p = q_0q_0p = q_0p$ and q_0p is also an idempotent. It satisfies $pq_0p = q_0pp = q_0p$, and hence $q_0p \leq p$. But p was assumed to be minimal with respect to the order $\leq$, and so $q_0p = p$. This gives $pG = q_0pG \subseteq q_0G$ and shows that pG is itself a minimal right ideal.

(ii) Assume that p is an idempotent in a minimal right ideal R. Note that pG is a right ideal and since $pG \subset R$ we have $pG = R$. Assume now that for some idempotent q one has $q \leq p$ and show that $p \leq q$. Since $q \leq p$ then we have also $qG \subset R$. It follows that for some $g \in G$, $p = qg$. Hence $qp = qqg = qg = p$. But $qp = q$ (because $q \leq p$). So $p \leq q$ and we are done. $\square$

Remark. It follows from the proof of part (i) above that for any idempotent q, there exists a minimal idempotent p with $p \leq q$.

Before moving to some applications of minimal idempotents in Ramsey theory, we want to introduce one more useful algebraic-topological notion.

Definition 2.8. *Let $(G, \cdot)$ be a compact left-topological semigroup. Then $K(G) = \bigcup\{R : R$ is a minimal right ideal $\}$.*

Note that in view Exercise 6, $K(G) \neq \emptyset$.

Theorem 2.9. *Let $(G, \cdot)$ be a compact left topological semigroup. Then $K(G)$ is a two-sided ideal, and, in fact, the smallest two-sided ideal.*

Proof. Being the union of right ideals, $K(G)$ is trivially a right ideal. We note also that if I is a two-sided ideal of G then $K(G) \subseteq I$. Indeed, for any minimal right ideal R of G one has $R \cap I \neq \emptyset$ (since for any $x \in R$ and $y \in I$ one has $xy \in R \cap I$) and hence $R \cap I$ is a right ideal which, in view of minimality of R implies $R \cap I = R$. Hence, $K(G) \subseteq I$.

It remains to show that $K(G)$ is a left ideal of G, i.e. $G \cdot K(G) \subseteq K(G)$. Let $x \in K(G)$ and let R be a minimal right ideal of G such that $x \in R$. For arbitrary $y \in G$ one has: $yx \in yR$ and, since yR is a right ideal, it remains to show that it is a minimal right ideal. Indeed, then one has $yx \in yR \subseteq K(G)$ which clearly implies $G \cdot K(G) \subseteq K(G)$.

So let J be a right ideal of G satisfying $J \subseteq yR$ and let $C = \{z \in R : yz \in J\}$. Then C is a right ideal of G which is contained in R and so $C = R$ and $J = yR$. $\qquad\square$

We are going to present now a proof, via the minimal idempotents, of the celebrated van der Waerden theorem on arithmetic progressions, which states that for any finite partition $\mathbb{N} = \bigcup_{i=1}^{r} C_i$, one C_i contains arbitrarily long arithmetic progressions. (This proof is a slight modification of the proof from [BFHK], which, in its turn, was a modification of an argument due to Furstenberg and Katznelson that first appeared, in the framework of an Ellis enveloping semigroup, in [FK2]).

Since for any minimal idempotent $p \in (\beta\mathbb{N}, +)$ and any partition $\mathbb{N} = \bigcup_{i=1}^{r} C_i$, one C_i belongs to p, van der Warden's theorem clearly follows from the following result.

Theorem 2.10. *Let $p \in (\beta\mathbb{N}, +)$ be a minimal idempotent and let $A \in p$. Then A contains arbitrarily long arithmetic progressions.*

Proof. Fix $k \in \mathbb{N}$ and let $G = (\beta\mathbb{N})^k$. Clearly, G is a compact left topological semigroup with respect to the product topology and coordinatewise addition. Let

$$E_0 = \{(a, a+d, ..., a+(k-1)d) : a \in \mathbb{N}, \ d \in \mathbb{N} \cup \{0\}\},$$

$$I_0 = \{(a, a+d, ..., a+(k-1)d) : a, d \in \mathbb{N}\}.$$

Clearly, E_0 is a semigroup in $\mathbb{N}^k$ and I_0 is an ideal of E_0. Let $E = cl_G E_0$ and $I = cl_G I_0$ be, respectively, the closures of E_0 and I_0 in G. It follows by an easy argument, which we leave to the reader, that E is a compact subsemigroup of G and I is a two-sided ideal of E. Let now $p \in K(\beta\mathbb{N}, +)$ be a minimal idempotent and let $\tilde{p} = (p, p, ..., p) \in G$. We claim that $\tilde{p} \in I$ and that this implies that each member of p contains a length k arithmetic progression. Indeed, assume that $\tilde{p} \in I$ and let $A \in p$. Then $\overline{A} \times ... \times \overline{A} = (\overline{A})^k$ is a neighborhood of $\tilde{p}$. Hence $\tilde{p} \in (\overline{A})^k \cap cl_G I_0 = cl_G(A^k \cap I_0)$, which implies

$A^k \cap I_0 \neq \emptyset$. It follows that for some $a, d \in \mathbb{N}$ $(a, a+d, ..., a+(k-1)d) \in A^k$ which finally implies $\{a, a+d, ..., a+(k-1)d\} \subset A$.

So it remains to show that $\tilde{p} \in I$. We check first that $\tilde{p} \in E$. Let $A_1, A_2, ..., A_k \in p$. Then $\overline{A_1} \times \overline{A_2} \times ... \times \overline{A_k} \ni \tilde{p}$. If $a \in \bigcap_{i=1}^k A_i$ then $(a, a, ..., a) \in (\overline{A_1} \times \overline{A_2} \times ... \times \overline{A_k}) \cap E_0$ which implies $\tilde{p} \in E$.

Now, since $p \in K((\beta\mathbb{N}, +))$, there is a minimal right ideal R of $(\beta\mathbb{N}, +)$ such that $p \in R$. Since $\tilde{p} \in E$, $\tilde{p} + E$ is a right ideal of E and there is a minimal right ideal $\tilde{R}$ of E such that $\tilde{R} \subseteq \tilde{p} + E$. Let $\tilde{q} = (q_1, q_2, ..., q_k)$ be an idempotent in $\tilde{R}$. Then $\tilde{q} \in \tilde{p} + E$ and for some $\tilde{s} = (s_1, s_2, ..., s_k)$ in E we get $\tilde{q} = \tilde{p} + \tilde{s}$. We shall show now that $\tilde{p} = \tilde{q} + \tilde{p}$. Indeed, from $\tilde{q} = \tilde{p} + \tilde{s}$ we get, for each $i = 1, 2, ..., k$, $q_i = p + s_i$. This implies $q_i \in R$ and since R is minimal, $q_i + \beta\mathbb{N} = R$. Hence $p \in q_i + \beta\mathbb{N}$. Let , for each $i = 1, 2, ..., k$, $t_i \in \beta\mathbb{N}$ be such that $p = q_i + t_i$. Then $q_i + p = q_i + q_i + t_i = q_i + t_i = p$ and so we obtained $\tilde{p} = \tilde{q} + \tilde{p}$.

To finish the proof, we observe that $\tilde{p} = \tilde{q} + \tilde{p}$ implies $\tilde{p} \in \tilde{q} + E = \tilde{R}$ which, in its turn, implies $\tilde{p} \in K(E) \subseteq I$ (since $K(E)$ is the smallest ideal in E). We are done.   □

Exercise 9. Show that there is an idempotent p in $(\beta\mathbb{N}, +)$ with the property that not every member of p contains a 3-term arithmetic progression.
Hint. Consider $FS(10^n)_{n=1}^\infty$ and utilize Theorem 1.5.

Definition 2.11. *Let $(G, \cdot)$ be a discrete semigroup. A set $A \subseteq G$ is called central if there exists a minimal idempotent $p \in (\beta G, \cdot)$ such that $A \in p$.*

Exercise 10. Prove that any multiplicatively central set in $\mathbb{N}$ (namely, a member of a minimal idempotent in $(\beta\mathbb{N}, \cdot)$) contains arbitrarily long geometric progressions.

Exercise 11. (i) Let S be a central set in $(\mathbb{N}, +)$ and let $d \in \mathbb{N}$. Let $S/d = \{n : nd \in S\}$ and $dS = \{n : n/d \in S\}$. Prove that S/d and dS are central in $(\mathbb{N}, +)$.

(ii) Use (i) (for $n = 2$) to show that each of the following sets is additively central:

$$C_1 = \{3^k(3m+1) : k, m \in \mathbb{N} \cup \{0\}\},$$

$$C_2 = \{3^k(3m+2) : k, m \in \mathbb{N} \cup \{0\}\}.$$

As theorems above indicate, central sets are an ideal object for Ramsey-theoretical applications. For example, central sets in $(\mathbb{N}, +)$ not only are large (i.e. piecewise syndetic) but also are combinatorially rich and, in particular, contain IP sets and arbitrarily long arithmetic progressions. Similarly, the multiplicative central sets in $(\mathbb{N}, \cdot)$ (namely, the members of minimal idempotents in $(\beta\mathbb{N}, \cdot)$) are multiplicatively piecewise syndetic, contain finite products sets (i.e. the multiplicative IP sets), arbitrarily long geometric progressions etc.

The following theorem obtained in collaboration with N.Hindman may be viewed as enhancement of Theorem 1.8 above.

Theorem 2.12. ([BH1], p.312) *For any finite partition* $\mathbb{N} = \bigcup_{i=1}^{r} C_i$ *one of C_i is both additively and multiplicatively central.*

Sketch of the proof. Let $M = cl\{p : \ p$ is a minimal idempotent in $(\beta\mathbb{N}, +)\}$. Then one can show that M is a right ideal in $(\beta\mathbb{N}, \cdot)$ (see [BH1], Theorem 5.4, p.311). Let $R \subseteq M$ be a minimal right ideal and pick an idempotent $q = q \cdot q$ in R. Let $i \in \{1, 2, ..., r\}$ be such that $C_i \in q$. Since q is a minimal idempotent in $(\beta\mathbb{N}, \cdot)$, C_i is central in $(\mathbb{N}, \cdot)$. Since $C_i \in q$ and $q \in M$, there is some minimal idempotent p in $(\beta\mathbb{N}, +)$ with $C_i \in p$. Hence C_i is also central in $(\mathbb{N}, +)$. $\qquad\square$

The following theorem supplies a useful family of examples of additively and multiplicatively central sets in $\mathbb{N}$.

Theorem 2.13. (cf. [BH3], Lemma 3.3) *For any sequence $(a_n)_{n=1}^{\infty}$ and an increasing sequence $(b_n)_{n=1}^{\infty}$ in $\mathbb{N}$, $\bigcup_{n=1}^{\infty}\{a_n, a_n + 1, a_n + 2, ..., a_n + b_n\}$ is additively central and $\bigcup_{n=1}^{\infty}\{a_n \cdot 1, a_n \cdot 2, ..., a_n \cdot b_n\}$ is multiplicatively central.*

As we shall see in the subsequent sections, central sets play also an important role in the study of recurrence in topological dynamics and ergodic theory. The original definition of central sets in $(\mathbb{N}, +)$, due to H. Furstenberg, was made in the language of topological dynamics. Before introducing Furstenberg's definition of centrality, we want first to recall some relevant dynamical notions.

Given a compact metric space (X, d), a continuous map $T : X \to X$ and not necessarily distinct points $x_1, x_2 \in X$, one says that x_1, x_2 are *proximal*, if for some sequence $n_k \to \infty$ one has $d(T^{n_k}x_1, T^{n_k}x_2) \to 0$.

A point which is proximal only to itself is called distal. In case all the points of X are distal T is called a distal transformation and (X, T) is called a distal system.

Exercise 12. (i) Show that any isometry is a distal transformation.

(ii) Let α be an irrational number. Show that the *skew product* map defined on 2-torus $\mathbb{T}^2$ by $T(x, y) = (x + \alpha, y + x)$ is distal but cannot be made an isometry in any equivalent metric on $\mathbb{T}^2$.

Remark. The skew product map featured in the Exercise 12 (ii) is an example of an *isometric extension*. A deep structure theorem of distal systems proved by Furstenberg in [F2] states that, for a sufficiently broadly interpreted notion of an isometric extension, any minimal distal system is a (potentially transfinite) tower of successive isometric extensions. We shall see in the next section that the notion of distality is intrinsically linked with the idempotent ultrafilters.

Recall that a point x in a dynamical system (X, T) is called *uniformly recurrent* if for any neighborhood V of x the set $\{n : T^n x \in V\}$ is syndetic.

Exercise 13. Prove that if (X, T) is minimal (i.e. every point has dense orbit) then each point $x \in X$ is uniformly recurrent.

One can show (see for example [A1], [E2], [F4, p.160]) that in a dynamical system on a compact metric space every point is proximal to a uniformly recurrent point. In particular, this implies that any distal point is uniformly recurrent. (See Theorem 3.9 below for a proof of an enhanced version of this fact.)

We are now ready to formulate Furstenberg's original definition of central sets in $(\mathbb{N}, +)$. For the proof of the equivalence of this definition to Definition 2.11 above, see Theorem 3.6

Definition 2.14. (see [F4], p.161) *A subset $S \subseteq \mathbb{N}$ is a central set if there exists a system (X, T), a point $x \in X$, a uniformly recurrent point y proximal to x, and a neighborhood U_y of y such that $S = \{n : T^n x \in U_y\}$.*

We shall conclude this section by introducing and discussing various notions of largeness for subsets of $\mathbb{N}$, which will be utilized in the subsequent sections.

Following the terminology introduced in [F4], given a family $\mathcal{M}$ of nonempty sets in $\mathbb{N}$, let us call a set $E \subseteq \mathbb{N}$ an $\mathcal{M}^*$ set if for any $M \in \mathcal{M}$ one has $E \cap M \neq \emptyset$.

For example, if $\mathcal{M}$ consists of sets containing arbitrarily long blocks of consecutive integers (these sets are called *replete* in [GH] and *thick* in [F4]), the family $\mathcal{M}^*$ consists of syndetic sets. If $\mathcal{M}$ is the collection of all IP sets in $(\mathbb{N}, +)$, then the elements of $\mathcal{M}^*$ will be called IP* sets. Similarly, central* sets, or, simply C* sets are defined as the sets which have nontrivial intersection with any central set.

Theorem 2.15. (i) *A set $E \subseteq \mathbb{N}$ is an IP* set if and only if E is a member of any idempotent $p \in (\beta\mathbb{N}, +)$.*

(ii) *A set E is C* set if and only if it is a member of any minimal idempotent $p \in (\beta\mathbb{N}, +)$.*

Proof. (i) Let p be an idempotent in $(\beta\mathbb{N}, +)$. If $E \notin p$, then $E^c \in p$ and, by Hindman's theorem (Theorem 1.2 above), there exists an IP set in E^c which fails to have a non-trivial intersection with E, which contradicts the assumption that E is an IP* set.

(ii) The proof is similar to that of (i). Assume that E is C* set. If p is minimal idempotent and $E \notin p$, then $E^c \in p$ and hence E^c is a central set which has an empty intersection with E. This contradicts the assumption that E is C^*.   □

Corollary 2.16. *C^* sets and IP^* sets have finite intersection property.*

Remark. One can easily check that the definitions of IP^* and C^* sets, as well as Theorem 2.15 and Corollary 2.16, extend naturally to general semigroups. We record here the following trivial extension of Theorem 2.15 (ii), that will be used in the course of the proof of Theorem 4.4 below:

A set E in a countable semigroup G is C^* if and only if it is a member of every minimal idempotent $p \in \beta G$.

It immediately follows from Theorem 2.15 that every IP^* set is C^*. On the other hand, since, by Theorem 2.10, every central set contains arbitrarily long arithmetic progressions, the complement of $FS(10^n)_{n=1}^{\infty}$ is, in view of Exercise 9, a C^* set which is not IP^*. We shall indicate now still another possibility of making the distinction between C^* and IP^* sets.

Definition 2.17. *For a set $E \subseteq \mathbb{N}$ the upper Banach density, $d^*(E)$, is defined by the formula*

$$d^*(E) = \limsup_{N-M \to \infty} \frac{|E \cap [M, N]|}{N - M + 1}$$

Proposition 2.18. *If $E \subseteq \mathbb{N}$ is a C^* set and $P \subseteq \mathbb{N}$ satisfies $d^*(P) = 0$, then $E \setminus P$ is a C^* set.*

Proof. It is obvious that a set having zero upper Banach density cannot be piecewise syndetic. It follows from Theorem 2.4, (ii) that $d^*(P) = 0$ implies that $\mathbb{N} \setminus P$ is a member of any minimal idempotent and hence is a C^* set. The claim of the lemma follows now from Corollary 2.16.   □

Remark. Proposition 2.18 implies that if E is a C^* set and P is an IP set satisfying $d^*(P) = 0$, then $E \setminus P$ is a C^* set which is not IP^*.

Exercise 14. (i) Show that $2\mathbb{N} + 1$ is a syndetic set which is not C^*.
(ii) Show that every C^* set is syndetic.

It follows from the above exercise that, unlike the syndetic sets, IP^* and C^* sets are not stable under a shift: $2\mathbb{N}$ is a C^* set but $2\mathbb{N} + 1$ is not. This motivates the following definition.

Definition 2.19. *A set E is called C_+^* if for some $k \in \mathbb{Z}$, $E - k$ is a C^* set. Similarly, if for some $k \in \mathbb{Z}$, $E - k$ is IP^* set then E is called IP_+^* set.*

It is an immediate observation that a set E is IP_+^* (C_+^*) if and only if for some $k \in \mathbb{Z}$, $E - k$ is a member of any idempotent (any minimal idempotent) in $(\beta \mathbb{N}, +)$. It follows that any IP_+^* set is C_+^* and that any C_+^* set is syndetic. The following result shows that these inclusions are strict.

Theorem 2.20. (i) *Not every syndetic set is* C_+^**;* (ii) *not every* C_+^* *set is* IP_+^*.

Proof. (i) Let us call any set of the form $S = \bigcup_{i=1}^{\infty}[a_i, b_i]$, where $b_i - a_i \to \infty$, a T-set (T stands for Thick). By Theorem 2.13 and Exercise 11(i), any T-set S is central as well as is the set $2S$. Note now that one can easily construct T-sets S_i, $i \geq 0$, so that the sets $2S_0$, $2S_1 - 1$, $2S_2 + 1$, $2S_3 - 2$, $2S_4 + 2$, ..., $2S_{2n-1} - n$, $2S_{2n} + n$, ... are all disjoint. Let U be the union of these sets. Then $V = \mathbb{N} \setminus U$ is certainly syndetic. At the same time, for any $k \in \mathbb{Z}$, V misses a k-shift of a central set and hence is not C_+^*.

(ii) The proof is similar to that of (i). It is not hard to construct "thin" IP sets $(S_n)_{n=0}^{\infty}$ such that $S_0, S_1 - 1, S_1 + 1, S_2 - 2, S_2 + 2, ..., S_n - n, S_n + n, ...$ are all disjoint and their union U has zero upper Banach density. Then, by Proposition 2.18, $V = \mathbb{N} \setminus U$ is C^* but not IP^*, since for every $k \in \mathbb{Z}$, V misses a k-shift of an IP set. □

3. Convergence along ultrafilters, topological dynamics, and some Diophantine applications

In this section we shall introduce and exploit the notion of convergence along ultrafilters. As we shall see, this notion, especially the convergence along minimal idempotents, allows one to better understand distality, proximality and reccurence in topological dynamical systems. We want to point out that many proofs to be given below are similar to known proofs which utilize the so called Ellis enveloping semigroup (see [E3], [A2]). This is not too surprising since the Ellis semigroup is a particular type of compactification and is in many respects similar to the universal object, the Stone-Čech compactification. On the other hand, the usage of ultrafilters, especially the idempotent ones, has at least two advantages. First, one can, on many occasions, replace in the proofs the convergence along idempotent ultrafilters by the notion of IP convergence which is based on Hindman's theorem, and hereby eliminate the usage of nonmetrisable objects in metrisable dynamics (there are people who care about such things...). See [B2, p.34] for more discussion. Second, the usage of ultrafilters allows one to much more easily deal with combinatorial applications of topological dynamics. For example, we shall show in this section that the two different notions of central discussed in Section 2 are equivalent.

To keep the presentation more accessible and in order to be able to better emphasize the main ideas we shall be dealing in this section mostly with the topological systems of the form (X, T) where X is a compact metric space and T is a (not necessarily invertible) continuous selfmap of X. All the definitions and results below are more or less routinely transferable to actions of general (countably) infinite semigroups of continuous mappings of compact metric spaces. Also, many results in this section can be extended to the case where

X is not metrisable compact Hausdorff space. We leave all these extensions as an exercise to the reader.

Throughout this and the subsequent sections $\beta\mathbb{N}$ stands for $(\beta\mathbb{N}, +)$.

Given an ultrafilter $p \in \beta\mathbb{N}$ one writes p-$\lim_{n \in \mathbb{N}} x_n = y$ if for every neighbourhood U of y one has $\{n : x_n \in U\} \in p$.

Exercise 15. (i) Check that p-$\lim_{n \in \mathbb{N}} x_n$ exists and is unique in any compact Hausdorff space.

(ii) Fix a sequence $(x_n)_{n \in \mathbb{N}}$ in a compact Hausdorff space X. Prove that the map $F : \beta\mathbb{N} \to X$, defined by $F(p) = p$-$\lim_{n \in \mathbb{N}} x_n$ is continuous.

(iii) Prove that if x_1, x_2 are proximal points in a topological system (X, T), then there exists $p \in \beta\mathbb{N}$ such that p-$\lim_{n \in \mathbb{N}} T^n x_1 = p$-$\lim_{n \in \mathbb{N}} T^n x_2$.

Theorem 3.1. *Let X be a compact Hausdorff space and let $p, q \in \beta\mathbb{N}$. Then for any sequence $(x_n)_{n \in \mathbb{N}}$ in X one has*

$$(3.1) \qquad (q + p)\text{-}\lim_{r \in \mathbb{N}} x_r = p\text{-}\lim_{t \in \mathbb{N}} q\text{-}\lim_{s \in \mathbb{N}} x_{s+t}.$$

In particular, if p is an idempotent, and $q = p$, one has

$$p\text{-}\lim_{r \in \mathbb{N}} x_r = p\text{-}\lim_{t \in \mathbb{N}} p\text{-}\lim_{s \in \mathbb{N}} x_{s+t}.$$

Proof. Note that by Exercise 15 both sides of equation (3.1) are well defined. Let $x = (q + p)$-$\lim_{r \in \mathbb{N}} x_r$. Given a neighborhood U of x we have $\{r : x_r \in U\} \in q + p$. Recalling that a set $A \subseteq \mathbb{N}$ is a member of ultrafilter $q + p$ if and only if $\{n \in \mathbb{N} : (A - n) \in q\} \in p$, we get

$$\{t : (\{s : x_s \in U\} - t) \in q\} = \{t : \{s : x_{s+t} \in U\} \in q\} \in p$$

This means that, for p-many t, q-$\lim_{s \in \mathbb{N}} x_{s+t} \in U$ and we are done. $\qquad\square$

Exercise 16. Let R be a minimal right ideal in $\beta\mathbb{N}$. Recall that (R, σ), where $\sigma : p \mapsto p + 1$, is a minimal (non-metrizable) system (see Theorem 2.3 above). Given a topological system (X, T) and a point $x \in X$ consider the mapping $\varphi : R \to X$, defined by $\varphi(p) = p$-$\lim_{n \in \mathbb{N}} T^n x$. Denote by Y the set $\{p\text{-}\lim_{n \in \mathbb{N}} T^n x : p \in R\}$. Prove that (Y, T) is a minimal system by checking that the following diagram is commutative:

$$
\begin{array}{ccc}
R & \xrightarrow{\;\sigma\;} & R \\
\varphi \downarrow & & \downarrow \varphi \\
Y & \xrightarrow{\;T\;} & Y
\end{array}
$$

Proposition 3.2. *Let (X, T) be a topological system and let $x \in X$ be an arbitrary point. Given an idempotent ultrafilter $p \in \beta\mathbb{N}$, let p-$\lim_{n \in \mathbb{N}} T^n x = y$. Then p-$\lim_{n \in \mathbb{N}} T^n y = y$. If x is a distal point (i.e. x is proximal only to itself) then p-$\lim_{n \in \mathbb{N}} T^n x = x$.*

Proof. Applying Theorem 3.1 (and the fact that $p + p = p$), we have

$$p\text{-}\lim_{n \in \mathbb{N}} T^n y = p\text{-}\lim_{n \in \mathbb{N}} T^n \ p\text{-}\lim_{m \in \mathbb{N}} T^m x$$

$$= p\text{-}\lim_{n \in \mathbb{N}} \ p\text{-}\lim_{m \in \mathbb{N}} T^{m+n} x = p\text{-}\lim_{n \in \mathbb{N}} T^n x = y.$$

If x is a distal point, then the relations $p\text{-}\lim_{n \in \mathbb{N}} T^n x = y = p\text{-}\lim_{n \in \mathbb{N}} T^n y$ clearly imply $x = y$ and we are done. $\square$

Exercise 17. Show that if T is a continuous distal selfmap of a compact metric space then T is invertible and T^{-1} is also distal. *Hint.* It follows from Proposition 3.2 that T is *onto*.

Proposition 3.3. *If (X, T) is a minimal system then for any $x \in X$ and any minimal right ideal R in $\beta\mathbb{N}$ there exists a minimal idempotent $p \in R$ such that $p\text{-}\lim T^n x = x$.*

Proof. By Exercise 16, $X = \{p\text{-}\lim_{n \in \mathbb{N}} T^n x, \ p \in R\}$. It follows that the set $\Gamma = \{p \in R : p\text{-}\lim_{n \in R} T^n x = x\}$ is non-empty and closed. We claim that Γ is a semigroup. Indeed, if $p, q \in \Gamma$, one has :

$$(p + q)\text{-}\lim_{n \in \mathbb{N}} T^n x = q\text{-}\lim_{n \in \mathbb{N}} T^n p\text{-}\lim_{m \in \mathbb{N}} T^m x = x.$$

By Ellis theorem Γ contains an idempotent which has to be minimal since it belongs to R. We are done. $\square$

Exircise 18. Let (X, T) be a topological system, R a minimal right ideal in $\beta\mathbb{N}$, and let $x \in X$ be a point in X. Prove that the following are equivalent:
   (i) x is uniformly reccurent;
   (ii) there exists a minimal idempotent $p \in R$ such that $p\text{-}\lim_{n \in \mathbb{N}} T^n x = x$.

It follows from Proposition 3.2 that for any topological system (X, T), any $x \in X$, and any idempotent ultrafilter p, the points x and $y = p\text{-}\lim_{n \in \mathbb{N}} T^n x$ are proximal. (If (X, T) is a distal system then $y = x$). The following theorem gives a partial converse of Proposition 3.3.

Theorem 3.4. *If (X, T) is a topological system and x_1, x_2 are proximal, not necessarily distinct points and if x_2 is uniformly reccurent, then there exists a minimal idempotent $p \in \beta\mathbb{N}$ such that $p\text{-}\lim_{n \in \mathbb{N}} T^n x_1 = x_2$.*

Proof. Let $I = \{p \in \beta\mathbb{N} : p\text{-}\lim_{n \in \mathbb{N}} T^n x_1 = p\text{-}\lim_{n \in \mathbb{N}} T^n x_2\}$. By Exercise 15 (iii), I is a non-empty closed subset of $\beta\mathbb{N}$. One immediately checks that I is a right ideal. Let R be a minimal right ideal in I. Since x_2 is uniformly recurrent, its orbital closure is a minimal system. By Proposition 3.3 there exists a minimal idempotent $p \in R$ such that $p\text{-}\lim T^n x_2 = x_2$. Then $p\text{-}\lim_{n \in \mathbb{N}} T^n x_1 = p\text{-}\lim_{n \in \mathbb{N}} T^n x_2 = x_2$ and we are done. $\square$

One can give a similar proof to the following classical result due to J. Auslander [A1] and Ellis [E2].

Theorem 3.5. *Let (X, T) be a topological system. For any $x \in X$ there exists a uniformly recurrent point y in the orbital closure $\overline{\{T^n x\}}_{n \in \mathbb{N}}$, such that x is proximal to y. Moreover, for any minimal right ideal $R \subset \beta\mathbb{N}$ there exists a minimal idempotent $p \in R$ such that p-$\lim_{n \in \mathbb{N}} T^n x = y$.*

Proof. Let R be a minimal ideal in $\beta\mathbb{N}$ and let p be a (minimal) idempotent in R. Let $y = p$-$\lim_{n \in \mathbb{N}} T^n x$. Clearly, y belongs to the orbital closure of x. By Proposition 3.2, x and y are proximal. By Exercise 18, y is uniformly recurrent. We are done. ☐

We are in position now to establish the equivalence of two notions of central that were discussed in Section 2.

Theorem 3.6. *The following properties of a set $A \subseteq \mathbb{N}$ are equivalent:*

(i) (cf. [F4], Definition 8.3) *There exists a topological system (X, T), and a pair of (not necesserily distinct) points $x, y \in X$ where y is uniformly recurrent and proximal to x, such that for some neighborhood U of y one has:*

$$A = \{n \in \mathbb{N} : T^n x \in U\}$$

(ii) ([BH1], Definition 3.1) *There exists a minimal idempotent $p \in \beta\mathbb{N}$ such that $A \in p$.*

Proof. (i) $\Rightarrow$ (ii) By Theorem 3.5 there exists a minimal idempotent p, such that p-$\lim_{n \in \mathbb{N}} T^n x = y$. This implies that for any neighborhood U of y the set $\{n \in \mathbb{N} : T^n x \in U\}$ belongs to p.

(ii) $\Rightarrow$ (i) The idea of the following proof is due to B. Weiss. Let A be a member of a minimal idempotent $p \in \beta\mathbb{N}$. Let $X = \{0, 1\}$, the space of bilateral 0-1 sequences. Endow X with the standard metric:

$$d(\omega_1, \omega_2) = \min\{\frac{1}{n+1} : \omega_1(i) = \omega_2(i) \text{ for } |i| < n\}$$

It is easy to check that (X, d) is a compact metric space. Let $T : X \to X$ be the shift operator: $T(\omega)(n) = \omega(n+1)$. Then T is a homeomorphism of X and (X, T) is a topological dynamical system. Viewing A as a subset of $\mathbb{Z}$, let $x = 1_A \in X$. Finally, let $y = p$-$\lim_{n \in \mathbb{N}} T^n x$. By Proposition 3.2, x and y are proximal. Also, since p is minimal, y is, by Exercise 18, a uniformly recurrent point. We claim that $y(0) = 1$. Indeed, define $U = \{z \in X : z(0) = y(0)\}$, and note that, since $y = p$-$\lim_{n \in \mathbb{N}} T^n x$ and $A \in p$, one can find $n \in A$ such that $T^n x \in U$. But since $x = 1_A$, $(T^n x)(0) = 1$. But then, given $n \in \mathbb{Z}$, we have: $T^n x \in U \Leftrightarrow (T^n x)(0) = 1 \Leftrightarrow x(n) = 1 \Leftrightarrow n \in A$. It follows that $A = \{n \in \mathbb{Z} : T^n x \in U\}$ and we are done. ☐

Remark. One can show that the characterization of central sets given in Theorem 3.6 extends to general semigroups. See [BH1] and [SY].

Let (X, T) be a topological system. In [F4], a point $x \in X$ is called IP* recurrent (C* recurrent) if for any neighborhood U of x, $\{n \in \mathbb{N} : T^n x \in U\}$ is an IP* set (C* set). The following straightforward exercise establishes the connection between these notions of recurrence and convergence along idempotent ultrafilters.

Exercise 19. Let (X, T) be a topological system and let $x \in X$. Prove:

(i) x is IP* recurrent if and only if for any idempotent $p \in \beta\mathbb{N}$ one has $p\text{-}\lim_{n \in \mathbb{N}} T^n x = x$;

(ii) x is C* recurrent if and only if for any minimal idempotent $p \in \beta\mathbb{N}$ one has $p\text{-}\lim_{n \in \mathbb{N}} T^n x = x$.

We saw in Section 2 that the family of C* sets is wider them that of IP* sets (see the discussion after Corollary 2.16 and Remark after Proposition 2.18). It turns out, however, that, somewhat surprisingly, the notions of IP* recurrence and C* recurrence coincide.

Theorem 3.7. (cf. [F4], Proposition 9.17) *Let (X, T) be a dynamical system. A point $x \in X$ is IP* recurrent if and only if it is C* recurrent.*

Proof. We need to show only that C* recurrence implies IP* recurrence. Let $q \in \beta\mathbb{N}$ be an arbitrary idempotent and let us show that $q\text{-}\lim_{n \in \mathbb{N}} T^n x = x$. By the remark after Theorem 2.7, there exists a minimal idempotent p, such that $p \leq q$. Then $p + q = p$ and we have: $x = p\text{-}\lim_{n \in \mathbb{N}} T^n x = (p + q)\text{-}\lim_{n \in \mathbb{N}} T^n x = q\text{-}\lim_{n \in \mathbb{N}} T^n (p\text{-}\lim_{m \in \mathbb{N}} T^m x) = q\text{-}\lim_{n \in \mathbb{N}} T^n x$. We are done. $\square$

We recall that a point is called distal if it is proximal only to itself.

Theorem 3.8. *Let (X, T) be a dynamical system and $x \in X$. The following are equivalent:*

(i) *x is a distal point;*

(ii) *x is IP* recurrent.*

Proof. (i) $\Rightarrow$ (ii). By Proposition 3.2, for any idempotent p, x and $p\text{-}\lim_{n \in \mathbb{N}} T^n x$ are proximal. Since x is distal, this may happen only if $x = p\text{-}\lim T^n x$. By Exercise 19 (i), this means that x is an IP* recurrent point.

(ii) $\Rightarrow$ (i). If x is not distal, then there exists $y \neq x$, such that x and y are proximal. But then, by Theorem 3.4, there exists an idemponent p such that $p\text{-}\lim T^n x = y$. Since $y \neq x$, this contradicts (ii). (We are using again the characterization of IP* recurrence given in Exercise 19). $\square$

Remark. The property of a point x to be IP* recurrent is much stronger than that of uniform recurrence (which, by Exercise 18, is equivalent to the fact that for *some* minimal idempotent $p\text{-}\lim_{n \in \mathbb{N}} T^n x = x$). While, in a minimal system, every point is uniformly recurrent, there are minimal systems having no distal points. In particular, any minimal topologically weakly mixing system is such. (See [F4, Theorem 9.12]).

We shall conclude this section with some diophantine applications which may be viewed as enhancements of classical results due to Hardy–Littlewood and Weyl. But first we need the following variation on the theme of Theorem 3.8.

Theorem 3.9. *Assume that (X, T) is a minimal system. Then it is distal if and only if for any $x \in X$ and any open set $U \subseteq X$ the set $\{n : T^n x \in U\}$ is IP_+^*.*

Proof. Assume that (X, T) is distal. By minimality, there exists $n_0 \in \mathbb{N}$ such that $T^{n_0} x \in U$. By Theorem 3.8, the set $\{n : T^n(T^{n_0} x) \in U\}$ is IP^* which, of course, implies that the set $\{n : T^n x \in U\}$ is IP_+^*.

Assume now that for any x_1, x_2 and a neighborhood U of x_2 the set $\{n : T^n x_1 \in U\}$ is IP_+^*. We will find it convenient to call an IP_+^* set $A \subseteq \mathbb{N}$ *proper* if A is not IP^* (i.e. A is a nontrivial shift of an IP^* set and, moreover, this shifted IP^* set is not IP^*). If T were not distal, then for some distinct points x_1, x_2 and idempotents p, q one would have: $p\text{-}\lim_{n \in \mathbb{N}} T^n x_1 = x_2$, $q\text{-}\lim_{n \in \mathbb{N}} T^n x_2 = x_1$ and also $p\text{-}\lim_{n \in \mathbb{N}} T^n x_2 = x_2$, $q\text{-}\lim_{n \in \mathbb{N}} T^n x_1 = x_1$ (see Theorem 3.4 and Proposition 3.2). Let U be a small enough neighborhood of x_2. Then, since $p\text{-}\lim_{n \in \mathbb{N}} T^n x_1 = x_2$, the set $S = \{n : T^n x_1 \in U\}$ is a member of p, and hence cannot be a proper IP_+^* set. But, since $q\text{-}\lim_{n \in \mathbb{N}} T^n x_1 = x_1$, the set S cannot be an improper IP_+^* set (that is, an IP^* set) either: if U is small enough, $S \notin q$. So T has to be distal. We are done.   □

We record the following immediate corollary (of the proof) of Theorem 3.9.

Corollary 3.10. *If (X, T) is distal and minimal, and x_1, x_2 are distinct points in X, then if U is a small enough neighbourhood of x_2, the set $\{n : T^n x_1 \in U\}$ is a proper IP_+^* set.*

We move now to diophantine applications. Our starting point is Kronecker's approximation theorem.

Theorem 3.11. *([Kro]) If the numbers $1, \alpha_1, \alpha_2, ..., \alpha_k$ are linearly independent over Q, then for any k subintervals $I_j = (a_j, b_j) \subset [0, 1]$ there exists $n \in \mathbb{N}$ such that one has simultaneously $n\alpha_j \bmod 1 \in I_j$, $j = 1, 2, ..., k$.*

In 1916 H. Weyl ([Weyl]) revolutionized the field of diophantine approximations by introducing the notion of uniform distribution in terms of exponential sums. One of his celebrated results dealing with the values of polynomials mod 1 will be discussed below. As for the Kronecker's theorem, Weyl's approach gives the fact that the set $\Gamma = \{n \in \mathbb{N} : n\alpha_j \bmod 1 \in I_j, j = 1, ..., k\}$

has positive density (which is equal $\prod_{j=1}^{k}(b_j - a_j)$). By considering the uniform Cesàro averages one can actually show that the set Γ is syndetic. Since, as we saw above, the property of a set to be IP_+^* is considerably stronger than syndeticity, the following refinement of Kronecker's theorem is of interest, especially, since it does not seem to follow from considerations involving the Cesàro averages.

Theorem 3.12. *Under the assumptions and notation of Theorem 3.11 the set $\Gamma = \{n \in \mathbb{N} : n\alpha_j \mod 1 \in I_j, \ j = 1, ..., k\}$ is IP_+^*.*

Proof. Let $T_j : \mathbb{T} \to \mathbb{T}$ be defined by $x \mapsto x + \alpha_j \mod 1$, $j = 1, ..., k$. Noticing that the product transformation $T_1 \times T_2 \times \cdots \times T_k$, acting on k-dimensional torus $\mathbb{T}^k$, is distal (this is obvious) and minimal (this follows from Kronecker's theorem, but may be proved in a variety of ways), we see that our result immediately follows from Theorem 3.9. □

We are going to discuss now similar refinements of some other classical results. In what follows the crucial role will be played by minimal distal systems. It was H. Furstenberg who introduced and applied the idea of using the unique ergodicity of a class of affine transformations of the torus to obtain a dynamical proof of Weyl's theorem on equidistribution of polynomials (see [F1] and [F4, p.69]). As we shall see, affine transformations of the kind treated by Furstenberg can also be utilized to obtain polynomial results in the spirit of Theorem 3.12.

The following extension of Kronecker's theorem was obtained by Hardy and Littlewood in [HaL].

Theorem 3.13. *If the numbers $1, \alpha_1, ..., \alpha_k$ are linearly independent over $\mathbb{Q}$, then for any $d \in \mathbb{N}$ and any kd intervals $I_{lj} \subset [0,1]$, $l = 1, ..., d; j = 1, ..., k$ the set*

$$\Gamma_{dk} = \{n \in \mathbb{N} : n^l \alpha_j \mod 1 \in I_{lj}, \ l = 1, ..., d; \ j = 1, ..., k\}$$

is infinite.

As with Kronecker's theorem, Weyl was able to show in [Weyl] that the set Γ_{dk} has positive density equal to the product of lengths of I_{lj}. In 1953 P. Szüsz [Sz] proved that the set Γ_{dk} is syndetic. The following theorem shows that Γ_{dk} is actually IP_+^*.

Theorem 3.14. *Under the assumptions and notation of Theorem 3.13, the set Γ_{dk} is IP_+^*.*

Proof. To make the formulas more transparent we shall put $d = 3$. It will be clear that the same proof gives the general case.

We start with easily checkable claim that if $T_\alpha : \mathbb{T}^3 \to \mathbb{T}^3$ is defined by $T_\alpha(x, y, z) = (x + \alpha, y + 2x + \alpha, z + 3x + 3y + \alpha)$ then $T_\alpha^n(0, 0, 0) = (n\alpha, n^2\alpha, n^3\alpha)$. This transformation T is distal (easy) and minimal. The last assertion can actually be derived from the case $k = 1$ of Hardy - Littlewood theorem above, but also can be proved directly. (For example, this fact is a special case of Lemma 1.25, p.36 in [F4]). Our next claim is that if the numbers $1, \alpha_1, \alpha_2, ..., \alpha_k$ are linearly independent over Q, then the product map $T = T_{\alpha_1} \times \cdots \times T_{\alpha_k}$ (acting on $\mathbb{T}^{3k}$) is distal and minimal as well. (The distality is obvious, the minimality follows, again, from an appropriately modified Lemma 1.25 in [F4]). By minimality of T, the orbit of zero in $\mathbb{T}^{3k}$ is dense which, together with Theorem 3.9, gives the desired result. □

Remark. In case the intervals I_{lj} contain 0, it can be shown that Γ_{dk} is IP*. This special case is also treated in [FW2].

Exercise 20. (i) Derive from Theorem 3.14 the following fact: for any irrational numbers $\alpha_1, ..., \alpha_k$ and any subintervals $I_j \subset [0, 1]$, $j = 1, ..., k$, the set

$$\{n \in \mathbb{N} : n^j \alpha_j \bmod 1 \in I_j, \ j = 1, ..., k\}$$

is IP$^*_+$.

(ii) Use (i) to obtain the following fact: for any real polynomial $p(t)$ having at least one coefficient other than the constant term irrational and for any subinterval $I \subset [0, 1]$ the set $\{n \in \mathbb{N} : p(n) \bmod 1 \in I\}$ is IP$^*_+$.

Remark. Another possibility of proving the statement (ii) is to apply Theorem 3.9 to the transformation $T : (x_1, x_2, ..., x_k) \mapsto (x_1 + \alpha, x_2 + x_1, ..., x_k + x_{k-1})$ which is used in [F4] to derive Weyl's theorem on uniform distribution.

We conclude this section by formulating a general result which may be proved by refining the techniques used above.

Theorem 3.15. *If real polynomials* $p_1(t), p_2(t), ..., p_k(t)$ *have the property that for any non-zero vector* $(h_1, h_2, ..., h_k) \in \mathbb{Z}^k$ *the linear combination* $\sum_{i=1}^{k} h_i p_i(t)$ *is a polynomial with at least one irrational coefficient other than the constant term then for any k subintervals* $I_j \subset [0, 1]$, $j = 1, ..., k$, *the set*

$$\{n \in \mathbb{N} : p_j(n) \bmod 1 \in I_j, \ j = 1, ..., k\}$$

is IP$^*_+$.

4. MINIMAL IDEMPOTENTS AND WEAK MIXING

In this chapter we shall connect convergence along minimal idempotents with the theory of weakly mixing unitary actions and weakly mixing measure preserving systems. While the customary definitions of weakly mixing $\mathbb{Z}$-actions can be more or less routinely extended to actions of abelian or even amenable (semi)groups (cf. [D]), the study of the weakly mixing actions of non-amenable groups necessitates introduction of new tools and ideas.

Before moving to a more advanced discussion, we want to illustrate the multifariousness of the notion of weak mixing by listing some equivalent conditions for a system to be weakly mixing. (In most books either (i) or (ii) below is taken as "official" definition of weak mixing).

Theorem 4.1. *Let T be an invertible measure-preserving transformation of a probability measure space $(X, \mathcal{B}, \mu)$. Let U_T denote the operator defined on measurable functions by $(U_T f)(x) = f(Tx)$. The following conditions are equivalent:*

(i) For any $A, B \in \mathcal{B}$

$$\lim_{N \to \infty} \frac{1}{N} \sum_{n=0}^{N-1} |\mu(A \cap T^{-n}B) - \mu(A)\mu(B)| = 0;$$

(ii) For any $A, B \in \mathcal{B}$ there is a set $P \subset \mathbb{N}$ of density zero such that

$$\lim_{n \to \infty, \, n \notin P} \mu(A \cap T^{-n}B) = \mu(A)\mu(B);$$

(iii) $T \times T$ is ergodic on the Cartesian square of $(X, \mathcal{B}, \mu)$;

(iv) For any ergodic probability measure preserving system $(Y, \mathcal{D}, \nu, S)$ the transformation $T \times S$ is ergodic on $X \times Y$;

(v) If f is a measurable function such that for some $\lambda \in \mathbb{C}$, $U_T f = \lambda f$ a.e., then $f = \text{const}$ a.e.;

(vi) For $f \in L^2(X, \mathcal{B}, \mu)$ with $\int f = 0$ consider the representation of the positive definite sequence $\langle U_T^n f, f \rangle, n \in \mathbb{Z}$, as a Fourier transform of a measure ν on $\mathbb{T}$:

$$\langle U_T^n f, f \rangle = \int_{\mathbb{T}} e^{2\pi i n x} d\nu, \quad n \in \mathbb{Z}$$

(this representation is guaranteed by Herglotz theorem, see [He]). Then ν has no atoms.

(vii) If for some $f \in L^2(X, \mathcal{B}, \mu)$ the set $\{U_T^n f, \ n \in \mathbb{Z}\}$ is totally bounded, then f is a constant a.e.;

(viii) Weakly independent sets are dense in $\mathcal{B}$. (A set $A \in \mathcal{B}$ is called weakly independent if there exists a sequence $n_1 < n_2 < \cdots$ such that the sets $T^{-n_i}A, \ i = 1, 2, ...,$ are mutually independent);

(ix) For any $A \in \mathcal{B}$ and $k \in \mathbb{N}, \ k \geq 2$, one has

$$\lim_{N \to \infty} \frac{1}{N} \sum_{n=0}^{N-1} \mu(A \cap T^{-n}A \cap T^{-2n}A \cap \cdots \cap T^{-kn}A) = (\mu(A))^{k+1}$$

(x) For any $k \in \mathbb{N}, \ k \geq 2$, any $f_1, f_2, ..., f_k \in L^\infty(X, \mathcal{B}, \mu)$ and any non-constant polynomials $p_1(n), p_2(n), ..., p_k(n) \in \mathbb{Z}[n]$ such that for all $i \neq j$, $\deg(p_i - p_j) > 0$, one has

$$\lim_{N \to \infty} \frac{1}{N} \sum_{n=0}^{N-1} f_1(T^{p_1(n)}x) f_2(T^{p_2(n)}x) \cdots f_k(T^{p_k(n)}x) = \int f_1 d\mu_1 \int f_2 d\mu_2 \cdots \int f_k d\mu_k$$

in L^2-norm.

Remark. Weakly mixing systems (for measure preserving $\mathbb{R}$-actions) were introduced under the name *dynamical systems of continuous spectra* in [KN]. See also [Hopf1] and [Hopf2]. These papers, as well as Hopf's book [Hopf3], already contain (versions of) the conditions (i) through (vii). Condition (viii) is due to Krengel (see [Kre] for this and other related results). The last two conditions, while being of interest in their own right, are strongly connected with combinatorial and number-theoretical applications. In particular, (ix) plays a crucial role in Furstenberg's ergodic proof of Szemerédi's theorem on arithmetic progressions (see [F3] and [F4]). Similarly, variations on condition (x) (see [B1]) are needed for proofs of polynomial extensions of Szemerédi's theorem (see, for example, [BL], [BM1], [BM2], [L]).

We are moving now to the discussion of weak mixing for general group actions. For sake of simplicity we shall be dealing with countably infinite (but not necessarily amenable) groups. We remark however that the results below can be extended to actions of general locally compact (semi)groups.

One possible approach to weak mixing for general groups is via the theory of invariant means. For example, if the acting group is amenable, one can replace the condition (i) in Theorem 4.1 by the assertion that the averages of the expressions $|\mu(A \cap T_g B) - \mu(A)\mu(B)|$ taken along any Følner sequence converge to zero. If the acting group G is noncommutative, one has, in addition, to replace conditions (v) and (vi) by the assertion that the only finite-dimensional subrepresentation of $(U_g)_{g \in G}$ (where U_g if defined by $(U_g f)(x) = f(T_g x)$) on $L^2(X, \mathcal{B}, \mu)$ is its restriction to the subspace of constant functions.

H. Dye has shown in [D] that under these modifications the conditions (i), (iii) and (v) in Theorem 4.1 are equivalent for measure-preserving actions of general amenable semigroups. By using the Ryll-Nardzewski theorem ([R-N]) which guarantees the existence of unique invariant mean on the space of weakly almost periodic functions on a group, one can extend the notion of weak mixing to actions of general locally compact groups. See [BR] for the details.

We are going now to indicate how to recover and refine some of the main results in [BR] by using minimal idempotents in βG.

We shall deal first with unitary representations of a group G on a Hilbert space $\mathcal{H}$ and will specialize to the case of measure preserving actions later. Since we work with countable groups only, we may and will always assume that the Hilbert spaces we work with are separable. To appreciate the terminology one should think of Hilbert space $\mathcal{H}$ as $L_0^2(X, \mathcal{B}, \mu) = \{f \in L^2(X, \mathcal{B}, \mu) : \int f d\mu = 0\}$.

Definition 4.2. *Let $(U_g)_{g \in G}$ be a unitary representation of a group G on a separable Hilbert space $\mathcal{H}$.*

(i) *A vector $\varphi \in \mathcal{H}$ is called* compact *if the set $\{U_g\varphi : g \in G\}$ is totally bounded in $\mathcal{H}$.*

(ii) *The representation $(U_g)_{g \in G}$ is called* weakly mixing *if there are no nonzero compact vectors.*

In the following theorem we are going to use expressions of the form $p\text{-}\lim_{g \in G} U_g\varphi$. Since the unit ball in $\mathcal{H}$ is a compact metrizable space with respect to the weak topology, these expressions have a well defined meaning. We remark also that if $p\text{-}\lim_{g \in G} U_g\varphi = \varphi$ weakly (which will be the case when φ is a compact vector and p a minimal idempotent), then one actually has $p\text{-}\lim_{g \in G} U_g\varphi = \varphi$ strongly. (The verification of this easy fact is left to the reader).

Theorem 4.3. *Let $(U_g)_{g \in G}$ be a unitary representation of a group G on a Hilbert space $\mathcal{H}$. For any $\varphi \in \mathcal{H}$ the following conditions are equivalent:*

(i) *There exists a minimal idempotent $p \in \beta G$ such that $p\text{-}\lim_{g \in G} U_g\varphi = \varphi$;*

(ii) *φ is a compact vector;*

(iii) *For any idempotent $p \in \beta G$ one has $p\text{-}\lim_{g \in G} U_g\varphi = \varphi$.*

Proof. (i) $\Rightarrow$ (ii). Let $\varepsilon > 0$ and consider the set

$$\Gamma = \{g \in G : \|U_g\varphi - \varphi\| < \varepsilon/2\} \in p.$$

It follows that for any $g_1, g_2 \in \Gamma$ one has

$$\|U_{g_1}\varphi - U_{g_2}\varphi\| < \varepsilon.$$

This implies that for any $g \in \Gamma^{-1}\Gamma$ one has

$$\|U_g\varphi - \varphi\| < \varepsilon.$$

By Exercise 7 (iii), the set $\Gamma^{-1}\Gamma$ is syndetic. This means that a union of finitely many shifts of $\Gamma^{-1}\Gamma$ gives G and, hence, a union of finitely many balls of radius ε cover the set $\{U_g\varphi\}_{g \in G}$, which, in its turn, means that φ is a compact vector.

(ii) $\Rightarrow$ (iii) If φ is a compact vector, the group G acts on compact space $X = \overline{\{U_g\varphi\}}_{g \in G}$ by isometries $U_g, g \in G$, and hence distally. By utilizing the same argument as in the proof of Proposition 3.2 we see that for any idempotent $p \in \beta G$ one has to have $p\text{-}\lim_{g \in G} U_g\varphi = \varphi$. Since the implication (iii) $\Rightarrow$ (i) is trivial, this concludes the proof of Theorem 4.3. □

Theorem 4.4. *Let $(U_g)_{g \in G}$ be a unitary representation of a group G on a Hilbert space $\mathcal{H}$. The following conditions are equivalent:*

(i) *$(U_g)_{g \in G}$ is weakly mixing;*

(ii) *For any minimal idempotent $p \in \beta G$ and any $\varphi \in \mathcal{H}$ one has $p\text{-}\lim_{g \in G} U_g\varphi = 0$;*

(iii) *For any $\varepsilon > 0$ and $\varphi_1, \varphi_2 \in \mathcal{H}$ the set $\{g : |\langle U_g\varphi_1, \varphi_2\rangle| < \varepsilon\}$ is C^*;*

(iv) *There exists a minimal idempotent $p \in \beta G$ such that for any $\varphi \in \mathcal{H}$ one has $p\text{-}\lim_{g \in G} U_g\varphi = 0$.*

Proof. (i) $\Rightarrow$ (ii) If for some minimal $p \in \beta G$ and $\varphi \in \mathcal{H}$ one has $p\text{-}\lim_{g \in G} U_g \varphi = \psi \neq 0$, then (by Proposition 3.2) one has $p\text{-}\lim_{g \in G} U_g \psi = \psi$. It follows now from Theorem 4.3 that ψ is a nontrivial compact vector, which contradicts (i).

(ii) $\Rightarrow$ (iii) The equivalence of these two statements follows immediately from the fact that a set $E \subseteq G$ is C* if and only if E is a member of every minimal idempotent $p \in \beta G$ (see Remark after Corollary 2.16).

(iii) $\Rightarrow$ (iv) Trivial.

(iv) $\Rightarrow$ (i) If $(U_g)_{g \in G}$ is not weakly mixing, then there exists a non-zero compact vector $\varphi \in \mathcal{H}$. But then, by Theorem 4.3, one has $p\text{-}\lim_{g \in G} U_g \varphi = \varphi$, which contradicts (iv). We are done. $\qquad\square$

Exercise 21. Show that each of the following properties of a unitary representation $(U_g)_{g \in G}$ on a Hilbert space $\mathcal{H}$ is equivalent to weak mixing.

(i) For any $\varepsilon > 0$, $\varphi_1, \varphi_2 \in \mathcal{H}$ the set $E = \{g : |\langle U_g \varphi_1, \varphi_2 \rangle| < \varepsilon\}$ has the following property: for any $n \in \mathbb{N}$ and any $g_1, g_2, ..., g_n \in G$, $\bigcap_{i=1}^{n} g_i E$ is syndetic;

(ii) For any $\varepsilon > 0, n \in \mathbb{N}$, and $\varphi_1, ..., \varphi_n \in \mathcal{H}$, there exists $g \in G$ such that $|\langle U_g \varphi_i, \varphi_i \rangle| < \varepsilon$ for $i = 1, ..., n$;

(iii) For any unitary representation $(V_g)_{g \in G}$ of G on a Hilbert space $\mathcal{K}$, the tensor product representation $(U_g \otimes V_g)_{g \in G}$ on $\mathcal{H} \otimes \mathcal{K}$ is weakly mixing.

Theorem 4.5. *Given a unitary representation $(U_g)_{g \in G}$ of a group G on a Hilbert space $\mathcal{H}$, let*

$$\mathcal{H}_c = \{f \in \mathcal{H} : f \text{ is compact with respect to } (U_g)_{g \in G}\}.$$

Then the restriction of $(U_g)_{g \in G}$ to the invariant space $\mathcal{H}_{wm} = \mathcal{H}_c^{\perp}$ is weakly mixing.

Proof. Let $\varphi \in \mathcal{H}$, $\varphi \perp \mathcal{H}_c$, and let $p \in \beta G$ be a minimal idempotent. Since $\mathcal{H}_c$ is an invariant subspace, the vector $\psi = p\text{-}\lim_{g \in G} U_g \varphi$ is in $\mathcal{H}_{wm}$. If $\psi \neq 0$, then since $\psi = p\text{-}\lim_{g \in G} U_g \psi$, ψ is a non-zero compact vector, which contradicts the fact that $\psi \in \mathcal{H}_c^{\perp}$. We are done. $\qquad\square$

Corollary 4.6. *Let $p \in \beta G$ be a minimal idempotent and for $\varphi \in \mathcal{H}$, let $P\varphi = p\text{-}\lim_{g \in G} U_g \varphi$. Then P is an orthogonal projection onto $\mathcal{H}_c$.*

We shall specialize now to the case of measure-preserving actions. Let us call a measure-preserving action $(T_g)_{g \in G}$ on a probability space $(X, \mathcal{B}, \mu)$ weakly mixing if the unitary action of G defined on $L^2(X, \mathcal{B}, \mu)$ by $(U_g f)(x) = f(T_g x)$ is weakly mixing on the space $L_0^2(X, \mathcal{B}, \mu) = \{f \in L^2(X, \mathcal{B}, \mu) : \int f d\mu = 0\}$.

The following result is an immediate corollary of Theorem 4.4(i) and is of interest even for $\mathbb{Z}$-actions.

Theorem 4.7. *Let* $(T_g)_{g \in G}$ *be a weakly mixing measure preserving action on a probability space* $(X, \mathcal{B}, \mu)$. *Then for any* $A, B \in \mathcal{B}$ *and any* $\varepsilon > 0$, *the set*

$$\{g \in G : |\mu(A \cap T_g B) - \mu(A)\mu(B)| < \varepsilon\}$$

is a C^*-*set.*

The following theorem should be juxtaposed with the polynomial result formulated above as condition (x) in Theorem 4.1.

Theorem 4.8. *Assume that* $(X, \mathcal{B}, \mu, T)$ *is a weakly mixing system. For any* $k \in \mathbb{N}$, $f_0, f_1, f_2, ..., f_k \in L^\infty(X, \mathcal{B}, \mu)$, *any non-constant polynomials* $p_1(n), p_2(n), ..., p_k(n) \in \mathbb{Z}[n]$ *such that for all* $i \neq j$, $\deg(p_i - p_j) > 0$ *and for any minimal idempotent* $p \in \beta\mathbb{N}$, *one has*

$$p\text{-}\lim_{n \in \mathbb{N}} \int f_0(x) f_1(T^{p_1(n)} x) \cdots f_k(T^{p_k(n)} x) d\mu = \int f_0 d\mu \int f_1 d\mu \cdots \int f_k d\mu.$$

The proof of Theorem 4.8 is too long to give here. However, it should be noted, that the proof has essentially the same structure as the proof of the parallel Cesàro version given in [B1]. The main and, practically, the only distinction with the proof in [B1] is the need to replace the Cesàro van der Corput trick utilized in [B1] by the following useful fact the proof of which is left to the reader.

Proposition 4.9. *Assume that* $(x_n)_{n \in \mathbb{N}}$ *is a bounded sequence in a Hilbert space* $\mathcal{H}$. *Let* $p \in \beta\mathbb{N}$ *be an idempotent. If* $p\text{-}\lim_{h \in \mathbb{N}} p\text{-}\lim_{n \in \mathbb{N}} \langle x_{n+h}, x_n \rangle = 0$ *then* $p\text{-}\lim_{n \in \mathbb{N}} x_n = 0$ *weakly.*

The following proposition is an immediate corollary of Theorem 4.8.

Theorem 4.10. *If* $(X, \mathcal{B}, \mu, T)$ *is weakly mixing system, then for any* $k \in \mathbb{N}$, *any sets* $A_0, A_1, ..., A_k \in \mathcal{B}$, *any non-constant polynomials* $p_1(n), p_2(n), ..., p_k(n)$ $\in \mathbb{Z}[n]$, *such that for all* $i \neq j$, $\deg(p_i - p_j) > 0$, *and any* $\varepsilon > 0$, *the set*

$$\{n : |\mu(A_0 \cap T^{p_1(n)} A_1 \cap ... \cap T^{p_k(n)} A_k) - \mu(A_0)\mu(A_1)...\mu(A_k)| < \varepsilon\}$$

is a C^* *set.*

We shall conclude this section by proving a refinement of the so called Khintchine's recurrence theorem under the assumption that our system is ergodic. Khintchine's theorem (proved in [Kh] for measure preserving $\mathbb{R}$-actions) states that for any measure preserving system $(X, \mathcal{B}, \mu, T)$, any $A \in \mathcal{B}$, and any $\varepsilon > 0$, the set $E = \{n : \mu(A \cap T^{-n} A) > (\mu(A))^2 - \varepsilon\}$ is syndetic. As a matter of fact, it is quite easy to show that the set E is always an IP* set and, moreover, this result holds for general semigroup actions. (See [B2], section 5 for details). One would like to extend Khintchine's theorem to sets of the form $A \cap T^{-n} B$. It is clear, however, that any such result can hold only under the additional assumption of ergodicity. We have the following theorem which, for simplicity, will be formulated and proved for $\mathbb{Z}$-actions.

Theorem 4.11. *Assume that $(X, \mathcal{B}, \mu, T)$ is an ergodic, invertible, probability measure preserving system. Then for any $\varepsilon > 0$ and any $A, B \in \mathcal{B}$ the set*

$$E = \{n \in \mathbb{Z} : \mu(A \cap T^n B) > \mu(A)\mu(B) - \varepsilon\}$$

is a C_+^* set.

Proof. We are going to utilize the splitting $L^2(X, \mathcal{B}, \mu) = \mathcal{H}_c \oplus \mathcal{H}_{wm}$ (cf. Theorem 4.5 above). Let $1_A = f = f_1 + f_2$, $1_B = g = g_1 + g_2$, where f_1, g_1 belong to the space of compact vectors $\mathcal{H}_c$ and $f_2, g_2 \in \mathcal{H}_c^\perp = \mathcal{H}_{wm}$. Note that since the constants belong to the space $\mathcal{H}_c$, one has $\int f_1 d\mu = \mu(A)$, $\int g_1 d\mu = \mu(B)$. By ergodicity of T one has

$$\frac{1}{N} \sum_{n=0}^{N-1} \int f_1(T^n x) g_1(x) d\mu \to \int f_1 d\mu \int g_1 d\mu = \mu(A)\mu(B)$$

as $N \to \infty$. It follows that for any $\varepsilon > 0$ one can find n_0 such that

$$\int f_1(T^{n_0} x) g_1(x) d\mu > \mu(A)\mu(B) - \varepsilon.$$

Let $p \in \beta\mathbb{N}$ be a minimal idempotent. Using the fact that $\mathcal{H}_c$ and $\mathcal{H}_{wm}$ are orthogonal spaces, invariant with respect to the unitary operator U defined by $(Uf)(x) = f(Tx)$, we have:

$$p\text{-}\lim_{n \in \mathbb{N}} \mu(T^{n_0} A \cap T^n B) = p\text{-}\lim_{n \in \mathbb{N}} \int f(T^{n_0} x) g(T^n x) d\mu$$

$$= p\text{-}\lim_{n \in \mathbb{N}} \int f_1(T^{n_0} x) g_1(T^n x) d\mu + p\text{-}\lim_{n \in \mathbb{N}} \int f_2(T^{n_0} x) g_2(T^n x) d\mu$$

$$= \int f_1(T^{n_0} x) g_1(x) d\mu > \mu(A)\mu(B) - \varepsilon.$$

(We also used the fact that $p\text{-}\lim_{n \in \mathbb{N}} g_1(T^n x) = g_1(x)$.) It follows that

$$F = \{n : \mu(T^{n_0} A \cap T^n B) > \mu(A)\mu(B) - \varepsilon\} \in p.$$

Since p was an arbitrary minimal idempotent, E is a C^* set and hence the set $\{n : \mu(A \cap T^n B) > \mu(A)\mu(B) - \varepsilon\}$ is C_+^*. We are done. $\square$

5. SOME CONCLUDING REMARKS

We started this essay with a question: "What is common between the invertibility of distal maps, partition regularity of the diophantine equation $x - y = z^2$, and the notion of mild mixing"? It was claimed in the Introduction that the answer is: idempotent ultrafilters; the purpose of this short concluding section is to convince the reader that this is indeed so.

As for the relevance of idempotents to the invertibility of distal maps, it is apparent from Proposition 3.2 and Exercise 17 above. So it remains to explain the connection of idempotent ultrafilters to the notion of mild mixing and to partition regularity of the equation $x - y = z^2$.

The notion of mild mixing, which lies between weak and strong mixing, was introduced by Walters in 1972 ([Wa1]) and rediscovered by Furstenberg and Weiss in 1978 ([FW1]). Not unlike the weak mixing, mild mixing admits many equivalent, yet diverse, formulations and plays a crucial role in ergodic proofs of some strong combinatorial results. (See, for instance, [FK2] and [BM2].) The following proposition, which we do not prove here, lists some interesting equivalent conditions, each of which may be taken for a definition of mild mixing. For sake of simplicity, we restrict ourself to measure preserving $\mathbb{Z}$-actions, but it should be mentioned that the notion of mild mixing makes perfect sense for unitary actions of locally compact groups and is of importance in the study of non-singular (i.e. not necessarily measure-preserving) actions as well. (See, for example, [ScW], [Sc] and [Wa2] for more information.) The connection of mild mixing to idempotent ultrafilters is perfectly clear from the items (iv) and (v) below.

Theorem 5.1. *Let T be an invertible measure-preserving transformation of a probability measure space $(X, \mathcal{B}, \mu)$. Let U_T denote the operator defined on measurable functions by $(U_T)f(x) = f(Tx)$. The following conditions are equivalent:*

(i) *For every $A \in \mathcal{B}$ with $0 < \mu(A) < 1$, one has*

$$\liminf_{|n| \to \infty} \mu(A \triangle T^n A) > 0.$$

(ii) *For any ergodic measure preserving system $(Y, \mathcal{D}, \nu, S)$, the transformation $T \times S$ is ergodic on $X \times Y$. (Note: it is not assumed that $\nu(Y)$ is finite.)*

(iii) *There are no non-constant rigid functions in $L^2(X, \mathcal{B}, \mu)$. (A function $f \in L^2(X, \mathcal{B}, \mu)$ is called rigid, if for some sequence $(n_k)_{k \in \mathbb{N}}$ one has $U_T^{n_k} f \to f$ in L^2.)*

(iv) *For any idempotent $p \in \beta\mathbb{N}$ and any $f \in L^2(X, \mathcal{B}, \mu)$ one has*

$$p\text{-}\lim_{n \in \mathbb{N}} U_T^n f = \int f \, d\mu \ \text{ (weakly)}.$$

(v) *For any $k \in \mathbb{N}$, any $f_0, f_1, \ldots, f_k \in L^\infty(X, \mathcal{B}, \mu)$ and any non-constant polynomials $p_1(n), p_2(n), \ldots, p_k(n) \in \mathbb{Z}[n]$, such that for all $i \neq j$, $\deg(p_i - p_j) > 0$, and any idempotent $p \in \beta\mathbb{N}$, one has*

$$p\text{-}\lim_{n \in \mathbb{N}} \int f_0(x) f_1(T^{p_1(n)}x) \ldots f_k(T^{p_k(n)}x) d\mu = \int f_0 \, d\mu \int f_1 \, d\mu \ldots \int f_k \, d\mu.$$

Finally, we shall explain briefly how one proves that for any finite partition of $\mathbb{N}$ there exist x, y, z in the same cell of partition, such that $x - y = z^2$. The reader will find the missing details and more discussion in [B2] and [B3]. The proof hinges in the following fact.

Proposition 5.2. *(Cf. [B2], Propositions 3.11 and 3.12; see also [BFM])*
For any probability measure preserving system $(X, \mathcal{B}, \mu, T)$, *any* $A \in \mathcal{B}$ *and*
any idempotent $p \in \beta \mathbb{N}$ *one has* $p\text{-}\lim_{n \in \mathbb{N}} \mu(A \cap T^{n^2} A) \geq \mu^2(A)$.

It follows (via Furstenberg's correspondence principle, see, for example, [F4], p.77, or [B3], Theorems 6.4.4 and 6.4.7), that for any set of positive upper density $E \subseteq \mathbb{N}$ and any IP set Γ, there exists $z \in \Gamma$ with $d^*(E \cap (E - z^2)) > 0$. This, in its turn, immediately implies that, for some $x, y \in E$ and $z \in \Gamma$, one has $x - y = z^2$.

Consider now an arbitrary finite partition $\mathbb{N} = \bigcup_{i=1}^{r} C_i$. Reindexing if necessary, we may assume that those C_i which have positive upper Banach density have indices $1, 2, \ldots, s$, where $s \leq r$. Let $U = \bigcup_{i=1}^{s} C_i$. It is not hard to see that U contains an IP set (this follows from the almost obvious fact that U has to contain arbitrarily long blocks of consecutive integers). It follows now from Hindman's theorem that there exist i_0, $1 \leq i_0 \leq s$, and an IP set Γ such that $\Gamma \subseteq C_{i_0}$. It follows now from the remarks above that for some $x, y \in C_{i_0}$ and $z \in \Gamma$ one has $x - y = z^2$. We are done.

Acknowledgment. I wish to thank Hillel Furstenberg, Neil Hindman, Sasha Leibman and Christian Schnell for their help in preparing these notes. I also would like to thank Sergey Bezuglyi and Sergiy Kolyada for organizing, under the aegis of INTAS, CRDF and ESF, the Katsiveli-2000 meeting and for their efforts to promote the theory of dynamical systems.

REFERENCES

[A1] J. Auslander, On the proximal relation in topological dynamics, *Proc. Amer. Math. Soc.*, **11** (1960), 890–895.

[A2] J. Auslander, *Minimal Flows and their Extensions*, North-Holland Mathematics Studies, 153, North-Holland Publishing Co., Amsterdam, 1988. xii+265 pp.

[B1] V. Bergelson, Weakly mixing PET, *Ergodic Theory and Dynamical Systems*, **7** (1987), 337-349.

[B2] V. Bergelson, Ergodic Ramsey theory–an update. *Ergodic theory of* $\mathbb{Z}^d$ *actions* (Warwick, 1993–1994), 1–61, London Math. Soc. Lecture Note Ser., 228, Cambridge Univ. Press, Cambridge, 1996.

[B3] V. Bergelson, Ergodic theory and Diophantine problems. *Topics in symbolic dynamics and applications*, 167–205, London Math. Soc. Lecture Note Ser., 279, Cambridge Univ. Press, Cambridge, 2000.

[BFM] V. Bergelson, H. Furstenberg and R. McCutcheon, IP-sets and polynomial recurrence, *Ergodic Theory Dynam. Systems*, **16** (1996), no. 5, 963–974.

[BFHK] V. Bergelson, H. Furstenberg, N. Hindman, and Y. Katznelson, An algebraic proof of van der Waerden's theorem, *Enseign. Math. (2)*, **35** (1989), no. 3-4, 209–215.

[BH1] V. Bergelson and N. Hindman, Nonmetrizable topological dynamics and Ramsey theory, *Trans. AMS*, **320** (1990), 293-320.

[BH2] V. Bergelson and N. Hindman, Additive and multiplicative Ramsey theorems in $\mathbb{N}$ – some elementary results, *Combin. Probab. Comput.*, **2** (1993), no. 3, 221–241.

[BH3] V. Bergelson and N. Hindman, On IP* sets and central sets, *Combinatorica*, **14** (1994), no. 3, 269–277.

[BL] V. Bergelson and A. Leibman, Polynomial extensions of van der Waerden's and Szemerédi's theorems, *Journal of AMS*, **9** (1996), 725 –753.

[BM1] V. Bergelson and R. McCutcheon, Uniformity in the polynomial Szemerédi theorem.
Ergodic theory of $\mathbb{Z}^d$ actions (Warwick, 1993–1994), 273–296, London Math. Soc.
Lecture Note Ser., 228, Cambridge Univ. Press, Cambridge, 1996.

[BM2] V. Bergelson and R. McCutcheon, An ergodic IP polynomial Szemerédi theorem,
Mem. Amer. Math. Soc., **146** (2000), no. 695, viii+106 pp.

[BR] V. Bergelson and J. Rosenblatt, Mixing actions of groups, *Illinois J. Math.*, **32** (1988),
no. 1, 65–80.

[D] H. Dye, On the ergodic mixing theorem, *Trans. AMS*, **118** (1965), 123– 30.

[E1] R. Ellis, Distal transformation groups, *Pacific J. Math.*, **8** (1958), 401–405.

[E2] R. Ellis, A semigroup associated with a transformation group, *Trans. AMS*, **94** (1960),
272–281.

[E3] R. Ellis, *Lectures on Topological Dynamics*, Benjamin, New York, 1969.

[F1] H. Furstenberg, Strict ergodicity and transformations of the torus, *Amer. J. Math.*,
83 (1961), 573–601.

[F2] H. Furstenberg, The structure of distal flows, *Amer. J. Math.*, **85** (1963), 477–515.

[F3] H. Furstenberg, Ergodic behavior of diagonal measures and a theorem of Szemerédi
on arithmetic progressions, *J. d'Analyse Math.*, **31** (1977), 204-256.

[F4] H. Furstenberg, *Recurrence in Ergodic Theory and Combinatorial Number Theory*,
Princeton University Press, 1981.

[FK1] H. Furstenberg and Y. Katznelson, An ergodic Szeméredi theorem for IP-systems
and combinatorial theory, *J. Analyse Math.* **45** (1985), 117–168.

[FK2] H. Furstenberg and Y. Katznelson, Idempotents in compact semigroups and Ramsey
theory, *Israel J. Math.*, **68** (1990), 257-270.

[FW1] H. Furstenberg and B. Weiss, The finite multipliers of infinite ergodic transforma-
tions. The structure of attractors in dynamical systems (Proc. Conf., North Dakota
State Univ., Fargo, N.D., 1977), 127–132, *Lecture Notes in Math.*, **668**, Springer,
Berlin, 1978.

[FW2] H. Furstenberg and B. Weiss, Simultaneous Diophantine approximation and IP-sets.
Acta Arith., **49** (1988), 413–426.

[GH] W. Gottschalk and G. Hedlund, *Topological Dynamics*, AMS Coll. Publ., vol. 36,
Providence, 1955.

[HaL] G. Hardy and J. Littlewood, Some problems of diophantine approximation, *Acta
Mathematica*, **37** (1914), 155–191.

[He] H. Helson, *Harmonic Analysis*, Addison-Wesley, Reading, MA, 1983.

[Hi1] N. Hindman, Finite sums from sequences within cells of a partition of $\mathbb{N}$, *J. Combinatorial Theory* (Series A), **17** (1974), 1–11.

[Hi2] N. Hindman, Partitions and sums and products of integers, *Trans. AMS*, **247** (1979),
227–245.

[HiS] N. Hindman and D. Strauss, *Algebra in the Stone-Čech Compactification. Theory
and Applications*, de Gruyter Expositions in Mathematics, 27. Walter de Gruyter &
Co., Berlin, 1998. xiv+485 pp.

[Hopf1] E. Hopf, Complete transitivity and the ergodic principle, *Proc. Nat. Acad. Sci.
U.S.A.*, **18** (1932), 204–209.

[Hopf2] E. Hopf, Proof of Gibbs' hypothesis on the tendency toward statistical equilibrium,
Proc. Nat. Acad. Sci. U.S.A., **18** (1932), 333–340.

[Hopf3] E. Hopf, *Ergodentheorie*, Chelsea, New York, 1948.

[Kh] A.Y. Khintchine, Eine Verschärfung des Poincaréschen Wiederkehrsatzes, *Comp.
Math.*, **1** (1934), 177–179.

[KN] B.O. Koopman and J. von Neumann, Dynamical systems of continuous spectra, *Proc. Nat. Acad. Sci. U.S.A.*, **18** (1932), 255–263.

[Kre] U. Krengel, Weakly wandering vectors and weakly independent partitions, *Trans. AMS*, **164** (1972), 199–226.

[Kro] L. Kronecker, Die Periodensysteme von Funktionen Reeller Variablen, *Berliner Sitzungsberichte* (1884), 1071–1080.

[L] A. Leibman, Multiple reccurence theorem for measure preserving actions of a nilpotent group, *Geom. and Funct. Anal.*, **8** (1998), 853–931.

[R-N] C. Ryll-Nardzewski, On fixed points of semigroups of endomorphisms of linear spaces, *Proc. Fifth Berkeley Sympos. Math. Statist. and Probability* (Berkeley, Calif., 1965/66), Vol. II: Contributions to Probability Theory, Part 1, 55–61.

[Sc] K. Schmidt, Asymptotic properties of unitary representations and mixing, *Proc. London Math. Soc.* (3) **48** (1984), no. 3, 445–460.

[Sz] P. Szüsz, Verschärfung eines Hardy-Littlewoodschen Satzes, *Acta Math. Acad. Sci. Hungar.*, **4** (1953), 115–118.

[SY] H. Shi and H. Yang, Nonmetrizable topological dynamical characterization of central sets, *Fund. Math.*, **150** (1996), 1–9.

[ScW] K. Schmidt and P. Walters, Mildly mixing actions of locally compact groups, *Proc. London Math. Soc.* (3) **45** (1982), 506–518.

[Wa1] P. Walters, Some invariant σ-algebras for measure-preserving transformations, *Trans. Amer. Math. Soc.* **163** (1972), 357–368.

[Wa2] P. Walters, A mixing property for nonsingular actions, 163–183, Aste'risque, 98-99, Soc. Math. France, Paris, 1982.

[Weyl] H. Weyl, Über die Gleichverteilung von Zahlen mod. Eins, *Math. Ann.* **77** (1916), 313–352.

DEPARTMENT OF MATHEMATICS, THE OHIO STATE UNIVERSITY, COLUMBUS, OH 43210

E-mail address: vitaly@math.ohio-state.edu

SYMBOLIC DYNAMICS AND TOPOLOGICAL MODELS IN DIMENSIONS 1 AND 2

ANDRÉ DE CARVALHO AND TOBY HALL

CONTENTS

1. INTRODUCTION

Symbolic dynamics has a long and distinguished history, going back to Hadamard's work on the geodesic flow on negatively curved surfaces. Because of its success in describing the dynamics of systems with Markov partitions, it is most closely associated with the study of such systems. However, in the 1970's, beginning with the work of Metropolis, Stein and Stein and culminating with the *kneading theory* of Milnor and Thurston, symbolic dynamics was used in a quite different way in the study of 1-dimensional discrete dynamics: the partition was fixed (given by the critical points) and the maps were allowed to vary in families, as long as the critical points remained the same. This is fundamentally different from the use of symbolic dynamics in the presence of Markov partitions. Not only does it not require the maps being studied to have such partitions, but also, by fixing the partition, it permits the description of all maps in a family in terms of the same symbols, thus allowing the comparison of different maps.

One of the conclusions of kneading theory is that every family of *unimodal maps* (i.e. continuous piecewise monotone self-maps of the unit interval with exactly two monotone pieces) presents essentially the same dynamical behaviour as it passes from trivial to chaotic dynamics. This is sometimes called *topological universality* for 1-dimensional dynamics. In dimension 2 the situation is more complicated. In this article we present some topological aspects of kneading theory together with an analogous 2-dimensional theory called *pruning*. The ideas are presented in such a way as to emphasize

40

Dimension 1	Dimension 2
$f_\mu(x) = \mu x(1-x), \mu > 4$	Horseshoe map
Kneading	Pruning
Kneading sequence	Pruning front
Unimodal order	Ordering along stable and unstable manifolds
0-entropy equivalence for endomorphisms	0-entropy equivalence for homeomorphisms
Piecewise affine maps with constant slope	Generalized pseudo-Anosov homeomorphisms
Permutation of a periodic orbit	Braid type of a periodic orbit

TABLE 1. Parallel concepts in the 1- and 2-dimensional theories

the parallels between the two theories. The concept of pruning was introduced by Cvitanović to describe families of 2-dimensional diffeomorphisms
(or homeomorphisms) such as the Hénon family. Unlike their 1-dimensional
counterparts, such families do not have critical points and usually do not
have any obvious substitutes (much of the hard work in some papers about
the Hénon family (e.g. [BC91, WY01]) consists in finding parameter values for
which there are such substitutes.) This absence of critical points is one of the
main reasons for the additional complexity of the theory in dimension 2. The
intention of this article is to give an outline of both theories, omitting many
of the technical details, in order to bring out those analogies which do exist.
There are many good detailed accounts of kneading theory in the literature
(e.g. [MT88, dMvS93]): the approach based on abstract models taken in this
article, though folkloric, does not seem to have appeared in print before. For
more detail on pruning, see [Cvi91, CGP88, dCH, dCH01, dCH02]

Table 1 indicates the correspondence between the relevant parallel concepts in dimensions 1 and 2.

2. KNEADING THEORY

2.1. Unimodal maps.
In this section the basic concepts of kneading theory
for unimodal maps are presented. We omit those details which are not essential to the present discussion. More complete accounts of kneading theory
can be found in [MT88, Dev89, dMvS93, KH95].

Definitions 1. Let $I = [a, b]$. A *unimodal map* on I is a continuous map
$f: I \to \mathbb{R}$ with $f(a) = f(b) = a$, for which there exists $c \in (a, b)$ such that f
is increasing on $[a, c]$ and decreasing on $[c, b]$ (not necessarily strictly). The
point c is the *critical* or *turning point* of f, and $f(c)$ is its *critical value*.

Given a unimodal map $f \colon I \to \mathbb{R}$, we are interested from a dynamical point of view in the set

$$\Lambda^+ = \Lambda_f^+ = \{x \in I;\ f^n(x) \in I \text{ for all } n \in \mathbb{N}\}$$

of points whose orbits remain within I. Clearly f restricts to a continuous map $f \colon \Lambda^+ \circlearrowleft$.

Example 1. The best-known example of a family of unimodal maps is the *logistic family* $\{f_\mu\}_{\mu>0}$, defined on $I = [0,1]$ by

$$f_\mu(x) = \mu x (1 - x).$$

Here $c = 1/2$ for all μ, and $f_\mu(I) \subseteq I$ if and only if $\mu \le 4$.

Symbolic dynamics is introduced by defining the *itinerary* of a point $x \in \Lambda^+$ to be the sequence

$$i(x) = s_0 s_1 s_2 \ldots \in \{0, C, 1\}^{\mathbb{N}}$$

given by

$$s_j = \begin{cases} 0 & \text{if } f^j(x) < c, \\ C & \text{if } f^j(x) = c, \\ 1 & \text{if } f^j(x) > c. \end{cases}$$

This defines a map $i \colon \Lambda^+ \to \{0, C, 1\}^{\mathbb{N}}$ with the property that $i \circ f = \sigma \circ i$, where $\sigma \colon \{0, C, 1\}^{\mathbb{N}} \circlearrowleft$ is the *shift map* given by $\sigma(s_0 s_1 s_2 \ldots) = s_1 s_2 s_3 \ldots$. The symbol C can only occur in itineraries if $f(I) \subseteq I$.

Example 2. Consider a logistic map f_μ with $\mu > 4$, so that $I = [0,1]$ is not invariant under f_μ: points of the interval $A = (1/2 - \epsilon, 1/2 + \epsilon)$, where $\epsilon = \sqrt{\frac{\mu-4}{4\mu}}$, have images outside of I. The preimage $f_\mu^{-1}(A)$ consists of two smaller intervals A_0 and A_1 to the left and right of A: points in these intervals are mapped out of I by f_μ^2. More generally, the set of points of I which are mapped out of I by f_μ^n but not by f_μ^{n-1} consists of 2^n open intervals with

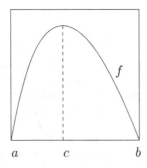

FIGURE 1. A unimodal map.

disjoint closures. This construction of 'excluded' intervals is similar to that of the middle-thirds Cantor set, and indeed it can be shown that $\Lambda_{f_\mu}^+$ is a Cantor set. (For μ large enough that $|f_\mu'(x)| > 1$ for $x \notin A$, it is straightforward to prove this statement. The proof in the general case is more subtle.) In this case the itinerary map $i\colon \Lambda_{f_\mu}^+ \to \{0,1\}^{\mathbb{N}}$ is a homeomorphism which conjugates f_μ and the shift map.

The itinerary of the critical value of a unimodal map $f\colon I \to \mathbb{R}$ with $f(I) \subseteq I$ plays a particularly important role in kneading theory.

Definition 2. The *kneading sequence* $\kappa(f)$ of a unimodal map $f\colon I \to \mathbb{R}$ with $f(I) \subseteq I$ is the itinerary $\kappa(f) = i(f(c))$ of its critical value.

Notice that $\kappa(f)$ describes the tail of the itinerary of any preimage of c: if $\sigma^n(i(x)) = C \ldots$ for any $n \in \mathbb{N}$, then $\sigma^{n+1}(i(x)) = \kappa(f)$.

In Section 2.3 it will be seen that the case where $\kappa(f)$ is *periodic* (i.e. $\kappa(f) = \overline{w}$, the infinite repetition of some word $w \in \{0, C, 1\}^n$) must be treated separately from the case where it is not periodic.

2.2. The unimodal order.

The unimodal order $\prec$ is a linear order on $\{0, C, 1\}^{\mathbb{N}}$ which reflects the usual ordering of points in I. A finite word $w \in \{0, 1\}^n$ is *even* or *odd* according as the sum of its entries is even or odd. Let $s = s_0 s_1 \ldots, t = t_0 t_1 \ldots \in \{0, C, 1\}^{\mathbb{N}}$ and assume that $s_j = t_j \neq C$ for all $j < n$ and that $s_n \neq t_n$ (note that if s and t are itineraries of points under a given unimodal map with $s_j = t_j = C$, then $s_n = t_n$ for all $n > j$). Set $0 < C < 1$ and define

$$s \prec t \text{ if } \begin{cases} s_0 s_1 \ldots s_{n-1} \text{ is even and } s_n < t_n \text{ or} \\ s_0 s_1 \ldots s_{n-1} \text{ is odd and } s_n > t_n. \end{cases}$$

The following straightforward lemma states that the unimodal order does indeed reflect the usual order on I:

Lemma 2.1. *Let $f\colon I \to \mathbb{R}$ be a unimodal map, and $x, y \in \Lambda_f^+$. If $i(x) \prec i(y)$ then $x < y$.*

Of course it follows immediately that if $x < y$ then $i(x) \preceq i(y)$: the possibility of equality in this partial converse cannot be excluded. Thus if we define

$$I(x) = \{y \in \Lambda_f^+;\ i(y) = i(x)\},$$

for $x \in \Lambda_f^+$, then $I(x)$ is an interval (possibly degenerate) containing x on which all iterates of f are monotone, and it is maximal with respect to these properties.

The next step is to describe the set of all itineraries of a given unimodal map $f\colon I \circlearrowleft$ (for the remainder of this section, we restrict to the case $f(I) \subseteq I$). Notice that if $s = s_0 s_1 \ldots$ is the itinerary of some point $x \in I$ then it must satisfy the following conditions:

1) If $s_n = C$ then $\sigma^{n+1}(s) = \kappa(f)$
2) $\sigma^n(s) \preceq \kappa(f)$ for all $n \geq 1$.

Condition 2) is a consequence of the observation that, for a unimodal map, the image of the critical point is greater than or equal to the image of any other point in the interval.

If the inequality in 2) is strict, these two conditions are also sufficient for the existence of a point $x \in I$ with itinerary s:

Theorem 2.2. *Let $f : I \circlearrowleft$ be a unimodal map, and suppose that $\kappa(f)$ is not periodic. Suppose $s = s_0 s_1 \ldots \in \{0, C, 1\}^{\mathbb{N}}$ satisfies*

1) If $s_n = C$ then $\sigma^{n+1}(s) = \kappa(f)$, and
2) $\sigma^n(s) \prec \kappa(f)$ for all $n \geq 1$.
Then there exists $x \in I$ such that $i(x) = s$.

Remarks.

a) Inequality in 2) is necessary: if $f(x) = f_4(x) = 4x(1-x)$, the sequences $s = 01000\ldots$ and $s = 11000\ldots$ both satisfy $\sigma(s) = \kappa(f_4) = 1000\ldots$, but neither is the itinerary of a point under f_4.
b) With some small provisos, the statement also holds for periodic kneading sequences (see [Dev89]).

For a proof, see for example [Dev89].

2.3. Abstract synthetic models. In this section we develop an abstract symbolic discussion and use it to construct synthetic models for unimodal maps.

Definitions 3. Let Σ^+ denote the 1-sided 2-shift space $\{0, 1\}^{\mathbb{N}}$. An element $\kappa \in \Sigma^+$ is a *kneading sequence* if it satisfies

$$\sigma^n(\kappa) \preceq \kappa \text{ for all } n \geq 0.$$

If κ is a kneading sequence, define

$$\Sigma_\kappa^+ = \{s \in \Sigma^+; \sigma^n(s) \preceq \kappa \text{ for all } n \geq 1\}.$$

Σ_κ^+ is clearly σ-invariant. We also denote by σ the restriction of the shift map to Σ_κ^+.

We now introduce an equivalence relation $\sim_\kappa$ on Σ_κ^+ by setting

$$s_0 \ldots s_{n-1} 0\kappa \sim s_0 \ldots s_{n-1} 1\kappa$$

for all $s_0 \ldots s_{n-1}$ for which these sequences lie in Σ_κ^+, and extending to the largest equivalence relation (i.e. with smallest equivalence classes) $\sim_\kappa$ with closed equivalence classes. Let I_κ denote the quotient space $\Sigma_\kappa^+ / \sim_\kappa$. If $\kappa = \overline{s_0 \ldots s_n}$ is periodic, define $\kappa(C) = \overline{s_0 \ldots s_{n-1} C}$ and

$$\tilde{\kappa} = \max\{\gamma \in \Sigma^+; \gamma \text{ is a periodic even kneading sequence, } \gamma \preceq \kappa\}$$

(where the maximum is taken with respect to the unimodal order).

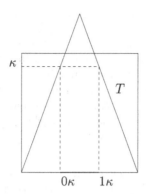

FIGURE 2. The first collapsed interval $A_0 = (0\kappa, 1\kappa)$.

Theorem 2.3. I_κ *is either a point or an interval. If it is an interval, then* $\sigma \colon \Sigma_\kappa^+ \circlearrowright$ *projects to a unimodal map* $\phi_\kappa \colon I_\kappa \circlearrowright$. *Moreover, the kneading sequence* $\kappa(\phi_\kappa)$ *of* ϕ_κ *is given by*

$$
\kappa(\phi_\kappa) = \begin{cases} \kappa & \text{if } \kappa \text{ is not periodic} \\ \kappa(C) & \text{if } \kappa \text{ is periodic even} \\ \tilde{\kappa}(C) & \text{if } \kappa \text{ is periodic odd.} \end{cases}
$$

Sketch Proof. Using the itinerary map, identify $\sigma \colon \Sigma^+ \circlearrowright$ with the restriction of the tent map $T \colon I \to \mathbb{R}$ defined by

$$
T(x) = \begin{cases} 3x & 0 \le x \le 1/2 \\ 3 - 3x & 1/2 \le x \le 1 \end{cases}
$$

to its invariant Cantor set Λ_T^+ (which is precisely the standard middle-thirds Cantor set). From now on we will not distinguish between σ and T, or between points of Σ^+ and points of $\Lambda_T^+ \subseteq I$.

By definition, Σ_κ^+ is obtained from Σ^+ by removing $A_0 = T^{-1}\left((\kappa, \infty)\right)$ and all of its preimages. Notice that $A_0 = (0\kappa, 1\kappa)$ (Figure 2). We can thus think of the equivalence relation $\sim_\kappa$ as an equivalence relation on the interval I which collapses to points the closures of the connected components of A_0 and its preimages. It is therefore a monotone decomposition of the interval and, since it has closed equivalence classes (by definition), the quotient space I_κ is either a point or an interval. Since the equivalence is σ-invariant, σ induces a continuous map $\phi_\kappa \colon I_\kappa \circlearrowright$. That ϕ_κ is unimodal follows from the monotonicity of the canonical projection $\pi \colon I \to I_\kappa$ (i.e. preimages of points are connected), and the unimodality of the tent map T. Moreover, denoting by $[s]$ the $\sim_\kappa$ equivalence class of $s \in \Sigma_\kappa^+$, the critical point of ϕ_κ is $[0\kappa] = [1\kappa]$, and its critical value is $[\kappa]$.

In order to determine the kneading sequence of ϕ_κ, it is necessary to consider the way in which Σ_κ^+ depends on κ:

Lemma 2.4. Σ_κ^+ *has isolated points if and only if κ is periodic odd.*

Sketch Proof. Because κ is a kneading sequence, connected components of preimages of A_0 under T are either contained in or disjoint from A_0. Since these components are open intervals, it follows that Σ_κ^+ has isolated points if and only if A_0 shares an endpoint with some component, not contained in A_0, of one of its preimages. This can happen if and only if κ is periodic (with period n, say) and has the property that nearby points exchange sides under the n-th iterate of T. This is equivalent to saying κ is periodic odd. ∎

Assume first that Σ_κ^+ has no isolated points. Then it is a Cantor set (since it is certainly compact and totally disconnected), and $\sim_\kappa$ is the equivalence which identifies two points if and only if they are the endpoints of a connected component of $[0,1] \setminus \Sigma_\kappa^+$, i.e. points of the form

$$s_0 \ldots s_{n-1}0\kappa \sim_\kappa s_0 \ldots s_{n-1}1\kappa.$$

Thus equivalence classes consist either of a single point (if the point is not a preimage of κ), or of exactly two points of the above form. In this situation, it is straightforward to describe the itinerary map i associated with $\phi_\kappa : I_\kappa \circlearrowleft$. If s is not a preimage of κ under σ, then $i([s]) = s$. Otherwise, there are two possibilities: if κ is not periodic, then

$$i([s_0 \ldots s_n\kappa]) = s_0 \ldots s_{n-1}C\kappa;$$

while if $\kappa = \overline{t_0 \ldots t_k}$, then

$$i([s_0 \ldots s_n\kappa]) = i([s_0 \ldots s_n\overline{t_0 \ldots t_k}]) = s_0 \ldots s_{n-1}\overline{Ct_0 \ldots t_{k-1}}$$

(since the critical point $[0\kappa] = [1\kappa]$ of ϕ_κ is periodic).

In particular, the itinerary of the critical value $[\kappa]$ is either κ (if κ is not periodic), or $\overline{t_0 \ldots t_{k-1}C}$ if $\kappa = \overline{t_0 \ldots t_k}$ is periodic even. This deals with the first two possibilities for $\kappa(\phi_\kappa)$ in the statement of the theorem.

The final case is where κ is periodic odd: it is illustrated by Examples 3–5, and the reader is encouraged to refer to them while reading the following argument. Suppose then that $\kappa = \overline{s_1 \ldots s_n}$ is periodic odd. Then $\kappa \succ (s_1 \ldots \hat{s}_n)^k\overline{s_1 \ldots s_n}$, for any $k > 0$ (where we write $\hat{0} = 1$ and $\hat{1} = 0$). Therefore

$$
\begin{aligned}
\kappa &= s_1 \ldots s_n\overline{s_1 \ldots s_n} \\
&\sim_\kappa s_1 \ldots \hat{s}_n\overline{s_1 \ldots s_n} \\
&\quad\vdots \\
&\sim_\kappa (s_1 \ldots \hat{s}_n)^k\overline{s_1 \ldots s_n} \\
&\quad\vdots \\
&\sim_\kappa \overline{s_1 \ldots \hat{s}_n}
\end{aligned}
$$

where the final equivalence is due to the equivalence classes being closed and the fact that

$$(s_1 \ldots \hat{s}_n)^k \overline{s_1 \ldots s_n} \to \overline{s_1 \ldots \hat{s}_n}, \text{ as } k \to \infty.$$

The sequence $\kappa_1 = \overline{s_1 \ldots \hat{s}_n}$ is clearly periodic even and it is also a kneading sequence. It may be the case, however, that $s_1 \ldots \hat{s}_n = (s_1 \ldots s_{j_1})^{n/j_1}$ (see examples below). If this is the case and $\kappa_1 = \overline{s_1 \ldots s_{j_1}}$ is odd, the same argument shows that $\kappa_1 \sim \overline{s_1 \ldots \hat{s}_{j_1}}$. Continuing this process produces finitely many periodic odd kneading sequences $\kappa_1, \ldots, \kappa_{m-1}$, with (minimal) periods satisfying $j_\ell \,|\, j_{\ell-1}$ for $\ell = 1, \ldots, m - 1$, until we arrive at a periodic even kneading sequence $\kappa_m = \tilde{\kappa}$ (which is not a repetition of a shorter word) and

$$\kappa = \kappa_0 \sim_\kappa \kappa_1 \sim_\kappa \ldots \sim_\kappa \kappa_{m-1} \sim_\kappa \kappa_m = \tilde{\kappa}.$$

That $\tilde{\kappa}$ is the largest periodic even kneading sequence smaller than κ follows from the observation that there exists no kneading sequence strictly between two kneading sequences of the forms $\overline{t_1 \ldots t_n}$ and $\overline{t_1 \ldots \hat{t}_n}$.

This argument shows that the $\sim_\kappa$ equivalence class of κ consists of the points

$$\kappa, \kappa_1, \ldots, \kappa_{m-1}, \tilde{\kappa}$$

together with some transitional orbits (see examples below). The points in this equivalence class are precisely those $s \in \Sigma_\kappa^+$ for which $s \succeq \tilde{\kappa}$. It follows that every $\sim_\kappa$ equivalence class contains a point of $\Sigma_{\tilde{\kappa}}^+$: $\Sigma_\kappa^+ \setminus \Sigma_{\tilde{\kappa}}^+$ consists of a countable collection of points contained in the complementary components of $\Sigma_{\tilde{\kappa}}^+$ in I; such points are exactly the preimages under T of those $s \in \Sigma_\kappa^+$ for which $s \succeq \tilde{\kappa}$; if A is an open interval in $I \setminus \Sigma_{\tilde{\kappa}}^+$, then all points in the closure of $A \cap \Sigma_\kappa^+$ are $\sim_\kappa$-equivalent and this closure contains both endpoints of A. Therefore, $\sim_\kappa$ 'collapses' the difference between Σ_κ^+ and $\Sigma_{\tilde{\kappa}}^+$. From this it follows that I_κ and $I_{\tilde{\kappa}}$ are homeomorphic, that ϕ_κ and $\phi_{\tilde{\kappa}}$ are topologically conjugate and, therefore, that the kneading sequences of both are equal to $\tilde{\kappa}(C)$.   ∎

The examples that follow clarify the last possibility in the statement of Theorem 2.3.

Example 3. Consider first some simple examples:

  i) If $\kappa = \overline{0}$ then $\Sigma_\kappa^+ = \{\overline{0}\}$.
  ii) If $\kappa = \overline{1}$ then $\Sigma_\kappa^+ = \{\overline{0}, 0^k\overline{1}; k \geq 0\} = \Sigma_{\overline{0}}^+ \cup \{0^k\overline{1}; k \geq 0\}$.
  iii) If $\kappa = \overline{10}$ then $\Sigma_\kappa^+ = \{\overline{0}, 0^k\overline{1}, 0^k 1^\ell \overline{10}; k, l \geq 0\} = \Sigma_{\overline{1}}^+ \cup \{0^k 1^\ell \overline{10}; k, l \geq 0\}$.

Although these are trivial examples from the point of view of the theorem — the quotient spaces I_κ are single points — they are representative of the general case: as κ increases from $\overline{0}$ to $\overline{1}$ to $\overline{10}$, the orbits which existed in a previous stage (e.g. $\overline{0}$) cease to be isolated, a new isolated periodic orbit (e.g. $\overline{1}$) is born, and transitional orbits (e.g. $0^k\overline{1}$) are created.

Example 4. For a more complicated example, in which I_κ is a nondegenerate interval, note that $\Sigma^+_{\overline{101}}$ is a Cantor set (since $\overline{101}$ is periodic even). It is easy to check that the sequences $s \in \Sigma^+_{\overline{100}}$ satisfying $\overline{101} \prec s \prec \overline{100}$ are of the form $(101)^k \overline{100}$, for $k \geq 1$. Notice that such points form a discrete set in $\Sigma^+_{\overline{100}}$ accumulating on $\overline{101}$ as $k \to \infty$. From this it follows that $\Sigma^+_{\overline{100}} \setminus \Sigma^+_{\overline{101}}$ is discrete.

Consider the equivalence relation $\sim \, = \, \sim_{\overline{100}}$. By definition, it satisfies

$$\overline{100} = 100\,\overline{100} \sim 101\,\overline{100} = 101\,100\,\overline{100} \sim 101\,101\,\overline{100} \sim \cdots \sim (101)^k\,\overline{100}.$$

Since the points of the form $(101)^k \overline{100}$ accumulate on $\overline{101}$ as $k \to \infty$ and the equivalence classes are closed, $\overline{101}$ is also equivalent to all of the points of this form. Since $\overline{101}$ is alone in its $\sim_{\overline{101}}$ equivalence class, and $\Sigma^+_{\overline{100}} \supset \Sigma^+_{\overline{101}}$ it follows that

$$\left\{ (101)^k\,\overline{100}, k \geq 0 \right\} \cup \left\{ \overline{101} \right\}$$

is a complete equivalence class and that $\phi_{\overline{100}}$ is topologically conjugate to $\phi_{\overline{101}}$: their kneading sequence is $\overline{10C}$.

Example 5. The final example is one in which two steps taking closures are involved. Let $\kappa = \overline{100101}$. Then the points $s \in \Sigma^+_\kappa$ with $s \succeq \overline{101}$ are of the form

$$(101)^k\overline{100} \quad \text{and} \quad (101)^k(100)^\ell\overline{100101} \quad \text{for } k, \ell \geq 0.$$

Here points of the form $(101)^k(100)^\ell\overline{100101}$ are isolated for all $k, \ell \geq 0$, but those of the form $(101)^k\overline{100}$ are accumulation points of $(101)^k(100)^\ell\overline{100101}$ as $\ell \to \infty$. Reasoning as in the previous example, we see that

$$\overline{100101} \sim (100)^\ell\overline{100101} \sim (101)^k(100)^\ell\overline{100101} \sim \overline{100} \sim \overline{101},$$

so that $\phi_{\overline{100101}}$ is also conjugate to $\phi_{\overline{101}}$.

Remark. While the symbolic approach taken above is rather messy, the results are beautifully explained by renormalization theory [CE80].

We can now compare these abstractly constructed maps to unimodal maps in general. Let $f \colon I \circlearrowleft$ be a unimodal map and assume first that its turning point is not periodic. Then the kneading sequence $\kappa = \kappa(f)$ is an element of Σ^+ and we can go through the construction above to obtain a unimodal map $\phi_\kappa \colon I_\kappa \circlearrowleft$. If the turning point of f is periodic, then $\kappa = s_0 \ldots s_{n-1}C$. The discussion above illustrates that changing the last symbol of κ to 0 or 1 produces the same quotient map: $\phi_{\kappa(0)} = \phi_{\kappa(1)}$ (where $\kappa(0), \kappa(1)$ denote the kneading sequences obtained changing the last symbol of κ to 0, 1 respectively). Thus the theorem below is quite general.

Theorem 2.5. *Let $f \colon I \circlearrowleft$ be a unimodal map with turning point c. If c is not periodic then let $\kappa = \kappa(f)$, and if c is periodic then let κ be the periodic sequence obtained by changing the symbol C in $\kappa(f)$ to either 0 or 1. Then* *the composition $I \to \Sigma_\kappa^+ \to I_\kappa$ of the itinerary map with the quotient map is a semi-conjugacy from $f \colon I \circlearrowleft$ to $\phi_\kappa \colon I_\kappa \circlearrowleft$.*

Thus every unimodal map $f \colon I \circlearrowleft$ can be modelled by a quotient of a subshift of the 1-sided 2-shift. The reader who has followed the arguments presented so far should have no difficulty providing a proof of this theorem. Observe that the fibres of the semi-conjugacy $I \to \Sigma_\kappa^+ \to I_\kappa$ in the statement are the equivalence classes of the largest equivalence relation on the interval whose equivalence classes are closed and either contain or are disjoint from intervals $I(x)$ of points with the same itinerary as x.

3. Pruning

We now start to consider 2-dimensional analogues of the theory presented in Section 2. There are, however, some fundamental differences between the two cases and it is not possible to define exact analogues of several concepts. In particular, the very definition of a '2-dimensional unimodal map' is lacking. The Hénon family $\{H_{a,b}\}_{a,b \in \mathbb{R}}$ defined by

$$H_{a,b}(x,y) = (f_\mu(x) - by, x)$$

is a family of diffeomorphisms of $\mathbb{R}^2$ which is a 2-dimensional counterpart of the logistic family; however, since the Hénon maps are invertible, they do not have any defining topological trait which entitles them to be called 'unimodal'. Nonetheless, it is possible to develop a number of analogies despite such basic difficulties. We will concentrate on orientation-preserving Hénon maps (i.e. we assume that $b > 0$).

3.1. The horseshoe.

The horsehoe map was introduced by Smale [Sma67]. In the context of this article, it should be thought of as a 2-dimensional analogue of the logistic map f_μ with $\mu > 4$. The horseshoe is a homeomorphism $F \colon S^2 \circlearrowleft$ of the sphere, all of whose interesting dynamics is contained in the stadium-shaped region $\mathbb{I}$ which is the union of a central square S and two half-disks as shown in Figure 3. It maps $\mathbb{I}$ into itself in a horseshoe shape, mapping vertical and horizontal line segments in $S \cap F^{-1}(S)$ into vertical and horizontal line segments respectively. In each component V_0 and V_1 of $S \cap F^{-1}(S)$ (Figure 4) the horseshoe is linear, contracting vertical segments by the factor $1/3$ and expanding horizontal segments by the factor 3. There is an attracting fixed point in the half-disk on the left denoted by x (shown as ■) and the point at infinity in S^2 is a repeller whose basin contains all points outside $\mathbb{I}$.

Notice that collapsing vertical segments and the two half-disks of $\mathbb{I}$ to points we obtain an interval and the horseshoe projects to a 'flat top' tent map $f \colon I \circlearrowleft$. This is also shown in Figure 3: the image $f(I)$ is shown slightly separated so it is possible to see what f does to the interval.

We now describe the dynamics of F inside the square S: see for example [Dev89] for a more detailed description. In Figure 5, $F^n(S) \cup F^{-n}(S)$ are

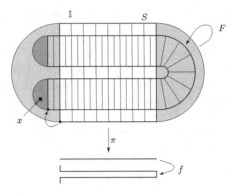

FIGURE 3. The horseshoe map.

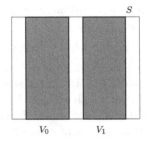

FIGURE 4. The vertical strips of $S \cap F^{-1}(S)$.

shown for $n = 1, 2$. The intersection $F^n(S) \cap F^{-n}(S)$ consists precisely of the set of points which remain inside S for n iterates of both F and F^{-1}. In general, $F^n(S) \cap F^{-n}(S)$ consists of 2^{2n} squares of side length 3^{-n}. The set of points all of whose iterates stay in the square is

$$\Lambda = \bigcap_{n=0}^{\infty} F^n(S) \cap F^{-n}(S).$$

Λ is a Cantor set and, as in Section 2, we define an itinerary map $i \colon \Lambda \to \{0,1\}^{\mathbb{Z}}$ by setting

$$i(x) = \begin{cases} 0 & \text{if } F(x) \in V_0 \\ 1 & \text{if } F(x) \in V_1 \end{cases}$$

As in Example 2, the itinerary map conjugates the restriction of F to Λ with the shift map $\sigma \colon \{0,1\}^{\mathbb{Z}} \circlearrowleft$, defined by

$$\sigma(\ldots s_{-2} s_{-1} \cdot s_0 s_1 \ldots) = \ldots s_{-1} s_0 \cdot s_1 s_2 \ldots$$

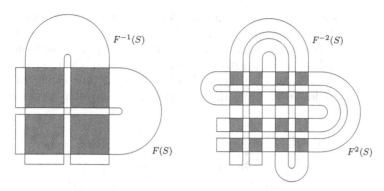

FIGURE 5. First steps in the construction of the horseshoe-invariant Cantor set

3.2. Models abstract and concrete.

The models in Section 2.3 were constructed starting from symbolic dynamics. It was then shown that they model actual unimodal maps. In dimension 2, it is necessary to build models which are homeomorphisms of the sphere. Once this is done, it is possible to describe them using symbolic dynamics just as in dimension 1. Because there are more technical problems in dimension 2, symbolic dynamics will be relegated to a secondary role. We first state the pruning theorem [dC99], which describes how it is possible to destroy the dynamics of a surface homeomorphism in a controlled way by isotopy. Our models will consist of all such *prunings* of the horseshoe and will form a family depending on infinitely many parameters.

Let $F: S \circlearrowleft$ be an orientation-preserving homeomorphism of an orientable surface S (which in this paper will always be taken to be $\mathbb{R}^2$ or S^2). The pruning theorem 3.1 gives conditions on a topological disk $D \subseteq S$ (the *pruning conditions*) which ensure that it is possible to perform an isotopy of F which destroys all of the dynamics in the (orbit of the) interior of D, while leaving the dynamics unchanged elsewhere. If this theorem is to be applied repeatedly then it is essential that it should only require continuity, since it is impossible in general for such pruning isotopies to preserve differentiability. Despite this, we start with an intuitive motivation for one of the pruning conditions in the context of diffeomorphisms.

Suppose for the time being, then, that F is a diffeomorphism. Roughly speaking, the first pruning condition for a disk D states that ∂D is the union of a segment C of stable manifold and a segment E of unstable manifold (perhaps of some periodic point of F). If this is the case, then a pruning isotopy can be thought of as uncrossing the stable and unstable manifolds within D (Figure 6), since if any such crossings remain then there are local horseshoes, and hence non-trivial dynamics, inside D. The second pruning condition, which should be thought of as a 2-dimensional analogue of the requirement that a sequence be a kneading sequence, ensures that such uncrossing is possible. To understand this condition on an intuitive level, consider the disk D depicted in Figure 7, together with an iterate $F^n(D)$. Uncrossing stable and unstable manifolds within D means uncrossing them within all of its images and preimages, including $F^n(D)$. However, the configuration of unstable manifolds in Figure 7 means that it is not possible to remove the circled crossings.

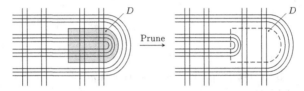

FIGURE 6. Uncrossing stable and unstable manifolds inside a pruning disk

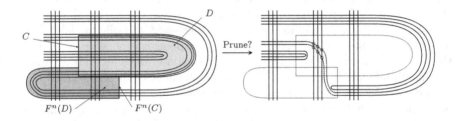

FIGURE 7. A disk which is not a pruning disk

After a little thought it can be seen that the problem in this figure arises because there is an integer $n > 0$ such that $F^n(C)$ intersects both $\mathrm{Int}(D)$ and $S \setminus D$, and by symmetry the same problem arises if $F^{-n}(E)$ intersects both $\mathrm{Int}(D)$ and $S \setminus D$ for some $n > 0$. The purpose of the second pruning condition is to prohibit such intersections. In fact, it goes further, and prohibits *any* intersection of $F^n(C)$ or $F^{-n}(E)$ with $\mathrm{Int}(D)$ for $n > 0$. Thus, in addition to bad intersections such as that of Figure 7, it also prohibits intersections where $F^n(C) \subseteq D$ or $F^{-n}(E) \subseteq D$. The reason for this is not that such intersections make it impossible to prune the dynamics in D, but rather that if they occur then D can be replaced by an equivalent disk D' for which there are no such intersections.

A sketch of the construction of such a disk D' is given in Figure 8. Here $F^n(C) \subseteq \mathrm{Int}(D)$: in particular, the circled intersections to the left of C are mapped by F^n to the circled intersections to the right of $F^n(C)$, which are contained in D. Hence any isotopy which removes all of the intersections in D also removes those immediately to the left of D. The disk D can therefore be enlarged along the unstable segment of its boundary to a new disk D', without changing the set of intersections to be removed. Such an enlargement can be continued until such a time as $F^n(C') \subseteq C'$, when C' is disjoint from $\text{Int}(D)$) and contains a fixed point p of F^n.

Hence all intersections of $F^n(C)$ with $\text{Int}(D)$ (and analogously all those of $F^{-n}(E)$ with $\text{Int}(D)$) can be prohibited by the conditions of the pruning theorem.

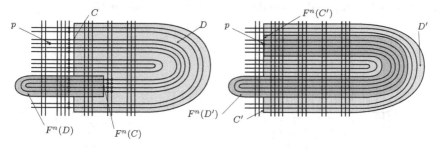

FIGURE 8. Equivalent pruning disks

Definition 4. A *pruning disk* for F is a closed topological disk D whose boundary can be written as the union of two arcs intersecting only at their endpoints, $\partial D = C \cup E$, satisfying the following *pruning conditions*:

i)

$$\text{diam}(F^n(C)) \to 0 \quad \text{as } n \to \infty$$
$$\text{diam}(F^{-n}(E)) \to 0 \quad \text{as } n \to \infty.$$

ii)

$$F^n(C) \cap \text{Int}(D) = \emptyset \quad \text{for all } n > 0$$
$$F^{-n}(E) \cap \text{Int}(D) = \emptyset \quad \text{for all } n > 0.$$

Remark. When dealing with non-differentiable maps in general, condition ii) above is more technical. The interested reader can see [dC99] for more details.

Example 6. In Figure 9, two pruning disks for Smale's horseshoe map $F\colon S^2 \to S^2$ are depicted. In a), the boundary of the disk is made up of stable and unstable segments of the fixed point with itinerary $\overline{0}$, while in b) the other fixed point, with itinerary $\overline{1}$, is used. In both cases, pruning condition i) is trivially satisfied (where C is the segment of stable manifold in ∂D, and E is the segment of unstable manifold). Straightforward arguments suffice to show that pruning condition ii) is also satisfied: in a), it can be seen by a simple calculation that the forward iterates $F^n(C)$ of C (depicted with bolder lines) are all disjoint from D (since they are attracted to the fixed point, only finitely many need be considered) — a similar argument applies to the backward iterates $F^{-n}(E)$ of E; in b), all iterates $F^n(C)$ are contained in the horizontal segment through the fixed point, and all iterates $F^n(C)$ (respectively $F^{-n}(E)$) are contained in the vertical (respectively horizontal) segment through the fixed point, and hence cannot intersect $\text{Int}(D)$. This argument generalizes readily to any configuration where D is bounded by segments C and E of the stable and unstable manifolds of a fixed point p, and where both the segment of stable manifold between p and C and the segment of unstable manifold between p and E are disjoint from the interior of D.

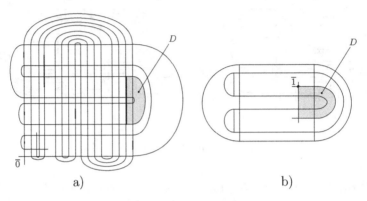

FIGURE 9. Pruning disks for the horseshoe

Theorem 3.1. *Let $F\colon S \circlearrowleft$ be a uniformly continuous orientation-preserving homeomorphism of an orientable surface, and let D be a pruning disk for F. Then there is an isotopy, supported on $U = \bigcup_{n \in \mathbb{Z}} F^n(\text{Int}(D))$, from F to a homeomorphism $F_D\colon S \circlearrowleft$ such that the non-wandering set of F_D is disjoint from U.*

The fact that the isotopy is supported in U means that the dynamics outside of U is unchanged by it. On the other hand, at the end of the isotopy there are no non-wandering points left inside U.[1] For example, in each of the two examples of Figure 9, an isotopy can be performed on the horseshoe which destroys precisely those orbits which enter the interior of the pruning disk.

Remark. A suitably generalized version of the pruning theorem holds for finite unions of pruning disks, which we refer to as *pruning fronts* (see [dC99] for details). In the language of Cvitanović [Cvi91], it would be more appropriate to call them *primary pruned regions.*

Let F be the horseshoe and D be a pruning disk for F. Denote by Δ the image of $\Lambda \cap D$ under the itinerary map. The set of nonwandering points of

---

[1]The two papers [dC99] and [dCH01] have slightly different definitions of pruning fronts and of the construction of the pruning isotopy. This statement is true for the formulation of [dC99]. For that of [dCH01], all of the dynamics is destroyed with the possible exception of a single periodic orbit.

F_D in S is given symbolically by

$$\Sigma_D = \{s \in \{0,1\}^{\mathbb{Z}}; \ \sigma^n(s) \notin \Delta \text{ for all } n \in \mathbb{Z}\}.$$

Clearly, Σ_D is σ-invariant and the map $\sigma: \Sigma_D \circlearrowleft$ is a direct analogue of the maps constructed in Section 2.3 using kneading sequences. Proving that an appropriately defined quotient of such a subshift is a surface homeomorphism is much harder than in dimension 1. For this reason, we construct the 2-dimensional models directly from the pruned maps F_D: this is an equivalent construction.

Definition 5. Let $\mathcal{D}$ denote the set of finite sequences $(D_1, \ldots, D_n)$ of disks with the property that D_1 is a pruning disk for F, and D_j is a pruning disk for $F_{D_1 \ldots D_{j-1}}$ for each j with $2 \leq j \leq n$. The *pruning family* associated to the horseshoe map F is

$$\mathcal{P} = \overline{\{F_{D_1 \ldots D_n}; \ (D_1, \ldots, D_n) \in \mathcal{D}\}},$$

where the closure is taken in the uniform topology.

We now define an equivalence relation for surface homeomorphisms, the 0-*entropy equivalence relation*: it plays a role analogous to that of $\sim_\kappa$ in Section 2.3. The quotient of an interval under a closed equivalence relation with connected classes is either a point or an interval, but in dimension 2 the situation is more complex, requiring more careful definitions and more elaborate proofs.

3.2.1. *The 0-entropy equivalence relation.* We start by recalling Bowen's definition of topological entropy [Bow71]. If (X,d) is a metric space and $F: X \circlearrowleft$ is uniformly continuous, we say that a subset S of X is (n, ϵ)-*separated* if it is possible to distinguish between the orbits of any two points $x, y \in S$ up to $n-1$ iterates with precision ϵ: $d(F^j(x), F^j(y)) > \epsilon$ for some $0 \leq j < n$. If X is compact, the topological entropy $h(F)$ of F is defined by

$$h(F) = \lim_{\epsilon \to 0} \limsup_{n \to \infty} \frac{1}{n} \ln N(n, \epsilon),$$

where $N(n, \epsilon)$ denotes the maximum cardinality of an (n, ϵ)-separated subset of X. For general X and $K \subset X$ compact, counting only those orbits which start in K, we obtain the entropy of F *in* K, denoted $h_F(K)$. Thus if we denote by $N(n, \epsilon, K)$ the maximum cardinality of an (n, ϵ)-separated subset of K, then

$$h_F(K) = \lim_{\epsilon \to 0} \limsup_{n \to \infty} \frac{1}{n} \ln N(n, \epsilon, K).$$

The topological entropy of F in X can then be defined as

$$h(F) = \sup\{h_F(K); \ K \subseteq X, K \text{ compact}\}.$$

Definition 6. Let $F\colon X \circlearrowleft$ be a homeomorphism. Then two points $x, y \in X$ are 0-*entropy equivalent* if there exists a continuum (i.e. a compact connected set) $K \subseteq X$ containing both x and y with

$$h_F(K) = 0 = h_{F^{-1}}(K).$$

Remarks.

a) That this indeed defines an equivalence relation follows from two facts: 1) $h_F(K \cup K') = \max\{h_F(K), h_F(K')\}$ and 2) the union of two continua containing a point in common is also a continuum.

b) If K is a proper subset of X, then it is not necessarily the case that $h_F(K) = h_{F^{-1}}(K)$.

c) In general, we can consider the family of equivalence relations $\sim_\alpha$, indexed by positive reals α, declaring two points to be $\sim_\alpha$-equivalent if there is a continuum containing both and carrying entropy smaller than α.

d) If $f\colon X \circlearrowleft$ is not invertible, we can define 0-entropy equivalence by omitting the condition on f^{-1}. It can be shown that the fibres of the semi-conjugacy given by Theorem 2.5 are exactly the 0-entropy equivalence classes under the unimodal map $f\colon I \circlearrowleft$. The equivalence classes of the α equivalence relations mentioned in the previous item can also be understood for unimodal maps in terms of renormalization.

The 0-entropy equivalence relation is most interesting for two-dimensional systems.

Example 7. In this example we describe the equivalence classes for the horseshoe map F. Denote by $\mathcal{H}^u$ and $\mathcal{H}^s$ the closures of the unstable and stable manifolds of the fixed point with itinerary $\overline{0}$ (or indeed of any other periodic point, since their closures coincide) and let $\mathcal{H} = \mathcal{H}^s \cup \mathcal{H}^u$. Equivalence classes are of four kinds:

a) Closures of connected components of $S^2 \setminus \mathcal{H}$.

b) Closures of connected components of $\mathcal{H}^u \setminus \mathcal{H}^s$ (not already contained in sets in a)).

c) Closures of connected components of $\mathcal{H}^s \setminus \mathcal{H}^u$ (not already contained in sets in a)).

d) Single points which are in none of the sets in a), b) or c).

To see that these sets do not carry entropy, notice that all points in any connected component of $S^2 \setminus \mathcal{H}$ (before taking the closure) converge to the attracting fixed point x. It is not hard to see that nothing significant happens on taking the closure, so that the sets in a) indeed carry no entropy. The same holds for sets of types b) and c). To see that any larger continuum must contain entropy, notice that if C is a connected set that contains two distinct sets among the ones described above, then it must intersect a Cantor set's worth of invariant manifolds, either stable or unstable (or both). It follows that either its ω- or its α-limit set contains the whole nonwandering set of the horseshoe, so that either $h_F(C) = \ln 2$ or $h_{F^{-1}}(C) = \ln 2$.

If F is a $C^{1+\epsilon}$ diffeomorphism of a surface S, then its 0-entropy equivalence classes form an upper semi-continuous decomposition of S into continua [dCP]. The main idea of the proof is that, by results of Katok and Pesin theory [Kat80, KH95], any continuum C carrying positive entropy must intersect a Cantor set of segments of stable or unstable manifolds of a periodic orbit which contains transverse homoclinic intersections. In particular, carrying positive entropy is stable under small perturbations of C, and so if a sequence (C_n) of continua carrying 0 entropy converges in the Hausdorff topology, then the limiting continuum also carries 0 entropy. This is equivalent to saying that the equivalence classes form an upper semi-continuous decomposition of S.

By results of Moore [Moo62] and Roberts-Steenrod [RS38], it follows immediately that $S/\sim$ is a *cactoidal surface* (intuitively, a surface with nodes). Since the equivalence is dynamically defined, F induces a homeomorphism $F/\sim: S/\sim \circlearrowleft$, which has the same topological entropy as F (by a result of Bowen [Bow71], using the fact that the equivalence classes carry 0 entropy). Moreover, any non-trivial continuum in $S/\sim$ must carry entropy of either $F/\sim$ or its inverse.

Intuitively, we can think of $F/\sim$ as a 'tight' version of F, in which all of the wandering domains have been collapsed to points. As an example, the quotient of the plane by the 0-entropy equivalence relation of the horseshoe (Example 7) is depicted in Figure 10. The quotient space is a sphere (since everything outside the homoclinic tangle is collapsed to a point), obtained by identifying the solid boundary in the figure along the dotted semi-circular arcs from the mid-point at the top to the corner point on the lower left. The stable and unstable manifolds of the horseshoe project to a transverse pair foliations with singularities, represented by solid and dashed lines, respectively. These foliations carry holonomy-invariant measures on transverse arcs, whose product gives a measure on the sphere. The quotient map preserves both foliations, dividing one of the measures by 2 and multiplying the other by 2, so that the product measure is invariant. This map is a *generalized pseudo-Anosov map*: it has all the defining characteristics of a pseudo-Anosov map, except that it has infinitely many singularities (and an accumulation point of singularities).

We finish with a conjectural analogue of Theorem 2.5. This is a precise statement of a conjecture due to Cvitanović [Cvi91].

Pruning Front Conjecture. *Let F be the horseshoe and $H_{a,b}$ denote the Hénon family. Then, for each choice of parameters a, b there exists a map $F_{a,b} \in \mathcal{P}$ such that $F_{a,b}/\sim$ and $H_{a,b}/\sim$ are topologically conjugate, where $\sim$ is the 0-entropy equivalence relation.*

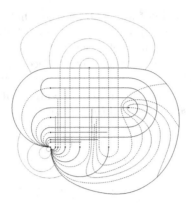

FIGURE 10. The quotient of the plane under the 0-entropy equivalence relation for the horseshoe

In other words, up to taking quotients by the 0-entropy equivalence relation, the Hénon family is contained in the pruning family of the horseshoe.

A similar conjecture has been proved by Ishii [Ish97] for the Lozi family

$$L_{a,b}(x,y) = (1 - a|x| - by, x),$$

which is more tractable than the Hénon family: it is piecewise affine and its non-differentiability locus $x = 0$ divides the plane naturally into two parts which can be used to introduce symbolic dynamics.

REFERENCES

[BC91]   M. Benedicks and L. Carleson. The dynamics of the Hénon map. *Ann. of Math. (2)*, 133(1):73–169, 1991.

[Bow71] R. Bowen. Entropy for group endomorphisms and homogeneous spaces. *Trans. Amer. Math. Soc.*, 153:401–414, 1971.

[CE80]   P. Collet and J.-P. Eckmann. *Iterated maps on the interval as dynamical systems.* Birkhäuser Boston, Mass., 1980.

[CGP88] P. Cvitanović, G. Gunaratne, and I. Procaccia. Topological and metric properties of Hénon-type strange attractors. *Phys. Rev. A (3)*, 38(3):1503–1520, 1988.

[Cvi91] P. Cvitanović. Periodic orbits as the skeleton of classical and quantum chaos. *Phys. D*, 51(1-3):138–151, 1991. Nonlinear science: the next decade (Los Alamos, NM, 1990).

[dC99]   A. de Carvalho. Pruning fronts and the formation of horseshoes. *Ergodic Theory Dynam. Systems*, 19(4):851–894, 1999.

[dCH]   A. de Carvalho and T. Hall. The forcing relation for horseshoe braid types. To appear in Experiment. Math.

[dCH01] A. de Carvalho and T. Hall. Pruning theory and Thurston's classification of surface homeomorphisms. *J. Eur. Math. Soc. (JEMS)*, 3(4):287–333, 2001.

[dCH02] A. de Carvalho and T. Hall. How to prune a horseshoe. *Nonlinearity*, 15(3):R19–R68, May 2002.

[dCP]   A. de Carvalho and M. Paternain. Monotone quotients of surface diffeomorphisms. In preparation.

[Dev89] R. Devaney. *An introduction to chaotic dynamical systems*. Addison-Wesley Publishing Company Advanced Book Program, Redwood City, CA, second edition, 1989.

[dMvS93] W. de Melo and S. van Strien. *One-dimensional dynamics*. Springer-Verlag, Berlin, 1993.

[Ish97] Y. Ishii. Towards a kneading theory for Lozi mappings. I. A solution of the pruning front conjecture and the first tangency problem. *Nonlinearity*, 10(3):731–747, 1997.

[Kat80] A. Katok. Lyapunov exponents, entropy and periodic orbits for diffeomorphisms. *Inst. Hautes Études Sci. Publ. Math.*, (51):137–173, 1980.

[KH95] A. Katok and B. Hasselblatt. *Introduction to the modern theory of dynamical systems*. Cambridge University Press, Cambridge, 1995. With a supplementary chapter by Katok and Leonardo Mendoza.

[Moo62] R. Moore. *Foundations of point set theory*. American Mathematical Society, Providence, R.I., 1962.

[MT88] J. Milnor and W. Thurston. On iterated maps of the interval. In *Dynamical systems (College Park, MD, 1986–87)*, pages 465–563. Springer, Berlin, 1988.

[RS38] J. Roberts and N. Steenrod. Monotone transformations of two-dimensional manifolds. *Ann. of Math.*, 39:851–862, 1938.

[Sma67] S. Smale. Differentiable dynamical systems. *Bull. Amer. Math. Soc.*, 73:747–817, 1967.

[WY01] Q. Wang and L-S. Young. Strange attractors with one direction of instability. *Comm. Math. Phys.*, 218(1):1–97, 2001.

INSTITUTE FOR MATHEMATICAL SCIENCES, STATE UNIVERSITY OF NEW YORK, STONY BROOK, NY 11794-3660, USA
E-mail address: andre@math.sunysb.edu

DEPARTMENT OF MATHEMATICAL SCIENCES, UNIVERSITY OF LIVERPOOL, LIVERPOOL L69 7ZL, UNITED KINGDOM
E-mail address: T.Hall@liverpool.ac.uk

MARKOV ODOMETERS

ANTHONY H. DOOLEY

ABSTRACT. We set up the theory of G-measures on Bratteli-Vershik systems and show how these correspond to induced transformations of odometers. Further, we show that each non-singular system corresponds to a Markov odometer which is uniquely ergodic as a G-measure.

CONTENTS

1. INTRODUCTION

By a **measurable dynamical system** we shall mean a quadruple $(X, \mathcal{B}, \mu, T)$ where X is a set, $\mathcal{B}$ a σ-algebra on X, μ a measure on $\mathcal{B}$ and T is a bimeasurable invertible transormation on X with the property that $\mu \circ T \sim \mu$. The system is called measure-preserving if $\mu \circ T = \mu$.

Actually, such a quadruple leads to an action of $\mathbb{Z}$ on X via $m \cdot x = T^m x$, $m \in \mathbb{Z}$. We shall sometimes also use the term measurable dynamical system to mean a quadruple $(X, \mathcal{B}, \mu, \Gamma)$ where $(X, \mathcal{B}, \mu)$ is as above, and Γ is a group of bimeasurable invertible transformations of X.

A system is **ergodic** if $E \in \mathcal{B}$ and $E = \gamma E$ for all $\gamma \in \Gamma \Rightarrow \mu(E) = 0$ or $\mu(E^c) = 0$.

Perhaps the best-known examples of measurable dynamical systems is the **odometer** defined on $X = \prod_{i=1}^{\infty} \mathbb{Z}_{l(i)}$ by $T(s, *) = (s+1, *)$ if $s+1 < l(0) - 1$ and $T(l(0) - 1, l(1) - 1, \ldots, l(j-1) - 1, s*) = (0, 0, \ldots 0, s+1, *)$ if $s + 1 < l(j) - 1$. Note that we do not define $T(l(0) - 1, l(1) - 1, (2) - 1, \ldots)$.

In this example, the odometer may act on any measure on X with the property that $\mu \circ T \sim \mu$. We shall look at some examples in section 3.

One of the major preoccupations of last century in this area was a classification of measurable dynamical systems.

What do we mean by classification? We have to define some kind of equivalence. For measure-preserving systems, the natural equivalence is **metric isomorphism**.

Definition 1.1. $(X, \mathcal{B}, \mu, T)$ *is* **metrically isomorphic** *to* $(Y, \mathcal{C}, \nu, S)$ *if there exists* $\phi : X \to Y$, *invertible and bi-measurable such that*

$$\phi \circ T = S \circ \phi \qquad \text{and} \qquad \mu \circ \phi^{-1} = \nu.$$

A looser type of equivalence is orbit equivalence.

Definition 1.2. *Let* Γ *act on* X *and* Δ *act on* Y. *Then* $(X, \mathcal{B}, \mu, \Gamma)$ *is orbit equivalent to* $(Y, \mathcal{C}, \nu, \Delta)$ *if there exists* $\phi : X \to Y$, *invertible and bi-measurable such that* $\mu \circ \phi^{-1} \sim \nu$, *and for all* $\gamma \in \Gamma$ *and for almost all* $x \in X$ *there exists* $\delta \in \Delta$ *with* $(\phi \circ \gamma)(x) = (\delta \circ \phi)(x)$.

The first classification of dynamical systems into types goes back to von Neumann.

Definition 1.3. *A non-singular system is of type:*

I_n *if* X *is a finite set of* n *elements and* μ *atomic*

I_∞ *if* X *is an infinite countable set and* μ *atomic*

II_1 *if there is a continuous measure* ν *equivalent to* μ *which is an invariant probability measure*

II_∞ *if there is a continuous invariant measure* ν *equivalent to* μ *with* $\nu(X) = \infty$

III *if* X *has no atoms and there is no equivalent invariant measure.*

It is easy to see that every ergodic non-singular system is of one of these types. Without loss of generality a system of type III may be considered to have a probability measure, as we may replace any measure by an equivalent probability measure.

A major step forward was taken by Henry Dye [9] in 1963 — actually, he only stated the following result for invariant-systems, though his techniques work just as well for non-singular systems. (See [23] for the details.)

Theorem 1.1 (Dye) *If* μ *is a continuous probability measure, then* $(X, \mathcal{B}, \mu, T)$ *is orbit equivalent to the odometer.*

Actually, Krieger [17] showed that the same is true for $(X, \mathcal{B}, \mu, \Gamma)$, Γ any discrete amenable group.

In fact, Dye showed more: there is just one orbit equivalence class of type II_1 and of type II_∞ systems. (The same statement for types I_n and I_∞ is trivial.) The task of sub-dividing and classifying measure-preserving systems into metric isomorphism classes, using entropy is another chapter of ergodic theory which I shall not discuss here.

There remained the problem of type III systems. A beautiful exposition of the classification of these was given in [21]. It is not my intention to reproduce that here. However, I should like to recall the definition of the Krieger-Araki-Woods ratio set. It is a subset of $[0, \infty]$ whose intersection with $(0, \infty)$ is a closed subgroup.

Definition 1.4. *We say that $r \in r(X, \mathcal{B}, \mu, \Gamma)$ if for all $\epsilon > 0$ and for every set A of positive measure there exists $\gamma \in \Gamma$ and there exists $B \subseteq A$, and*

$$\left| \frac{d\mu \circ \gamma}{d\mu}(x) - r \right| < \epsilon \qquad \text{for all } x \in B.$$

[Actually, the definition needs a slight (but obvious) modification for $r = \infty$.]

Since there are not too many closed subgroups of $(0, \infty)$, we see that one of the following cases must hold:

- the ratio set is $\{1\}$ — in this case we have a system of type II
- the ratio set is $[0, \infty]$ — the system is said to be of type III_1
- the ratio set is $\{\lambda^n : n \in \mathbb{Z}\}$ for some $0 < \lambda < 1$ — the system is said to be of type III_λ
- the ratio set is $\{0, 1, \infty\}$ — the system is said to be of type III_0

The basic properties of the ratio set are described in [19]. The celebrated Connes-Krieger theorem [7] states

Theorem 1.2 (Connes-Krieger) *There is a unique orbit equivalence class of sets of type III_λ for $0 < \lambda \leq 1$.*

It is known that this statement is false for the type III_0 systems — and in fact, there are uncountably many different orbit equivalence classes within type III_0. Paradoxically, it is quite hard to write down concrete examples of type III_0 measures. (For some examples, see [14], [12].)

The classification of type III_0 measures into orbit equivalence classes remains as a tantalizing problem. By Dye's Theorem, it suffices to describe all measures on an infinite product space X which are ergodic for the odometer and of type III_0.

2. BRATTELI-VERSHIK DIAGRAMS

Bratteli introduced these to study the spectral space of AFC^*-algebras, and Vershik to study topological dynamics. As we shall point out below, they are also useful in measurable dynamics.

A nice exposition of the basics of Bratteli-Vershik diagrams is given in [15] — I shall summarize what I need from this source.

Let $V = \cup_{k=1}^\infty V_i$ be a union of finite sets of vertices, and let

$$E = \bigcup_{i=1}^\infty E_i$$

be a union of finite sets of edges with given range and source maps $r(E_n) \subseteq V_n$ and $s(E_n) \subseteq V_{n-1}$.

V_{n-1}

E_n

V_n

Figure 1

The sets V and E together with the maps r and s define a **Bratteli diagram**.

For an **ordered Bratteli diagram**, we assume in addition an order $\geq$ on E, so that e and e' are comparable if and only if $r(e) = r(e')$.

For fixed $k, l \in \mathbb{N}$, let $P_{k,l}$ be the set of all paths from V_k to V_l , $P_{k,l} = \{(e_{k+1}, \ldots, e_l) : e_i \in E_i$ for $i = k+1, \ldots, l$ and $r(e_i) = s(e_{i+1})$ for $i = k+1, \ldots, l-1\}$.

We define

$$r(e_{k+1}, \ldots, e_l) = r(e_l)$$

and

$$s(e_{k+1}, \ldots, l_l) = s(e_{k+1})$$

and an order

$$(e_{k+1}, \ldots, e_l) > (f_{k+1}, \ldots, f_l)$$

if and only if there is an

$$i \in \{k+1, \ldots, l\} \text{ with } e_i > f_i$$

and

$$e_j = f_j \text{ for } i < j \leq l.$$

For any sequence $m_0 = 0 < m_1 < m_2 < \ldots$, a contraction of (V, E) by (m_j) is the (ordered) Bratteli-Vershik diagram with edges $E'_n = P_{m_{j-1}, m_j}$ and $r' = r$, $s' = s$ (as above), and order as above.

As is well-known, the set

$$X = \varprojlim P_{0,k}$$

is a compact Hausdorff space — it is the set of infinite paths

$$X = \{\mathbf{e} = (e_1, e_2, \ldots) : e_i \in E_i \quad \text{and} \quad r(e_i) = s(e_{i+1}) \ \forall\, i \geq 0\}.$$

It is clear that this space is homeomorphic for contracted diagrams.

Vershik observed that X is a compact Hausdorff space when equipped with the topology generated by the open set

$$U(f_1, \ldots, f_k) = \{\mathbf{e} : e_i = f_i \quad \text{for } 1 \leq i \leq k\}$$

and that the Vershik transformation $\varphi : X \to X$ defined by

$$\varphi(e_1, e_2, \ldots) = (f_1, f_2 \ldots f_k, e_{k+1}, \ldots),$$

where $(f_1, f_2 \ldots f_k)$ is the successor of $(e_1 \ldots e_k)$ in $P_{0,k}$ is a homeomorphism of X.

If (V, E) has a unique maximal path $e_{\max}$, this defines φ on every non-maximal path. We then take $\varphi(e_{\max}) = e_{\min}$ — the unique minimal path.

It is then known that simple ordered Bratteli diagrams are in one-to-one correspondence with essentially minimal pointed topological systems (X, φ, y), where the right-hand side is up to homeomorphism and the left-hand side up to contraction and isomorphism of Bratteli diagrams.

Of course, (X, φ) also generates a measurable dynamical system. We shall see that it is closely related to the odometer. My basic question is: how can we put measures on this space, and how can we analyse them?

To recover the odometer picture, one takes every vertex set to be of cardinality 1. We allow the edge E_i to have cardinality l_i — and index it by $\mathbb{Z}_{l_{(i)}}$, with the usual order $0 \leq 1 \leq \ldots \leq l_i - 1$. Then X corresponds to $\prod_{i=1}^{\infty} \mathbb{Z}_{l_{(i)}}$, and φ corresponds to the odometer T introduced above.

$$\textit{The odometer on } \prod_{i=1}^{\infty} \mathbb{Z}_2.$$

Figure 2

Roughly speaking, a measurable dynamical system arising from a Bratteli diagram may be thought of as an odometer where not all transformations are permitted. More precisely, we will see that these correspond to partially defined odometers.

Given a Bratteli-Vershik diagram, there are two Bratteli-Vershik diagrams associated to it which are in some sense, extreme cases: the **acyclic cover** X_1, and the **projected odometer** X_0.

The first of these, X_1 is obtained by increasing the number of vertices at level n to equal the number of paths in $P_{0,n}$. The diagram is a tree graph.

The second is obtained by identifying all vertices at one level, but retaining the number of edges. The number of paths $P_{0,n}(X_0)$ is equal to $P_{0,n}(X)$.

The situation is illustrated below (figure 3).

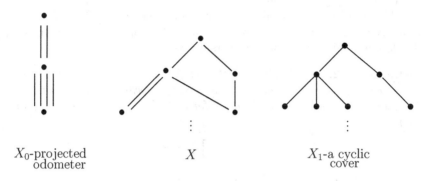

X_0-projected
odometer

X

X_1-a cyclic
cover

Figure 3

There are mappings $q_1 : X_1 \to X$ and $q_0 : X \to X_0$. It can easily be seen that both q_0 and q_1 are continuous maps of topological spaces.

Let us note that there is a natural identification of the paths of X as a subset of the paths of X_0. We may identify

$$X = \{(e_1, e_2, \ldots) \in X_0 : r_X(e_i) = s_X(e_{i+1}) \ \forall i\}.$$

This is a Borel set in X_0, as it is an infinite intersection of open sets.

We may without loss of generality, assume that the order on X (as a subset of X_0) is compatible with the order on X_0 in the sense that $e < f \implies e <_{X_0} f$.

Definition 2.1. *Suppose that the orders on X_0 and X are compatible. Let φ_0 be the Vershik transformation of X_0 and φ the Vershik transformation of X. Then φ is the induced transform on X, considered as a subset of X_0.*

Proof. We have to show that for $x \in X \subseteq X_0$

$$\varphi(x) = \varphi_0^n(x),$$

where $n = n(x)$ is the least value of $\{m \in \mathbb{Z} : \varphi_0^m(x) \in X\}$. But this is clear from the fact that the orders are compatible. $\qquad\square$

We should like to make some brief commentary on the acyclic cover — introduced in [10]. For this Bratteli diagram, the Vershik transform is trivial — and the associated AF algebra is abelian. The space of paths of X contains the space of paths of X_1 as a Borel subset, in exactly the same way as $X_0 \supseteq X$. The natural map $q_1 : X_1 \to X$ is a homeomorphism. X_1 may be thought of as a natural totally disconnected space which contains X.

There is a natural analogue of the finite coordinate change group on an infinite product space which can be found acting in the space X associated to any Bratteli diagram. In the case of an odometer on $\prod_{i=1}^{\infty} \mathbb{Z}_{l(i)}$, we set

$\Gamma = \coprod_{i=l}^{\infty} \mathbb{Z}_{l(i)}$ where $\Gamma = \{ \mathbf{x} \in X : x_i = 0 \text{ for } i \text{ sufficiently large} \}$ and Γ acts on X by $(\gamma.x)_i \geq \gamma_i + x_i (\operatorname{mod} l(i))$.

This group has the same orbits as the odometer, and has proved useful in analysing non-singular odometer actions ([2], [3], [4]). It is interesting to develop an analogue for Bratteli-Vershik systems.

Definition 2.2. *Let (V, E) be a Bratteli-Vershik diagram, and for each n, let Θ_n be the subgroup of the permutations of E_n which preserves the set with common range and source. That is, for $e \in E_n$, $\theta \in \Theta_n$, $r(e) = r(\theta e)$ and $s(e) = s(\theta e)$.*

Now let
$$\Gamma_n = \prod_{k=1}^{n} \Theta_k.$$

This group acts on $P_{0,n}$ by $\gamma(e)_i = (\gamma_i e_i)$, where $\gamma = (\gamma_1, \gamma_2, \gamma_3, \ldots, \gamma_n)$, $\gamma_i \in \Theta_i$ and $e = (e_1, e_2, e_3, \ldots, e_n) \in P_{0,n}$. In fact, G_n acts naturally on X and $\Gamma = \coprod_{k=1}^{\infty} \Theta_k$ acts on X.

This action of Γ is continuous in the topology on X. Γ is a discrete abelian group, given the normal product topology. As in the case of the odometer, let us note that the Γ-orbits coincide with the φ-orbits — in each case the orbit of a sequence $e \in X$ is the set of sequences $(f_1, f_2, \ldots)$ such that for i sufficiently large (depending on e) $e_i = f_i$, and for all values of i, $r(e_i) = r(f_i)$ and $s(e_i) = s(f_i)$.

3. G-MEASURES, MARKOV MEASURES AND UNIQUE ERGODICITY

In this section, we will describe the set of measures on X via the G-measure formalism introduced in [3] for the odometer. The idea of this formalism is to simultaneously generalize Mike Keane's g-measures [16] and the Riesz products of classical harmonic analysis.

Throughout this section, (V, E) will be a Bratteli-Vershik diagram, (X, φ) the associated dynamical system. Let $\mathcal{B}$ be the Borel σ-algebra. Here, we shall treat it as a non-singular measurable dynamical system — the problem being to put measures μ on $(X, \mathcal{B})$ such that $\mu \circ \varphi \sim \mu$.

Such a measure is called **ergodic** if for all $B \in \mathcal{B}$, $\varphi(B) = B \implies \mu(B) = 0$ or $\mu(B^c) = 0$. These measures are the building blocks of all the measures.

Now, by the remarks at the end of the previous section, it is clear that a measure μ is quasi-invariant (and ergodic) for φ, if and only if it is quasi-invariant (and ergodic) for Γ.

Definition 3.1. *Given a probability measure on X, we will consider the **tail measure***

$$\mu^{(n)} = \frac{1}{|\Gamma_n|} \sum_{\gamma \in \Gamma_n} \mu \circ \gamma.$$

Note that μ is one of the summands here and so $\mu < \mu^{(n)}$ (and indeed, if μ is Γ-quasi-invariant, $\mu \sim \mu^{(n)}$). Thus, the function $G_n(x) = \frac{d\mu}{d\mu^{(n)}}(x)$ is well-defined (at least, almost everywhere) and satisfies the following two conditions

1. $\dfrac{1}{|\Gamma_n|} \displaystyle\sum_{\gamma \in \Gamma_n} G_n(\gamma x) = 1$ (normalization), and

2. If $\gamma \in \Gamma_k$ and $n \geq k$, then

$$G_n(\gamma x) G_k(x) = G_k(\gamma x) G_n(x) \qquad \text{(compatibility)}$$

(The compatibility comes simply from the fact that $\Gamma_k \subseteq \Gamma_n$.)

We are going to turn this definition around. Given a normalized compatible family $G = (G_n)_{n=1}^{\infty}$ of Borel functions, we say that μ is a **G-measure** if

(3.1) $$\frac{d\mu}{d\mu^{(n)}}(x) = G_n(x).$$

Actually, there is an equivalent formulation of this notion in terms of a sequence $(g_n)_{n=1}^{\infty}$ satisfying

1. $\dfrac{1}{|\Theta_n|} \displaystyle\sum_{\theta \in \Theta_n} g_n(\theta x) = 1$ (normalization)

2. $g_n(\gamma x) = g_n(x)$ for all $\gamma \in \Gamma_{n-1}$ (invariance)

If one writes $G_n(x) = g_1(x) \ldots g_n(x)$, then the (G_n) is a normalized compatible family. Conversely, given (G_n), we can define g_n by $g_n(x) = \frac{G_n(x)}{G_{n-1}(x)}$. In terms of the measures μ, we have

$$g_n = \frac{d\mu^{(n-1)}}{d\mu^{(n)}}.$$

The functions G_n (and g_n) are defined by μ up to sets of measure zero only. However, it turns out [3] that we can, by moving to an equivalent measure, assume that they are continuous. We shall discuss this further later. Thus, we may turn the definition around and ask, for a given normalized compatible sequence $(G_n)_{n=1}^{\infty}$, to describe the set of G measures. Of particular interest is to know when there is a unique G-measure μ for a given sequence $(G_n)_{n=1}^{\infty}$. This property was studied for odometers in [3], where various criteria were given for uniqueness. In particular, if there is a unique G-measure, it is easy to see that it is ergodic for Γ (or equivalently for φ).

Let (V, E) be a Bratteli-Veshik diagram, and let

Definition 3.2. Let $G = (G_n)_{n=1}^{\infty}$ be a normalized compatible family of continuous functions on X. Then we say that we have a **uniquely ergodic G-measure** if there is a unique measure μ such that (3.1) holds.

It is of interest to determine uniqueness conditions on a Bratteli-Vershik diagram which extend those of [3].

Examples 3.1 (i) A G-measure on X is a **product measure** if for each n, $g_n(x) = g_n(x_n)$ is a function of x_n alone. In the case of an odometer, where $X = \prod_{n=1}^{\infty} \mathbb{Z}_{l_n}$, this corresponds exactly to an infinite product measure on X. To recover the measure μ_n on the factor $\mathbb{Z}_{l_n}$, one sets $\mu_n(\{i\}) = l_n g_n(i)$, $i \in \mathbb{Z}_{l_n}$.

(ii)   A G-measure on X is a **Markov measure** if for each n, $g_n(x) = g_n(x_n, x_{n+1})$ depends only on the two successive coordinates x_n, x_{n+1}.

A useful way to think about Markov odometers on a full odometer space $X = \prod_{n=1}^{\infty} \mathbb{Z}_{l_n}$ is that the n-th state has l_n elements and for each n, the $l_n \times l_{n+1}$ matrix $(\frac{1}{l_n} g(x_n, x_{n+1}))$ specifies the probabilities of transition to the $n + 1$-th state. The normalization condition then becomes the usual Markov condition. In the case of a Bratteli diagram, one can identify X with a subset of X_0 as in Definition 2.1. The X transitions may then be thought of as being zero if $r(x_n) \neq s(x_{n+1})$.

Let us note that the Vershik transformation φ is not necessarily non-singular for a given G-measure. Indeed, the Vershik transformation is not defined on any path $(e_1, e_2, \ldots)$ which is maximal. For non-singularity, we must insist that maximal paths have measure zero. In fact, an obvious argument shows that

Proposition 3.3. *The measure μ is non-singular for φ (or alternatively, for Γ) if and only if the following two conditions hold:*

(i) For μ-a.e. $x \in X$ $\exists y \in X$ such that $x < y$.
(ii) For μ-a.e. $x \in X$ $\exists y \in X$ such that $x > y$.

Proof. Suppose that (i) holds. Then we have

$$\varphi x = \min\{z \in X : x < z\}.$$

Similarly, if (ii) holds then

$$\varphi^{-1} x = \max\{z \in X : z < x\}$$

and these are both defined μ-almost everywhere. Conversely, if either (i) or (ii) fails, then there is a set E of positive measure on which either φ or φ^{-1} is not defined.   □

In the case of a product measure, unique ergodicity is actually equivalent to non-singularity. In fact, we have

Proposition 3.4. *Let μ be a product measure. The following conditions are equivalent*

(i) μ is non-singular for φ
(ii) μ is ergodic for φ
(iii) $g_n(x_n) > 0$ for all n.

Proof. If μ is non-singular for φ, it is also non-singular for Γ.

Let $e = (e_1, \ldots e_n) \in P_{0n}$. Then $\mu(U(e)) > 0$ if and only if for each $\gamma \in \Gamma_n$, $\mu(U(\gamma e)) > 0$ by Proposition 3.3. Thus

$$\mu(U(e)) > 0 \quad \Leftrightarrow \quad \mu\left(\bigcup_{\gamma \in \Gamma_n} U(\gamma e) \right) > 0.$$

Since μ is non-trivial, we have $\mu(U(e)) > 0$ for all $e \in P_{0,n}$. But

$$\mu(U(e)) = l_1 \ldots l_n \, g_1(e_1) g_2(e_2) \ldots g_n(e_n).$$

It follows that $\mu(U(e) \triangle E) < \epsilon$ for all $\gamma \in \Gamma_n$.

Thus, by Proposition 3.3, (i) and (iii) are equivalent.

The implication (ii) $\Rightarrow$ (i) is by definition. The following standard proof gives the opposite implication.

Let $\mathcal{B}_n$ be the σ-algebra on X generated by $\{U(e_1 \ldots e_n) : (e_1 \ldots e_n) \in P_{0,n}\}$.

Now, let E be a Γ invariant set in $\mathcal{B}$. Let $\epsilon > 0$. We may choose n and $A \in \mathcal{B}_n$ so that $\|\chi_E - \chi_A\|_1 < \epsilon$. Then $|\mu(E) - \int \chi_A \chi_E d\mu| < \epsilon$. Since μ is a product measure

$$\int \chi_A \chi_E d\mu = \int \lambda_A d\mu \int \chi_E d\mu = \mu(A)\mu(E).$$

Letting $\epsilon \to 0$, $\mu(A) \to \mu(E)$ and we obtain $\mu(E) = \mu(E)^2$, i.e., $\mu(E) = 0$ or 1. $\qquad \square$

It is an interesting matter to give easily computable criteria for unique ergodicity of Markov measures. Here is one of the uniqueness results from [3], specialized to Markov measures, and generalized to Bratteli-Vershik systems.

Proposition 3.5. *Suppose that* $\{g_i\}$ *is a sequence of Markov functions on* X*-normalized tail-dependent and* $g_i(x) = g_i(x_i, x_{i+1})$.

Suppose further that

(i) $\liminf_i g_i(x_i, x_{i+1}) > 0$ *for all* $x \in X$, *and*

(ii) there exists $0 < \alpha < 1$ *and* $K > 0$ *so that for all* $e \in E_i$, $f, g \in E_{i+1}$ *with* $r(e) = s(f) = s(g)$,

$$|g_i(e, f) - g_i(e, g)| < \frac{K}{l(e)}|f - g|^\alpha,$$

where $l(e) = \text{card} \{f : r(e) = s(f)\}$.

Then there exists a unique G-measure.

The proof of this proposition follows the same lines as that of Theorem 1 of [3].

Actually, it was shown in [13] that in some senses, every G-measure is a Markov measure in the case of an odometer. This also holds true for Bratteli-Vershik systems. To see this, there are two steps.

Lemma 3.6. *Let μ be a G-measure. Then there exists a normalized compatible sequence $H = (H_n)$ and a H-measure ν such that $\nu \sim \mu$ and for each n, $H_n(x)$ depends on finitely many coordinates.*

Proof. We give a sketch of a proof of this result. By Lusin's theorem, each function $g_n(x)$ may be approximated by a continuous function h_n on X off a set of measure less that $\frac{\epsilon}{2^n}$. The topology on X is such that each h_n depends only on finitely many coordinates $h_n(x) = h_n(x_n, x_{n+1}, \ldots, x_m)$. Let $H_n(x) = h_1(x) \ldots h_n(x)$. Letting ν be a weak*-limit point of $\frac{G_n(x)}{H_n(x)} \mu$, we see that ν is an H-measure. Further, $\frac{G_n(x)}{H_n(x)}$ converges outside a set of measure less than $\frac{\epsilon}{2} + \frac{\epsilon}{4} + \ldots = \epsilon$. Now an application of Borel-Cantelli tells us that ν is equivalent to μ. □

We may then contract our Bratteli diagram to obtain a Markov measure. More specifically, suppose that $g_n(x)$ depends on $(x_n, x_{n+1}, \ldots, x_{m(n)})$ for a suitable sequence of integers $m(n)$.

We define inductively $k(1) = 1$, $k(2) = m(1)$, and

$$k(j) = \max_{k_{(j-1)} \leq i \leq m(j)} m(i).$$

Then the sequence $\{k(i)\}$ defines a contraction of the Bratteli diagram, with new vertex sets $\widetilde{V}_j = V_{k(j)}$ and new edge sets $\widetilde{E}_j = P_{k(j),k(j+1)}$. The space $\widetilde{X}$ for $(\widetilde{V}, \widetilde{E})$ is homeomorphic to X.

Furthermore, we may define a sequence of g-functions on $\widetilde{X}$ by, for $(y_1, y_2, \ldots) \in \widetilde{X}$

$$\widetilde{g}_k(y_k, y_{k+1}) = \prod_{i=k(j-1)}^{m(j)} g_i(y_k, y_{k+1}).$$

The choice of the sequence $k(i)$ guarantees that $g_i(y_k, y_{k+1})$ is defined. The fact that $\{g_i\}$ is a normalized sequence for Γ guarantees that $\{\widetilde{g}_k\}$ is a normalized sequence for $\widetilde{\Gamma}$ — the finite coordinate change group of $\widetilde{X}$.

We have shown

Proposition 3.7. *Suppose that (V, E) is a Bratteli-Vershik diagram and X is the associated space. Suppose that $\{G_n\}$ is a normalized compatible family of functions on X, each depending on finitely many coordinates.*

Then there exists a contraction $\widetilde{X}$ of X so that G_n corresponds to $\widetilde{G}_n$, a normalized compatible family of Markov transition functions.

These two lemmas allow us to assert that every G-measures "is", up to equivalence and contraction, a Markov measure. However, this process does not preserve unique ergodicity.

4. THE MAIN THEOREM

The previous section concluded with the heuristic statement that every G-measure "is" a Markov measure. This, combined with another heuristic statement, that every (φ-quasi-invariant) measure on X "is" a G-measure, and Dye's theorem [9]: every measurable dynamical system is orbit equivalent to an odometer, means that up to orbit equivalence, every dynamical system "is" a Markov odometer.

Recently, the author and Hamachi made this statement more precise. In fact, we can show more: for every ergodic measurable dynamical system, we may pass to a suitable induced transformation, obtaining a Markov odometer which is uniquely ergodic as a G-measure. In this section, I would like to give a proof of this result. The proof is based on the multiple tower construction of [21]. Vershik ([24]) showed that in the measure-preserving case that $(X, \mathcal{B}, T, \mu)$ is orbit equivalent to a Markov odometer via an orbit equivalence which is almost an equivalence — it preserves many properties of μ — and it is uniquely ergodic. Thus, our result may be considered to be an extension of Vershik's to the non-singular case.

Theorem 4.1. *Every ergodic non-singular transformation is orbit equivalent to a Markov odometer on a Bratteli-Vershik system which is uniquely ergodic.*

Remark 4.2 Let me remark that the results of Connes-Krieger tell us that there is a unique orbit equivalence class of systems of type II_1, II_∞, III_1, and III_λ for each $0 < \lambda < 1$. Given that each of these may be realized explicitly as a product measure, which is *a fortiori* uniquely ergodic, our theorem has new content only for type III_0 systems. We shall discuss explicit constructions of the Markov type III_0 systems which are not orbit equivalent to products in the next section.

Proof of Theorem 4.1. Full details of the proof are given in [11]. Because the details are rather technical, it seems useful to give a rather full outline of the proof here.

Throughout, we take $(X, \mathcal{B}, \mu, T)$ to be an ergodic non-singular transformation. We shall assume that $\mu(X) < \infty$. This is not strictly necessary, but it will simplify our proof.

We shall carry out the following steps:

1. Define the notion of an ordered multiple tower with constant Jacobian — this will be a layer of a Bratteli-Vershik diagram.
2. Explain how to refine an ordered multiple tower, and see how passing from a tower to its refinement induces a stochastic matrix.
3. Show how to approximate X by a sequence of refining ordered multiple towers, so that μ is obtained as a uniquely ergodic Markov measure.

Suppose we are given a Bratteli-Vershik diagram (V, E), and a Markov sequence of positive G-functions (i.e., the G_n are normalized and compatible and g_n depends only on (e_n, e_{n+1}) with $r(e_n) = s(e_{n+1})$).

For $\epsilon \in E(n)$ and $i \in V(n-1)$, choose $e(i)$ so that $r(e(i)) = i$. Then set

$$P_{i,\epsilon}^{(n)} = \begin{cases} |\Theta_n| g_n(i, \epsilon) & \text{if } s(\epsilon) = i; \\ 0 & \text{if } s(\epsilon) \neq i. \end{cases}$$

The matrix

$$P_{i,\epsilon}^{(n)} = \left\{ P_{i,\epsilon}^{(n)} \right\}_{i \in V(n-1), \epsilon \in E(n)}$$

is then a stochastic matrix such that $P_{i,\epsilon}^{(n)} > 0$ if and only if $s(\epsilon) = i$.

We also choose an "initial state" — a measure ν on $V(0)$ with $\nu(\{i\}) > 0$ for all $i \in V(0)$.

A G-measure ν on X is then a Markov measure for the initial state ν and the sequence $P^{(n)}$ in the sense that

(4.2) $\mu[U(e_1, e_2, \ldots, e_n)] = \nu(s(e_1)) P_{s(e_1), e_1}^1 P_{s(e_2), e_2}^2 \cdots P_{(e_n), e_n}$.

This coincides with the usual definition of Markus measures. We shall also need condition (i) and (ii) of Proposition 3.3, in order to guarantee non-singularity of μ with respect to φ.

Definition 4.2. *Let $(X, \mathcal{B}, \mu, T)$ be a non-singular transformation system. We denote by $[T]_*$ the set of all **partially defined elements of the full group**, i.e. transformations $\phi : A \to B$, where $A = \operatorname{Dom} \phi$ and $B = \operatorname{Im} \phi$ have positive measure in X and such that for almost all $x \in A$, $\phi(x) \in \operatorname{Orb}_T(x)$.*

Definition 4.3. *If $A \subseteq X$ is a set of positive measure on **ordered tower of** A is:*

- *A partition of A, $\mathcal{P} = \{A_\alpha : \alpha \in \Lambda\}$ by measurable sets A_α, where Λ is a finite totally ordered set.*
- *A collection of maps $\xi_{\alpha,\beta} \in [T]_*$, for all $\alpha, \beta \in \Lambda$ such that*
  - *(i) $\operatorname{Dom} \xi_{\alpha,\beta} = A_\beta$*
  - *(ii) $\operatorname{Im} \xi_{\alpha,\beta} = A_\alpha$,*
  - *(iii) $\xi_{\alpha\alpha} = \operatorname{Id}\big|_{A_\alpha}$ and*
  - *(iv) $\xi_{\alpha\beta} \circ \xi_{\beta\gamma}$ for all $\alpha, \beta \in \Lambda = \xi_{\alpha,\gamma}$ for all triples $\alpha, \beta, \gamma \in \Lambda$.*

*The **support** of the tower is the union of the sets A_α.*

*Given an ordered tower, we define the **associated transformation** S by*

$$Sx = \xi_{\beta\alpha} x \quad \text{if } x \in A_\alpha,$$

where $\beta = \min_{\theta \in \Lambda}\{\theta : \theta > \alpha\}$. One sees easily that $s \in [T]_*$.

*A tower $\xi = (\xi_{\alpha,\beta})$ is of **constant Jacobian** if*

$$\frac{d\mu \circ \xi_{\alpha\beta}}{d\mu}(x) = \text{const} = \frac{\mu(A_\alpha)}{\mu(A_\beta)}$$

for all $\alpha, \beta \in \Lambda$ and for a.e. $x \in A_\beta$.

Our main tool — as in the proof of Dye's theorem — is the following version of Rokhlin's lemma. In this form, it goes back to Krieger at least.

Lemma 4.4. *Let $(X, \mathcal{B}, \mu, T)$ be a non-singular ergodic system. Let $A_1, A_2, \ldots, A_k \in \mathcal{B}$ be sets of positive measure, let $\epsilon > 0$ and let $N \geq 1$.*

Then there exists a measurable subset E of positive measure and a tower ξ of E such that

  (i) $\mu(E) > 1 - \epsilon$.
 (ii) $\forall x \in E$, $T^i x \in \mathrm{Orb}_S(x)$ for $-N \leq i \leq N$ (where S is the associated transformation of ξ).
(iii) Each set A_i is approximated by a union of levels of ξ, up to symmetric difference of measure less than ϵ.

We shall omit the proof of this (standard) result. Our strategy is to use this lemma to create a sequence of ordered towers of constant Jacobian which successively approximate X.

In order to carry out this strategy, we need to consider an ordered multiple tower, i.e. a set $\xi^{(i)} : i = 1, \ldots, m$ of towers with the sets $\mathrm{supp}\, \xi^{(i)}$ disjoint.

Given such a set of towers, let $\Lambda = \{\alpha i : 1 \leq i \leq m$ and $\alpha \in \Lambda^{(i)}\}$ and $\xi_{\alpha_i \beta_i}(x) = \xi_{\alpha\beta}^{(i)}$ if $x \in \mathrm{supp}\, \xi^{(i)}$ and define the associated transformation in the obvious way.

We then need to **refine** this ordered multiple tower. Choose and fix some index $\alpha^{(i)}$ in each $\Lambda^{(i)}$ and let $E = \cup_{i=1}^m P(\xi^{(i)})_{\alpha_i}$. The set E has a natural partition $E = \cup_{i=1}^m E_i$, as above. Now let us suppose that we have an ordered multiple tower of E whose partition is finer than this one. Call it $\eta = \{\eta^{(i)}, \ldots, \eta^{(m)}\}$, where $\eta^{(j)} = \{\eta_{\epsilon\delta}^{(j)} : \epsilon, \delta \in \Lambda\}$.

We will define an ordered multiple tower ζ of the ξ-invariant set $\cup_{i=1}^n \cup_{\beta \in \Lambda^{(i)}} \xi_{\beta, \alpha^{(i)}}(\mathrm{supp}(\eta))$. Put

$$\Lambda(\zeta^{(j)}) = \{\beta i \epsilon : 1 \leq i \leq n, \beta \in \Lambda(\xi^{(i)}), \epsilon \in \Lambda(\eta^{(j)}) \quad \text{and} \quad P(\eta^{(j)})_\epsilon \subseteq E_i\}.$$

On $\Lambda(\zeta^{(j)})$, define the lexicographic ordering $\beta i \epsilon < \beta' i' \epsilon'$ if

$$\text{either} \quad \begin{cases} \epsilon < \epsilon' & \text{or} \\ \epsilon = \epsilon', i = i' \quad \text{and} \quad \beta < \beta'. \end{cases}$$

Let

$$A_{\beta i \epsilon}^j = \xi_{\beta, \alpha^{(i)}}^{(i)}(P(\eta^j)_\epsilon) \leq P(\xi^i)_\beta$$

and

$$\zeta_{\alpha^{(i)} i \delta, \alpha^{(i')} i' \epsilon} = \eta_{\delta\epsilon}^{(j)} \quad \text{for} \quad \alpha^{(i)} i \delta, \alpha^{(i')} i' \epsilon \in \Lambda(\xi^{(i)})$$

$$\zeta_{\beta i' \epsilon, \alpha^{(i')} i' \epsilon}^{(j)} = \xi_{\beta, \alpha^{(i')}} \quad \text{for} \quad \beta i' \epsilon, \alpha^{(i')} i' \epsilon \in \Lambda(\zeta^{(j)})$$

and

$$\zeta^{(j)}_{\beta i\delta,\gamma i'\epsilon} = \zeta^{(j)}_{\beta i\delta,\alpha^{(j')}i\delta} \circ \zeta^{(j')}_{\alpha^{(i)}i\delta,\alpha^{(i')}i'\epsilon} \circ \left(\zeta^{(j)}_{\gamma i'\epsilon,\alpha^{(i')}i'\epsilon}\right)^{-1}$$

$$\text{for } \beta i\delta, \gamma i'\epsilon, \alpha^{(i)}i\delta, \alpha^{(i')}i'\epsilon \in \Lambda(\zeta^{(j)}).$$

This is exactly the transformation used in Vershik's lemma on essentially minimal topological spaces [15].

Definition 4.5. *We call ζ the **extension of** ξ **by** η.*

We are going to think of ξ as one layer of a Bratteli-Vershik diagram, and ζ as the next layer. The edges connecting these two layers are given by containments of sets. The set of edges joining $\xi^{(i)}$ and $\zeta^{(j)}$ being in one-to-one correspondence with $\{\beta i\epsilon : \beta i\epsilon \in \Lambda\xi^{(j)}\}$. For such an edge, we set $S(\beta i\epsilon) = i$ and $r(\beta i\epsilon) = j$.

The interesting (and novel) thing for us to understand is how a tower of constant Jacobian is transformed by this process.

In fact, suppose that ξ is of constant Jacobian with respect to μ, and that there is a measure $\nu \sim \mu$ on E so that η is of constant Jacobian with respect to ν. We define $\rho \sim \mu$ on X by setting

(1) $\rho = \nu$ on E.

(2) For $\beta i\delta, \gamma i'\epsilon \in \Lambda(\xi^{(j)})$, $x \in A^{(j)}_{\gamma i'\epsilon}$

$$\frac{d\rho \circ \zeta^{(i)}_{\beta i\delta,\gamma i'\epsilon}}{d\rho}(x) = \frac{\mu(P(\xi^{(i)})_\beta)}{\mu(E_i)} \frac{\nu(P(\eta^{(j)})_\delta)}{\nu(P(\eta^{(j)})_\epsilon)} \frac{\mu(E_{i'})}{\mu(P(\xi^{(i')})_\gamma)}.$$

In fact, it is easy to check that for $B \in \mathcal{P}(\xi^{(i)})\gamma$, we have

$$\rho(B) = \frac{\mu(\mathcal{P}(\xi^{(i)})_\gamma)}{\nu(P(\xi^{(i)})_{\alpha^{(i)}})}\nu(\xi^{(i)}_{\alpha^{(i)}_\gamma}(B))$$

and

$$\frac{d\rho}{d\mu}(x) = \frac{d\nu \circ \xi^{(i)}_{\alpha^{(i)},\gamma}}{d\mu}(x), \quad x \in \mathcal{P}(\xi^{(i)})_\gamma.$$

Definition 4.6. *Then $\mu_\nu = \rho$ is called the **Markov extension** of μ **by** ν.*

In fact, we can define a Markov transition matrix for the two layers of a Bratteli diagram defined above. Letting $S(\beta i\epsilon) = i$ we take

$$P_{i,\beta i\epsilon} = \frac{\mu_\nu(P(\xi^{(j)})_{\beta i\epsilon})}{\mu_\nu(\text{supp}(\xi^{(i)}))} = \frac{\mu(P(\xi^{(i)})_\beta)}{\mu(\text{supp}\,\xi^{(i)})} \frac{\nu(P(\eta^{(j)})_\epsilon)}{\nu(P(\xi^{(j)})_{\alpha^{(i)}})}.$$

Then $P_{i,\beta i\epsilon}$ is the stochastic matrix associated with the extension of μ by ν.

Now for the inductive step. Let $\epsilon > 0$ and use Lemma 4.4 to construct a tower ξ which covers nearly all of X. Partition each $P(\xi)_\alpha$ into a finite number of sets of positive measure $E(j)_\alpha$, $1 \leq j \leq n+1$, so that

$$\frac{\mu(E(j)_{\beta'})}{\mu(E(j)_\beta)} e^{-\epsilon} < \frac{d\mu \circ \xi_{\beta'\beta}}{d\mu}(x) < \frac{\mu(E(j)_{\beta'})}{\mu(E(j)_\beta)} e^{\epsilon}.$$

The tower obtained by this finer partition has approximately constant Jacobian.

Continuing in this way (for full details of the proof, see [11]) we get, for any (ϵ_n), $(\epsilon_n \to 0$ and $\Sigma\epsilon_n < \infty)$, a sequence ξ_n of ordered multiple towers, each of constant Jacobian for measures μ_n such that:

(1) ξ_{n+1} is an extension of ξ_n
(2) μ_{n+1} is a Markov extension of μ_n
(3) $\mu(\mathrm{supp}(\xi_{n+1})) > (1 - \epsilon_n)\mu(\mathrm{supp}(\xi_n))$
(4) $e^{-\epsilon_n} < \frac{d\mu_{n+1}}{d\mu_n}(x) < e^{\epsilon_n}$ for all $x \in \mathrm{supp}(\xi_{n+1})$
(5) $\mu\{x \in \mathrm{supp}(\xi_n) : S_n x$ or $S_n^{-1}x$ is not defined$\} \leq \epsilon_n\mu(\mathrm{supp}(\xi_n))$ (where S_n is the associated transformation of ξ_n)
(6) A_n is approximated by a union of levels of ξ_n, in symmetric difference of levels to within ξ_n.
(7) $\mu\{x \in \mathrm{supp}(\xi_n) : T^i x \in \mathrm{Orb}_{S_n}(x), -n \leq i \leq n\} > (1 - \epsilon_n)\mu(\mathrm{supp}\,\xi_n)$.

Using this inductive procedure, we set

$$d\nu = \prod_{n=1}^{\infty} \frac{d\mu_{n+1}}{d\mu_n}d\mu \text{ on } F = \cap_{n=1}^{\infty}\mathrm{supp}\,(\xi_n)$$

and use the Borel-Cantelli Lemma to see that (X, μ, T) is orbit equivalent to the Markov system associated to (ξ_n) and ν as above. Property (4) guarantees that ν is uniquely ergodic for the g-functions $\frac{d\mu_{n+1}}{d\mu_n}(x)$. This proves the Theorem.   □

Remarks 4.7. (i) Of course, Krieger showed that every ergodic system of type III_λ $(0 < \lambda \leq 1)$ is unique up to orbit equivalence — and they can be realized as product measures on full odometers ([21]). So our result is new only for systems of type III_0.

(ii) The result of Proposition 3.7 is refined by Theorem 4.1 with the extra conclusion that we have a uniquely ergodic G-measure.

(iii) In this context, it is natural to ask for an example of a Markov odometer (necessarily of type III_0) which is not orbit equivalent to a product. This is done in [11] and we will discuss this in section 6.

(iv) Another natural question is the study of non-uniqueness for Markov measures — how bad can the set of G-measures be? We shall discuss this in the next section.

5. NON-UNIQUENESS RESULTS

Keane's original g-measure paper [16] contains the proof that if $l_n = l$ for all n, if the function g is positive on $[0, 1]$ and C^1, and if we define $g_n(x) =$

$g(l^n x)$ for $n \in \mathbb{N}$, then the G-measure associated to (g_n) (he called it a g-measure) is uniquely ergodic. Keane's student B. Petit showed the result was still valid if $g \in Lip(\alpha)$ — and it was this result which inspired the uniqueness result for G-measures from [2] cited in Proposition 3.5.

However, the question remained as to whether positivity and continuity of g suffice to give uniqueness. Finally, Bramson and Kalikow [1] showed that there was a positive function g on $\prod^\infty \mathbb{Z}_l$, continuous in the product topology for which the corresponding g-measure was not uniquely ergodic. Their proof involved a clever use of a random joinings argument. Finally, Anthony Quas [22] found an example of a function continuous in the $[0,1]$ topology for which there is not a unique g-measure.

Given that we now know that every G-measure is a Markov measure, it is of some interest to ask how bad non-uniqueness can be for a set of Markov G functions. In a recent preprint [13], Dan Rudolph and I have shown that it can be quite arbitrary.

Theorem 5.1. *([13]) Let $\widetilde{\Lambda} \subseteq [0,1]$ be an arbitrary G_δ set. Then there exists a sequence $\{P_{i,\epsilon}^{(n)}\}$ of Markov transition matrices so that (4.1) holds for a set of probability measures S, where the Krieger types of the measures in S are exactly $\{\mathrm{III}_\lambda : \lambda \in \widetilde{\Lambda}\}$.*

Again, the details are quite technical, so I do not want to give full details here. However, I would like to sketch the ideas of the proof.

Suppose we take a full odometer with a fast-increasing number of coordinates, whose Bratteli diagram is like this

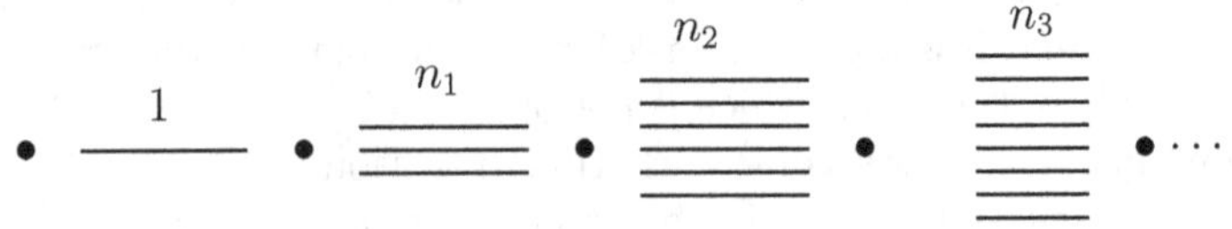

Figure 4

Let us also take a collection of numbers $\{\lambda_{i,j} : i, j \in \mathbb{N}\}$ the set of whose limit points is $\widetilde{\Lambda}$. We then make up an $n_i \times n_{i-1}$ stochastic matrix like this

$$
P^n = c \left(
\begin{array}{ccccc|ccc}
 & & & & & \vdots & \vdots & 1 \\
 & & * & & & \lambda_{i,2} & 1 & \lambda_{i,2}^{-1} \\
 & & & & & 1 & \lambda_{i,2}^{-1} & \lambda_{i,2}^{-2} \\
\hline
\cdots & \lambda_{i,1}^3 & \lambda_{i,1}^2 & \lambda_{i,1} & 1 & & & \\
 & \cdots & \lambda_{i,1} & 1 & \lambda_{i,1}^{-1} & & * & \\
 & & 1 & \lambda_{i,1}^{-1} & \lambda_{i,1}^{-2} & & &
\end{array}
\right),
$$

where the entries in $*$ are very small and the λ_i's in each block converge to a different element of $\widetilde{\Lambda}$, and c is organized so that the Markov condition holds.

We then partition the Bratteli diagram so that each "branch" has all its transitions "almost" in the group generated by λ, and there is a unique III$_\lambda$ G-measure on this branch. The full gory details are in [13]. Of course, the uniform Lipschitz condition (3.3) fails badly on X.

I believe that such a partitioning of X into uniquely ergodic branches should always be possible, and indeed, should give the ergodic decomposition of $(X, \mathcal{B}, T, \mu)$ in a natural way.

6. NON-PRODUCT MARKOV ODOMETERS

In this final section, I would like to explain some recent work with Hamachi [11] to produce an example of a Markov odometer which is not orbit equivalent to a product. Of course, by the Connes-Krieger theorem, each of the types III$_\lambda$, $0 < \lambda \leq 1$, and type II are unique up to orbit equivalence, and can be realized as product measures, so our example is necessarily of type III$_0$. Krieger [17] gave an example of a type III$_0$ system which is not orbit equivalent to a product, but it is not clearly an odometer (even though, by Dye's theorem it must be orbit equivalent to one), and it is certainly not Markov.

With Hamachi, I set out to give an explicit example of a Markov odometer which is not orbit equivalent to a product measure. The example is a Bratteli-Vershik system where the number of edges grows quite quickly. I will define the stochastic matrices below. As in the previous section, I shall leave it to the interested reader to find the details in the paper.

Example 6.1. Let us now define the explicit parameters for our example. Let $k_0 = 2$, $\nu(0) = \nu(1) = \frac{1}{2}$ and $r_0 = 1$. We will give an inductive definition of two sequences of positive integers, $\{k_n\}_{n \geq 0}$ and $\{r_n\}_{n \geq 0}$. Put $l_n = [\sqrt{n}] + 1$ (the integer part of $\sqrt{n}$). Suppose that we have defined $k_0, k_1, \ldots, k_{n-1}$ and $r_0, r_1, \ldots, r_{n-1}$, $n \geq 1$. We take an integer r_n satisfying the following conditions (C1)-(C3):

  : (C1) $\log r_n > 4 \sum_{i=1}^{n-1} k_{i-1} \log r_i$, $\quad n \geq 2$, and $r_1 = 2$.

: (C2) $2\sum_{i=1}^{n-1} k_{i-1}\log r_i\{\log r_{n-2} - 2\sum_{i=1}^{n-3} k_{i-1}\log r_i\} <$
$\log r_n - 4\sum_{i=1}^{n-1} k_{i-1}\log r_i, \quad n \geq 2.$

: (C3) $\log r_{n-1} - 2\sum_{i=1}^{n-2} k_{i-1}\log r_i < \log r_n - 2\sum_{i=1}^{n-1} k_{i-1}\log r_i, \quad n \geq 3.$

Obviously such a choice of r_n is possible. Now we set

$$r_{0,0} = 1,$$

$$r_{n,i} = r_n^{2k_{n-1}-i-1}, \quad 0 \leq i \leq 2k_{n-1} - 1, \quad (n \geq 1)$$

$$k_n = r_{n,0}(l_n + (l_n - 1)(k_{n-1} - r_{n-1,0})), \quad (n \geq 1).$$

Next we define the stochastic $k_{n-1} \times k_n$-matrix $M^{(n)}$. For this we set

$$P_0 = \frac{1}{r_{n,0}l_n}$$

$$P_i = \frac{1}{r_{n,i}l_n}, \quad i \in [r_{n-1,0}, k_{n-1}] \cup [k_{n-1} + r_{n-1,0}, 2k_{n-1} - 1].$$

We define $r_{n,0}$-dimensional row vectors $a_0, a_1, a_2, \ldots, a_{2k_{n-1}-1}$ by setting

$$a_i = \underbrace{(P_0, \ldots, P_0)}_{r_{n,0} \text{ times}} \qquad 0 \leq i < r_{n-1,0}$$

$$a_i = (\underbrace{P_i, \ldots, P_i}_{r_{n,i} \text{ times}}, \underbrace{0, \ldots, 0}_{r_{n,0}-r_{n,i} \text{ times}}) \quad i \in [r_{n-1,0}, k_{n-1}] \cup [k_{n-1} + r_{n-1,0}, 2k_{n-1} - 1].$$

We note that the sum of the entries of each vector a_i is $\dfrac{1}{l_n}$. The matrix $M^{(n)}$ is defined by

$$
\begin{pmatrix}
a_0 & 0\ldots0 & 0 & \ldots & 0\ldots0 & 0\ldots0 & a_k\ldots a_k \\
a_1 & 0\ldots0 & 0 & \ldots & 0\ldots0 & 0\ldots0 & a_k\ldots a_k \\
\vdots & \vdots & \vdots & \vdots & \vdots & \vdots & \vdots \quad \vdots \\
a_{j-1} & 0\ldots0 & 0 & \ldots & 0\ldots0 & 0\ldots0 & \underbrace{a_k\ldots a_k}_{l_n-1\text{times}} \\
a_j & 0\ldots0 & 0 & \ldots & 0\ldots0 & \underbrace{a_{j+k}\ldots a_{j+k}}_{l_n-1\text{times}} & 0\ldots0 \\
a_{j+1} & 0\ldots0 & 0 & \ldots & \underbrace{a_{j+1+k}\ldots a_{j+1+k}}_{l_n-1\text{times}} & 0\ldots0 & 0\ldots0 \\
\vdots & \vdots & \vdots & \vdots & \vdots & \vdots & \vdots \\
a_{k-1} & \underbrace{a_{2k-1}\ldots a_{2k-1}}_{l_n-1\text{times}} & 0 & \ldots & 0\ldots0 & 0\ldots0 & 0\ldots0
\end{pmatrix}
$$

where $j = r_{n-1,0}$ and $k = k_{n-1}$.

To show this is not a product, we used Krieger's Property A.

Theorem 6.1. [17] *Amongst the product type transformations, exactly the type III ones have Krieger's Property A.*

Here is Property A:

Definition 6.2. *Let* Γ *act as non-singular transformations on* $(X, \mathcal{B}, \mu)$. *We say that* Γ *has **Krieger's Property A** if there exist* $\delta, \eta > 0$ *and there exists a* σ-*finite measure* $\lambda \sim \mu$ *such that: every set* A *of positive measure contains*

a measureable subset B *of positive measure such that*

$$\limsup_{s \to \infty} \lambda \big\{ x \in B : \exists \gamma \in [\Gamma] \ \text{with} \ \gamma x \in B$$

and $\qquad \log \frac{d\mu \circ \gamma}{d\mu}(x) \ \in \ \left(e^{s-\delta}, e^{s+\delta} \right) \bigcup \left(-e^{s+\delta}, -e^{s-\delta} \right) \big\}$
$$> \ \eta \lambda(B)$$

In fact, we can show for our example that the lim sup above is zero on every set of positive measure, but that it is of type III. It follows from Theorem 6.1 that it must be not orbit equivalent to a product, yet of Type III.

Remarks 6.1 (i) Of course "orbit equivalent to a product" is the same as "the associated flow has the AT property", by the Theorem of Connes and Woods [8]. We are currently working on calculating the associated flow of our example in order to show directly that it is not AT. We would then like to generalize these techniques to other Markov measures.

(ii) It would be interesting to know if there are any non-product odometers which do have property A.

(iii) In our example, the number of coordinates increases very fast. It would be interesting to know if there are any non-product odometers of finite type. That is to say, Markov odometers with $\sup_{k \in \mathbb{N}} l_k < \infty$, which are not equivalent to products. In this connection, I note that the Markov odometers of finite type which were studied in [12] did in fact turn out to be of product type.

REFERENCES

[1] M. Bramson and S. Kalikov, Non-uniqueness in g-functions, *Israel J. Math.* **84** (1993), 153-160

[2] G. Brown and A.H. Dooley, Ergodic measures are of weak product type, *Math. Proc. Camb. Phil. Soc.* **98** (1985) 129-145

[3] G. Brown and A.H. Dooley, Odometer actions on G-measures, *Ergod. Th. Dyn. Systems* **11** (1991) 297-307

[4] G. Brown and A.H. Dooley, Dichotomy theorems for G-measures, *Internat. J. Math.* **5**(1994) 827-834

[5] G. Brown and A.H. Dooley, On G-measures and product measures, *Ergod. Th. Dyn. Systems* **18** (1998) 95-107

[6] G.Brown, A.H. Dooley and J. Lake, On the Krieger-Araki-Woods ratio set, *Tôhoku Math. J.* **47** (1995) 1-13

[7] A. Connes, Une classification des facteurs de Type III, *Ann. Sci. Ecole Norm. Sup.* **6** (1973), 133-252

[8] A. Connes and J. Woods, Approximately transitive flows and ITPFI factors, *Ergod. Jn. Dyn. Systems* **5** (1985) 203-236

[9] H. Dye, On groups of measure-preserving transformations I, *Trans. Amer. Math. Soc.* **85** (1963), 551-576

[10] A.H. Dooley, The spectral theory of posets and its applications to C^*-algebras, *Trans. Amer. Math. Soc.* **223** (1976) 143-156

[11] A.H. Dooley and T. Hamachi, Markov odometer actions not of product type, submitted

[12] A.H. Dooley, I. Klemeš and A. Quas, Product and Markov measures of type III, *J. Austral. Math. Soc. (Series A)* **64** (1998) 84-110

[13] A.H. Dooley and D. Rudolph, Non-uniqueness in G-measures, submitted

[14] T. Hamachi, Y. Oka and M. Osikawa, A classification of ergodic non-singular transformation groups, *Mem. Fac. Sci. Kyushu Univ. Ser. A* **28** (1974) 115-133

[15] R. Hermann, I. Putnam and C. Skau, Ordered Bratteli diagrams, dimension groups and topological dynamics, *Internat. J. Math.* **3** (1992) 827-864

[16] M. Keane, Strongly mixing g measures, *Invent. Math.* **16** (1972) 309-324

[17] W. Krieger, On non-singular transformations of measure, *Z. Wahrscheinlichkeitstheorie verw. Geb.*, **11** (1969) 83-97

[18] W. Krieger, II ibid **11** (1969) 98-119

[19] W. Krieger, On the Araki-Woods asymptotic ratio set and non-singular transformations of a measure space, *Lecture Notes in Math.* **160** (Springer, Berlin 1970) pp. 158-177

[20] W. Krieger, On the infinite product construction of non-singular transformations of a measure space, *Invent. Math.* **15** (1972) 144-163. Erratum ibid **26** (1974) 323-328

[21] Y. Katznelson and B. Weiss, The classification of non-singular actions revisited, *Ergod. Th. Dyn. Systems* **11** (1991), 333-348

[22] A. Quas, Non-ergodicity for C^1 expanding maps and g-measures, *Ergod. Th. Dyn. Systems* **16** (1996) 531-543

[23] C. Sutherland, Notes on orbit equivalence, Lecture note series 23 (Univ. i Oslo, 1976)

[24] A.M. Vershik, Uniform approximation of shift and multiplication operators, *Soviet Math. Doklady* **24** (1981) 97-100

SCHOOL OF MATHEMATICS, UNIVERSITY OF NEW SOUTH WALES, UNSW SYDNEY NSW 2052, AUSTRALIA

E-mail address: `A.Dooley@unsw.edu.au`

GEOMETRIC PROOFS OF MATHER'S CONNECTING AND ACCELERATING THEOREMS

V. KALOSHIN

ABSTRACT. In this paper we present simplified proofs of two important theorems of J.Mather. The first (connecting) theorem [Ma2] is about wandering trajectories of exact area-preserving twist maps naturally arising for Hamiltonian systems with 2 degrees of freedom. The second (accelerating) theorem is about dynamics of generic time-periodic Hamiltonian systems on two-torus (2.5-degrees of freedom). Mather [Ma6] proves that for a generic time-dependent mechanical Hamiltonian there are trajectories whose speed goes to infinity as time goes to infinity, in contrast to time-independent case, where there is a conservation of energy.

CONTENTS

The author is partially supported by AIM fellowship.

1. INTRODUCTION

The results of this paper are not new and the main purpose is to present simplified geometric proofs of two important theorems of J. Mather [Ma2, Ma6]. Both theorems are particular examples of instabilities in Hamiltonian systems or what is sometimes called *Arnold's diffusion*. Recently Mather [Ma7] announced a proof of existence of Arnold's diffusion for a generic nearly integrable Hamiltonian systems with 2.5 and 3-degrees of freedom using his variational approach developed in [Ma2]–[Ma6].

The first (connecting) Mather's theorem says that inside of a Birkhoff region of instability there are trajectories connecting any two Aubry-Mather sets, i.e. given any two Aubry-Mather sets Σ_ω and $\Sigma_{\omega'}$ inside of a Birkhoff region of instability there is a trajectory α-asymptotic to Σ_ω and ω-asymptotic to $\Sigma_{\omega'}$. Recently J. Xia [X] gave a simplified proof of the first result using the same variational approach as Mather. The second (accelerating) theorem says that a "generic" Hamiltonian time periodic system on the 2-torus $\mathbb{T}^2$ has trajectories whose speed goes to infinity as time goes to infinity. Different from [Ma6] proofs of this result are given by Bolotin-Treschev [BT] and Delshams-de la Llave-Seara [DLS1]. Our proof of the second theorem combines a geometric approach and Mather's variational approach. The second theorem is proved using ideas from the proof of the first theorem. Let's give the rigorous statement of both results.

First show that exact area-preserving twist maps naturally arise for Hamiltonian systems with two degrees of freedom. Indeed, let $H : \mathbb{R}^4 \to \mathbb{R}$ be a C^2-smooth function and consider the corresponding Hamiltonian system

$$(1) \qquad\qquad \begin{cases} \dot{x} = H_y(x,y) \\ \dot{y} = -H_x(x,y) \end{cases} \qquad x, y \in \mathbb{R}^2.$$

Denote by $\varphi_H^t(x_0, y_0) = (x_t, y_t)$ the time t map along the trajectories of (1). Each trajectory starting at some point $(x_0, y_0) \in \mathbb{R}^4$ belongs to its energy surface $E^3 = E_{(x_0, y_0)}^3 \subset \{H(x,y) = H(x_0, y_0) = \text{const}\} \subset \mathbb{R}^4$. Take a *Poincare section* $S^2 \subset E^3$ transverse to the vector field (1) and a trajectory $\{\varphi_H^t(p)\}_{t \geq 0}$ of a point $p \in S^2$. Let $\varphi_H^{\tau(p)}(p) \in S^2$ be the first return to S^2. For a nearby point $q \in S^2$ also there is a point $\varphi_H^{\tau(q)}(q) \in S^2$ of the first intersection with S^2 for $\tau(q)$ close to $\tau(p)$. This defines the *Poincare return map* which sends a point $q \in U \subset S^2$ into the point $\mathcal{P}(q) = \varphi_H^{\tau(q)}(q)$ of the first return of its forward trajectory to S^2 with U being the set where such a return exists. Given by the Hamiltonian flow φ_H^t the map $\mathcal{P} : U \to S^2$ preserves an area form on S^2. Such an area form is the restriction of the standard symplectic Darboux form $\omega = dx_1 \wedge dy_1 + dx_2 \wedge dy_2$ on $\mathbb{R}^4$ onto S and is non-degenerate, because (1) is transverse to S^2. The Poincare map $\mathcal{P}$ preserves this area form. Moreover, the regularity properties of $\mathcal{P}$ are the same as those of H, i.e. if H is C^r (with $R \geq 1$), then $\mathcal{P}$ is also C^r in the region where it is defined. To formulate the other two important properties of $\mathcal{P}$ we bring the domain of definition of $\mathcal{P}$ to the standard form.

Let $\mathbb{A} = \mathbb{S}^1 \times \mathbb{R} = \{(\theta, r) \in S^1 \times \mathbb{R}\}$ be the cylinder with the standard Lebesgue area form $d\theta \wedge dr$ and let $C \subset \mathbb{A}$ be an open which intersects every vertical line $\{\theta\} \times \mathbb{R}$ in an open interval. A non-contructable Jordan curve γ on $\mathbb{A}$ homeomorphic to a circle is called *rotational*. Consider a C^1-smooth orientation and area preserving map $\mathcal{P} : C \to \mathbb{A}$. $\mathcal{P}$ is called

• *exact* (or with no up/down drift) if $\mathcal{P}$ has zero flux, i.e. for any rotational curve $\gamma \subset C$ area of the regions above γ and below $\mathcal{P}(\gamma)$ equals area below γ and above $\mathcal{P}(\gamma)$[1];

• *monotone twist* (or simply twist) if for any vertical curve $l = \{\theta\} \times \mathbb{R}$ in $\mathbb{A}$ its image $\mathcal{P}(l)$ intersects every vertical line $\{\tilde{\theta}\} \times \mathbb{R}$ with a nonzero angle.

Assume also that $\mathcal{P}$ is homotopic to the inclusion map. We shall call a map with the above properties an *EAPT* (exact orientation and area preserving twist).

Important examples of EAPTs of a cylinder are *billiards* in convex bounded regions, *the plane restricted three-body* problem, the *standard map* of the 2-torus or Frenkel-Kontorova model, and etc (see e.g. [MF] and [Mo] II.4).

1.1. Mather's connecting theorem.

A compact region $C \subset \mathbb{A}$ is called *a Birkhoff region of instability (BRI)* if C is a compact $\mathcal{P}$-invariant set whose frontier consists of two components denoted by C_- and C_+ both rotational curves and no other rotational invariant curves in between. For convenience we call the upper frontier C_+— the top and the lower frontier C_- —the bottom. Let C be a BRI. Since both frontiers are invariant under $\mathcal{P}$ it induces two homeomorphisms of the circles. Therefore, there are two well-defined rotation numbers ω_- and ω_+ for $\mathcal{P}|_{C_-}$ and $\mathcal{P}|_{C_+}$ respectively. It follows from the twist condition that $\omega_- < \omega_+$.

Theorem 1. (Aubry [AL]-Mather [Ma1]) *For any rotation number $\omega \subset [\omega_-, \omega_+]$ there is an invariant set Σ_ω, called Aubry-Mather set, such that every orbit in Σ_ω has a rotation number of winding around the cylinder equal ω.*[2]

If a rotation number is rational $\omega = p/q \in \mathbb{Q}$, then generically Σ_ω is a finite union of period orbits of period q. In a highly degenerate case Σ_ω might be a rotational curve. If a rotation number is irrational, then Σ_ω is either a Denjoy-Cantor set or a rotational curve.

Mather's Connecting Theorem. *For any two rotation numbers $\omega, \omega' \in [\omega_-, \omega_+]$ inside of rotation interval of Birkhoff region of instability there is a point $p \in C$ such that its $\omega(\alpha)$-limit set of p is contained in Σ_ω (in $\Sigma_{\omega'}$) respectively.*

---

[1] Area preservation implies that this property is independent of the choice of γ and is necessary to have invariant curves at all

[2] We shall give a more precise statement of this theorem in Section 1.3

In this paper first we shall prove a weaker version of this theorem (see just below) and then for generic EAPTs extend it to a strong version.

Mather's Weak Connecting Theorem. *In the setting of the above theorem for any positive ε there is a point $p \in C$ such that for some positive n_+ and negative n_- we have $\mathcal{P}^{n_\pm}(p)$ belongs to the ε-neighborhood of Σ_ω and $\Sigma_{\omega'}$ respectively.*

The original proof of Mather [Ma2] using variational method is quite complicated and involved. Recently it was significantly simplified by J. Xia [X]. It might give some insight in how to estimate on diffusion time. Topological arguments presented here are sufficiently simple and based on Birkhoff's invariant set theorem. These arguments are qualitative and seem to give no insight *on diffusion time*. In the case $\Sigma_\omega = C_-$ and $\Sigma_{\omega'} = C_+$ Mather's Weak Connecting Theorem is the Theorem of Birkhoff [B1] and Mather's Connecting Theorem was also proved by Le Calvez [L1] using clever topological arguments.

1.2. Mather's accelerating theorem.

Let $\mathbb{T}^2$ be the two-torus with a C^2-smooth Riemannian metric ρ, defined on the tangent bundle $\mathbf{T}\mathbb{T}^2$ of $\mathbb{T}^2$ and by duality can be also defined on the cotangent bundle $\mathbf{T}^*\mathbb{T}^2$. Denote by $S\mathbb{T}^2$ ($S^*\mathbb{T}^2$) denotes the unit (co)tangent bundle of $\mathbb{T}^2$ and by $T_q(p)$ the associated kinetic energy to the metric ρ, i.e.

$$(2) \qquad T_q(p) = \rho_q(p, p)/2, \quad p \in \mathbf{T}_q^*\mathbb{T}^2,$$

where $\mathbf{T}_q^*\mathbb{T}^2$ denotes the cotangent bundle of $\mathbb{T}^2$ at q. Let $U : \mathbb{T}^2 \times \mathbb{T} \to \mathbb{R}$ be a C^2-smooth time periodic function on $\mathbb{T}^2$ which is the potential energy. We normalize period to be one. This defines the mechanical Hamiltonian system with

$$(3) \qquad H(q, p, t) = T_q(p) + U(q, t).$$

Mather's acceleration theorem I. *For a generic C^2-smooth metric ρ on $\mathbb{T}^2$ and a generic C^2-smooth function $U : \mathbb{T}^2 \times \mathbb{T} \to \mathbb{R}$, there exists a trajectory $q(t)$ of the Hamiltonian flow, defined by (3), with an unbounded speed $\dot{q}(t)$. Moreover, energy $E(t) = \rho_{q(t)}(p(t), p(t)) + U(q(t), t)$ tends to infinity as time goes to infinity.*

Remark 1. As the reader will see generic metrics on $\mathbb{T}^2$ and generic functions on $\mathbb{T}^2 \times \mathbb{T}$ form open dense sets in the corresponding spaces of C^2-metrics and C^2-functions.

Non-degeneracy Hypothesis on ρ and U.

Hypothesis 1. There is an indivisible homology class $h \in H_1(\mathbb{T}^2, \mathbb{Z})$ which has only one shortest hyperbolic periodic geodesic Γ;

Choose generators for the homology group $H_1(\mathbb{T}^2, \mathbb{Z})$. Let h_0 be the homology class corresponding to Γ. Lift $\mathbb{T}^2$ to the cylinder $\mathbb{A}$ with h_0 being the only nontrivial homology class of $\mathbb{A}$. Then Γ lifts to a countable collection of copies. Denote by Γ_0 and Γ_1 adjacent copies. Recall that a geodesic γ is called a *Morse Class A geodesic* if its lift $\hat{\gamma}$ to the universal cover $\mathbb{R}^2$ is globally length minimizing and γ is homoclinic to Γ, i.e. its α-limit set is Γ_0 and ω-limit set is Γ_1 (see [Ba1] (6.8) for more detailed discussion of Morse Class A geodesics).

Hypothesis 2. There is only one positive Morse Class A geodesic Λ.

Since Γ is hyperbolic it has stable and unstable manifolds $W^s(\Gamma)$ and $W^u(\Gamma)$ in $S\mathbb{T}^2$ respectively. Another way to view the geodesic Λ is as the intersection of invariant manifolds $W^s(\Gamma) \cap W^u(\Gamma)$.

Hypothesis 2′. Intersection of invariant manifolds $W^s(\Gamma) \cap W^u(\Gamma)$ is transversal.

This hypothesis is not necessary, but simplifies presentation. Let $s \mapsto \Gamma(s)$ and $s \mapsto \Lambda(s)$ be arclength parameterizations of both Γ and Λ with the same orientation. Then there are constants $a, b \in \mathbb{R}$ such that

$$
\begin{aligned}
\rho(\Gamma(s), \Lambda(s+a)) &\to 0 \quad \text{as} \quad s \to -\infty \\
\rho(\Gamma(s), \Lambda(s+b)) &\to 0 \quad \text{as} \quad s \to +\infty.
\end{aligned}
\tag{4}
$$

Both quantities converge to zero exponentially fast by hyperbolicity of Γ which implies that one can define the *Melnikov Integral*

$$
G(t) = \lim_{s \to \infty} \left[\int_{-s+a}^{s+b} U(\Gamma(\tau), t) d\tau - \int_{-s}^{s} U(\Lambda(\tau), t) d\tau \right].
\tag{5}
$$

Hypothesis 3. $G(t)$ is not constant.

Fact. *In the space of C^2-smooth Riemannian metrics on $\mathbb{T}^2$ and C^2-smooth time periodic functions on $\mathbb{T}^2$ there is an open dense set of those who satisfy Hypothesis 1-3.*

Mather's acceleration theorem II. *If Hypotheses 1-3 are satisfied, there exists a trajectory of the Hamiltonian system (3) whose energy goes to infinity as time goes to infinity.*

Remark 2. The fact stated above implies that Mather's acceleration theorem version II implies version I.

This last theorem was originally proved by Mather [Ma6] using variational method. Later Bolotin-Treschev [BT] and Delshams-de la Llave-Seara [DLS1] gave in a sense similar proofs analogous theorems using the standard geometric approach. In [DLS2] the second group of authors extended this result to some manifolds different from the two-torus with apriori unstable geodesic flow. The proof which we present here is a mixture of geometric and variational approaches. Accelerating trajectories constructed in the present proof slightly differ from both given in Mather's work [Ma6] by the variational method and in [BT] an [DLS1] by geometric methods, even though in the spirit our proof uses ideas from both approaches.

1.3. **Birkhoff's invariant set theorem and Aubry-Mather sets as action-minimizing sets.** We need Birkhoff's invariant set theorem.

Theorem 2. [B2], [MF] *Let $C \subset \mathbb{A}$ be a BRI of an EAPT $\mathcal{P} : C \to C$. Suppose that $V \subset C$ is a closed connected set separating the cylinder and invariant under $\mathcal{P}$. Then V is equal C^-, or C^+, or contains both frontiers.*

For the purpose of completeness we formally describe Aubry-Mather as action-minimizing sets. Denote by $\tilde{C}$ the natural lift of $C \subset \mathbb{A} = \mathbb{S}^1 \times \mathbb{R}$ to $\mathbb{R} \times \mathbb{R}$ and by $\tilde{\mathcal{P}} : \tilde{C} \to \tilde{C}$ the lift of the map $\mathcal{P}$. For a point $p \in C$ let $\tilde{p} \in \tilde{C}$ be its lift and let $(x_n(\tilde{p}), r_n(\tilde{p})) = \tilde{\mathcal{P}}^n \tilde{p}$. Then if the following limits

$$(6) \qquad \rho_\alpha(p) = \lim_{n \to +\infty} x_n(\tilde{p})/n \quad \text{and} \quad \rho_\omega(p) = \lim_{n \to -\infty} x_n(\tilde{p})/n$$

exist and equal $\rho(p) = \rho_\alpha(p) = \rho_\omega(p)$, then is is called a *rotation number*. Geometric meaning of rotation number is average amount of rotation of the trajectory of p around the cylinder $\mathbb{A}$.

A fundamental property of an EAPT $\mathcal{P} : C \to C$ is that it can be globally described by a "generating" function $h : D \to \mathbb{R}$, where $D \subset \mathbb{R}^2$ is the set of points $(x, x') \in \mathbb{R}^2$ such that there is $r, r' \in \mathbb{R}$ with $(x, r) \in \tilde{C}$ and $\tilde{P}(x, r) = (x', r')$. Clearly, D is open. By twist condition for each $(x, x') \in D$ the pair r, r' as above is unique. Moreover, (r, r') depends continuously on (x, x'). See [MF] §5 or [Ba1] sec. 7 for more. If $\mathcal{P}$ is C^r-smooth, then h is C^{r+1} on D. A generating function $h : D \to \mathbb{R}$ can also be defined by

$$(7) \qquad \begin{cases} r = -\partial_1 h(x, x') \\ r' = \partial_2 h(x, x'). \end{cases}$$

By agreement extend h from D to the whole plane $\mathbb{R}^2$ by $h(x, x') = +\infty$. Sometimes such a generating function h is called *variational principle* (see e.g. [MF] §5).

Denote $\mathbb{R}^{\mathbb{Z}} = \{x : x : \mathbb{Z} \to R\}$ the space of sequences. Given an arbitrary sequence $(x_j, \ldots, x_k), j < k$ of $x \in \mathbb{R}^{\mathbb{Z}}$ denote $h(x_j, \ldots, x_k) = \sum_{i=j}^{k-1} h(x_i, x_{i+1})$. A segment $(x_j, \ldots, x_k)$ is called *minimal* with respect to h if $h(x_j, \ldots, x_k) \leq h(x_j^*, \ldots, x_k^*)$ for all $(x_j^*, \ldots, x_k^*)$ with $x_j = x_j^*$ and $x_k = x_k^*$. A sequence $x \in \mathbb{R}^{\mathbb{Z}}$ is h-*minimal* if for every finite segment of x is h-minimal. Denote by $\mathcal{M}(h)$ the set of minimal sequences. If h is a generating function of an EAPT $\mathcal{P} : C \to C$, then using relation (7) each minimal sequence $x \in \mathbb{R}^{\mathbb{Z}}$ corresponds to a trajectory of $\mathcal{P}$. So the set $\mathcal{M}(h)$ corresponds to the set of points $\mathcal{M}(\mathcal{P}) \subset C$ (see [MF] §3 or [Ba1] sec. 7 for more). The set $\mathcal{M}(\mathcal{P})$ is called a set of *action-minimizing or h-minimal points* or, equivalently, $\mathcal{M}(\mathcal{P})$ the set of points whose trajectories are minimal.

Theorem 3. (Aubry [AL]-Mather [Ma1]) *With notations Theorem 1 for any rotation number $\omega \in [\omega_-, \omega_+]$ there is a nonempty invariant set $\Sigma_\omega \subset \mathcal{M}(\mathcal{P}) \subset$*

C of points whose trajectories are h-minimal and have rotation number $\rho(p) = \omega$.

Σ_ω is called an *action-minimizing* or *an Aubry-Mather set*.

1.4. Structure of Aubry-Mather sets. In this section we describe possible structures of Aubry-Mather sets. Recall that a point $p \in C$ is called *recurrent* with respect to $\mathcal{P}$ if its trajectory has p in the closure, i.e. $\overline{\{\mathcal{P}^n p\}}_{n \in \mathbb{Z}\backslash 0} \ni p$. Denote by $\Sigma_\omega^{rec} \subset \Sigma_\omega$ the set of minimal recurrent points with rotation number ω. Denote also by $\pi_1 : \mathbb{S}^1 \times \mathbb{R} \to \mathbb{S}^1$ the natural projection onto the first component. The theorem about structure of Aubry-Mather sets, given below, follows e.g. from thms (4.3) and (5.3) [Ba1] and the graph theorem (thm 14.1) [MF].

Structure Theorem. (*Irrational case $\omega \notin \mathbb{Q}$*) *The Aubry-Mather set Σ_ω is*

• *either an invariant curve or*
• Σ_ω^{rec} *and its projection $\pi_1(\Sigma_\omega^{rec})$ are Cantor sets. Then Σ_ω^{rec} consists of three different types of trajectories.*
There is the set of $p \in \Sigma_\omega^{rec}$ whose projection $\pi_1(p)$ can be approximated from above and below by $\pi_1(\Sigma_\omega^{rec})$. This set had power continuum and corresponds to those points which are not endpoints of $\mathbb{S}^1 \backslash \pi_1(\Sigma_\omega^{rec})$.
There is the set of $p \in \Sigma_\omega^{rec}$ whose projection $\pi_1(p)$ can be approximated either only from above or only from below by $\pi_1(\Sigma_\omega^{rec})$. These sets are countable and correspond to the right resp. left endpoints of components of $\mathbb{S}^1 \backslash \pi_1(\Sigma_\omega^{rec})$.[3]

To formulate the Structure Theorem in the case of rational rotation number introduce some sets. Let Σ_ω^{per} the set of periodic points. Two periodic points p^- and p^+ are neighboring elements of Σ_ω^{per} if projections $\pi_1(p^-)$ and $\pi_1(p^+)$ have a segment in $\mathbb{S}^1$ free from projection of all the other elements of $\pi_1(\Sigma_\omega^{per})$. For neighboring periodic points p^- and p^+ in Σ_ω^{per} let

$$
\begin{aligned}
\Sigma_\omega^+(p^-, p^+) &= \{p \subset \Sigma_\omega : \ p \text{ is } \alpha - \text{asymptotic to } p^- \\
&\qquad \text{and } \omega - \text{asymptotic to } p^+\}, \\
\Sigma_\omega^-(p^-, p^+) &= \{p \in \Sigma_\omega : \ p \text{ is } \alpha - \text{asymptotic to } p^+ \\
&\qquad \text{and } \omega - \text{asymptotic to } p^-\}
\end{aligned}
$$

(8)

Let $\Sigma_\omega^\pm$ be the union of $\Sigma_\omega^\pm(p^-, p^+)$ over all neighboring periodic points p^- and p^+ in Σ_ω^{per}.

Structure theorem. (*Rational case $\omega \in \mathbb{Q}$*) *The Aubry-Mather set Σ_ω is a disjoint union of Σ_ω^{per}, Σ_ω^+, and Σ_ω^-. Moreover, Σ_ω^{per} is always non-empty and if Σ_ω^{per} is not a curve, then Σ_ω^- and Σ_ω^+ are non-empty too.*

---

[3]Two points $p \in \Sigma_\omega^{rec}$ and $p' \in \Sigma_\omega^{rec}$ are asymptotic, i.e. $\text{dist}(\mathcal{P}^n p, \mathcal{P}^n p') \to 0$ as $|n| \to \infty$, if and only if $\pi_1(p)$ and $\pi_1(p')$ are endpoints of some component of $\mathbb{S}^1 \backslash \pi_1(\Sigma_\omega^{rec})$. In this case we have $\sum_{n \in \mathbb{Z}} |\pi_1(\mathcal{P}^n p) - \pi_1(\mathcal{P}^n p')| \leq 1$. Convergence in Σ_ω^{rec} is never uniform.

2. A PROOF OF MATHER'S WEAK CONNECTING THEOREM

The idea of the proof of Mather's weak connecting theorem is to choose two recurrent points p and p' in the starting Aubry-Mather sets Σ_ω and $\Sigma_{\omega'}$ respectively. Take an open ε-ball $V_\varepsilon(p)$ (resp. $V'_\varepsilon(p)$) about p (resp. p') and consider the union over forward (resp. backward) images of $V_\varepsilon(p)$ (resp. $V'_\varepsilon(p)$)[4]. It turns out that the following properties hold true.

Lemma 1. *With the notations above let ω^- and ω^+ be rotation numbers of the "top" C_+ and "bottom" C_- frontiers of a BRI C. Let $\omega \in [\omega^-, \omega^+]$ be a rotation number and $\Sigma_\omega \subset C$ be the corresponding Aubry-Mather set. Then for any recurrent point $p \in \Sigma_\omega^{rec}$ and any $\varepsilon > 0$ for the open ε-neighborhood $V_\varepsilon(p)$ of p the following holds true*
* *for some positive number $n_+ = n_+(p,\varepsilon)$ (resp. $n_- = n_-(p,\varepsilon)$) the union of images $\cup_{j=0}^{n_+} \mathcal{P}^j V_\varepsilon(p)$ (resp. $\cup_{j=0}^{n_+} \mathcal{P}^{-j} V_\varepsilon(p)$) separates the cylinder $\mathbb{A}$.*
* *the union over all forward (resp. backward) images $V_\varepsilon^+(p) = \cup_{j \in \mathbb{Z}_+} \mathcal{P}^j V_\varepsilon(p)$ (resp. $V_\varepsilon^-(p) = \cup_{j \in \mathbb{Z}_+} \mathcal{P}^{-j} V_\varepsilon(p)$) is connected and open.*
* *closure of $V_\varepsilon^+(p)$ (resp. $V_\varepsilon^-(p)$) contains both frontiers $C^\pm$ of C.*
* *let $V_\varepsilon^\infty(p) = \cup_{j \in \mathbb{Z}} \mathcal{P}^j V_\varepsilon(p)$. Then $V_\varepsilon^\infty(p)$ is invariant and both $V_\varepsilon^+(p)$ and $V_\varepsilon^-(p)$ are open dense in $V_\varepsilon^\infty(p)$.*

Let $V_\varepsilon^+(p)$ and $V_\varepsilon^-(p')$ be the union of one-sided iterates of ε-neighborhoods of $p \in \Sigma_\omega$ and $p' \in \Sigma_{\omega'}$. This lemma implies that $V_\varepsilon^+(p)$ and $V_\varepsilon^-(p')$ have to have nonempty intersection as connected open sets both separating the cylinder and having frontiers C_+ and C_- of a BRI C in its closure. Intersection of two open sets is open so there is an open subset $V_{\omega,\omega'}^\varepsilon$ inside of the ε-neighborhood $V_\varepsilon(p)$ of p such that after a number of forward iterations, say n, the set $V_{\omega,\omega'}^\varepsilon$ is mapped into $V_\varepsilon(p')$, i.e. $V_{\omega,\omega'}^\varepsilon \subset V_\varepsilon(p)$ and $\mathcal{P}^n V_{\omega,\omega'}^\varepsilon \subset V_\varepsilon(p')$ for some $n \in \mathbb{Z}_+$. This proves the required statement in Mather's Weak Connecting Theorem. Now we prove the lemma.

2.1. Union of iterates of a neighborhood of a recurrent point p separates the cylinder $\mathbb{A}$.

In this subsection we prove the first and the second parts of the lemma. Indeed, the set $V = \cup_{j \in \mathbb{Z}} \mathcal{P}^j V_\varepsilon(p)$ is invariant and open by definition. Show that V is connected.

Pick an arbitrary positive ε and consider an open ε-ball $V_\varepsilon(p)$ of p. Since p is recurrent, for an arbitrary ε there is $n \in \mathbb{Z}$ such that p and $\mathcal{P}^n p$ are $\varepsilon/2$-close. It follows from simple properties of Aubry graphs (see Lemma 4.5 [Ba1] or Theorem 11.3 [MF]) that any recurrent point p of an Aubry-Mather set can be approximated by a periodic point r from another Aubry-Mather set $\Sigma_{\tilde\omega}$, $\tilde\omega = s/q \in \mathbb{Q}$. Therefore, if approximation is close enough then r is $\varepsilon/4$-close to p and $\mathcal{P}^n r$ is $\varepsilon/4$-close to $\mathcal{P}^n p$. Thus, both $\mathcal{P}^n r$ and r are in $V_\varepsilon(p)$. We can also assume that q is prime. The projection onto the base circle of

---

[4]In [Ha] topological arguments of very different nature are used to prove existence of Aubry-Mather sets

points of an action-minimizing orbit are cyclically ordered (see (4.1) and (5.1) [Ba1] or Theorem 12.3 [MF]). Recall also that ω is with prime numerator. All these remarks imply that the set $\cup_{j=0}^{q}\mathcal{P}^{j}V_{\varepsilon}(p)$ separates the cylinder $\mathbb{A}$. This union of neighborhoods reminds a "bicycle chain". This proves the first claim of the lemma.

Connectivity of $\cup_{j\in\mathbb{Z}_+}\mathcal{P}^{j}\,V_{\varepsilon}(p)$ can be shown as follows. Notice that $V_{\varepsilon}(p)$ has a periodic point r of period q inside. Therefore, the set $\cup_{s\in\mathbb{Z}_+}\mathcal{P}^{sq}\,V_{\varepsilon}(p)$ is connected. The sets $\cup_{s\in\mathbb{Z}_+}\mathcal{P}^{sq+j}\,V_{\varepsilon}(p)$ for different j's with $0 \le j \le q-1$ are connected among each other because members of the orbit of r are connected by the "bicycle chain". This completes the proof of the second point.

The third point follows from Birkhoff's invariant set theorem. The last claim of s a direct corollary of area-preservation. This proves the lemma. Q.E.D.

3. Extension of Mather's Weak Connecting Theorem to Mather's Connecting Theorem for generic EAPTs

First we formulate genericity hypothesis for EAPTs:

Genericity Hypothesis: We say that an EAPT has KS property if linearization of any periodic point p has no eigenvalue 1 and transversal intersections of its stable and unstable manifolds.

We shall prove

Mather's Connecting Theorem for EAPTs with KS property. *With the notations of Mather's Connecting Theorem and an EAPT map $\mathcal{P}$: $\mathbb{A} \to \mathbb{A}$ satisfying KS property with a BRI $C \subset \mathbb{A}$ there is an open dense set $\mathcal{K} \subset [\omega_-, \omega_+]$ such that for any two pair $\omega, \omega' \in \mathcal{K}$ there is a heteroclinic trajectory with α-limit set in Σ_ω and ω-limit set in $\Sigma_{\omega'}$. Moreover, for any $\omega, \omega' \in [\omega_-, \omega_+]$ there is trajectory whose α-limit set contains Σ_ω and ω-limit set contains $\Sigma_{\omega'}$.*

Moreover, for any pair $\omega, \omega' \in [\omega_-, \omega_+]$ there is a heteroclinic trajectory whose α-limit set has nonempty intersection with Σ_ω and ω-limit set — with $\Sigma_{\omega'}$.

Remarks:

1. This property is an area-preserving analog of so called Kupka-Smale property that all periodic points are hyperbolic and their stable and unstable manifolds intersect transversally [Sm]. This property is generic in both sense topological (Baire residual) [Sm] and probabilistic/measure (prevalence) [Ka] for C^r-smooth (not necessarily area-preserving) diffeomorphisms of a compact manifold. It is also generic for area-preserving maps [Ro].

2. Direct computation [MF] show that an action-minimizing periodic point has linearization with trace at least two. Along with KS property and area-preservation this implies that *all* action-minimizing periodic points are saddles. So stable and unstable manifolds are 1-dimensional. By KS property intersection of stable and unstable manifolds of all periodic points transverse

which implies that any rotational invariant curve can't have rational rotation number, i.e. ω_- and ω_+ are in $\mathbb{R} \setminus \mathbb{Q}$.

3. For any rational rotation number $p/q \in (\omega_-, \omega_+)$ the corresponding Aubry-Mather set $\Sigma_{p/q}$ has at least one periodic point x of period q. Hyperbolicity of x implies that it is isolated.

4. Existence of bi-asymptotic trajectories to $\Sigma_{p/q}^{per}$, i.e. trajectories from $\Sigma_{p/q}^{\pm}$ (see Structure Theorem), implies that the union of stable and unstable manifolds of periodic points in $\Sigma_{p/q}^{per}$ separates the cylinder. Thus, their closure contains the "top" and the "bottom" frontiers $C_{\pm}$ of C. This, in particular, implies that for any pair of rational numbers p/q and p'/q' in (ω_-, ω_+) there is a trajectory whose α-limit set is contained in $\Sigma_{p/q}$ and ω-limit set is in $\Sigma_{p'/q'}$. So we have

Lemma 2. *Hyperbolicity of periodic points along with Structure Theorem (non-emptiness of $\Sigma_{p/q}^{per}$ for all $p/q \in (\omega_-, \omega_+)$) implies Mather's Connecting Theorem for rational rotation numbers.*

To prove Mather's Connecting Theorem for irrational or mixed rotation numbers we need to know additional structure of Aubry-Mather sets with irrational rotation numbers.

Definition 4. An Aubry-Mather set Σ_ω is called *hyperbolic* if it admits two continuous line fields one is contracting the other is expanding.

The standard Hadamard-Perron theorem implies that if an Aubry-Mather set Σ_ω is hyperbolic, then from every point p in Σ_ω there are two smooth curves which stable and unstable manifolds of p. Since Aubry-Mather sets depend continuously on rotation number and cone property is open, it is not difficult to prove that hyperbolicity of periodic points leads to hyperbolicity of most of Aubry-Mather sets.

Theorem 5. *(e.g. [L2] Thm. 1.10) If a C^1-smooth EAPT $\mathcal{P} : \mathbb{A} \to \mathbb{A}$ has KS property, then there is an open dense set $\mathcal{K}$ of rotation numbers in $\mathbb{R}$ such that any Aubry-Mather with its rotation number from $\mathcal{K}$ is hyperbolic.*

Lemma 3. *If an Aubry-Mather set Σ_ω is hyperbolic, then the union of stable and unstable manifolds of points in Σ_ω separates the cylinder.*

Proof: In the rational case $\omega = p/q \in \mathbb{Q}$ this is in Remark 4 above. In the irrational case $\omega \notin \mathbb{Q}$ the idea is analogous. Let ω be irrational and suppose that $\mathcal{P}$ has a right twist, if not take the inverse. By Structure Theorem Σ_ω^{rec} consists of points of three types. The first type of p's in Σ_ω^{rec} is when the projection $\pi_1(p)$ can be approximated from both sides by $\pi_1(\Sigma_\omega^{rec})$. The other two types is when the projection $\pi_1(p)$ can only be approximated from one side.

Since stable and unstable line field have to be transversal with a separated from zero angle and continiuosly chanimg from a point to a point, stable and unstable manifolds of points of the first type have to intersect with unstable and stable manifolds respectively or nearby points of the same type. We need only to connect through the "wholes", i.e. through preimages of open intervals which are in the complement of $\pi_1(\Sigma_\omega^{rec})$. To do that we need to construct connecting trajectories of neighboring points of the second and third type.

Let p_r and p_l be points in Σ_ω^{rec} whose projections into $\mathbb{S}^1$ are neighbors, i.e. $(\pi_1(p_l), \pi_1(p_r)) \subset \mathbb{S}^1 \setminus \pi_1(\Sigma_\omega^{rec})$. By the footnote to Structure Theorem we know that for any $\varepsilon > 0$ there is $n \in \mathbb{Z}$ such that $\mathcal{P}^n p_r$ and $\mathcal{P}^n p_l$ are ε-close. Since ε is arbitrary, an angle between stable and unstable line fields is separated from zero, and stable and unstable manifolds of points in Σ_ω^{rec} are C^1 by Hadamard-Perron Theorem, we have that $W^u(p_l)$ and $W^s(p_r)$ intersect and their projection contains $(\pi_1(p_l), \pi_1(p_r))$. This implies that for any two neighbors of the second and third type their stable and unstable manifolds do intersect and the union of stable and unstable manifolds of points in Σ_ω^{rec} separates the cylinder. Let $W^s(\Sigma_\omega)$ be the union of stable manifolds of all points in Σ_ω which is a stable lamination. Similarly, one can define unstable lamination $W^u(\Sigma_\omega)$.

In a view of the above arguments and Birkhoff invariant set theorem, if $\omega, \omega' \in \mathcal{K}$ from Theorem 5 or, equivalently, Aubry-Mather sets Σ_ω and $\Sigma_{\omega'}$ are hyperbolic, then their stable and unstable laminations $W^s(\Sigma_\omega)$ and $W^u(\Sigma_{\omega'})$ intersect and the intersection contains at least one trajectory whose α-limit set is contained in Σ_ω and ω-limit set is in $\Sigma_{\omega'}$. So we proved the following

Proposition 6. *If $\mathcal{P} : \mathbb{A} \to \mathbb{A}$ is a C^1-smooth EAPT with KS property, $C = \mathcal{P}(C)$ is a BRI, C_+ and C_- are "top" and "bottom" frontiers of C respectively, ω_+ and ω_- are rotation numbers of $\mathcal{P}$ restricted to C_+ and C_- respectively, and $\mathcal{K} \cap [\omega_-, \omega_+]$ be an open dense set of rotation numbers with hyperbolic Aubry-Mather sets (from Theorem 5). Then for any ω and ω' from $\mathcal{K} \cap [\omega_-, \omega_+]$ the stable and unstable laminations $W^s(\Sigma_\omega)$ and $W^u(\Sigma_{\omega'})$ of corresponding Aubry-Mather sets do intersect and the intersection contains a heteroclinic trajectory going from Σ_ω to $\Sigma_{\omega'}$.*

This proves the first part of Mather's Connecting Theorem for EAPTs with KS property. To prove the "moreover" part of the Mather's connecting theorem from this section we use standard arguments of Arnold [Ar] usually called whiskered tori. In our case tori are replaced by periodci points. In our case whiskered tori are 0-dimensional and correspond to action-minimizing periodic points. By Remark 2 above they are hyperbolic saddles. Chose a bi-infinite sequence of rational numbers $\{\omega_n\}_{n \in \mathbb{Z}} \subset \mathcal{K} \subset [\omega_-, \omega_+]$ so that corresponding Aubry-Mather sets are hyperbolic and $\lim \omega_n = \omega$ (resp. ω') if $n \to -\infty$ (resp. $+\infty$). This implies that each periodic points $\{p_n \in \Sigma_{\omega_n}\}_{n \in \mathbb{Z}}$ is hyperbolic. By Proposition 6 for any $n \in \mathbb{Z}$ their stable $W^s(p_n)$ and unstable $W^u(p_{n+1})$ manifolds consequently cross each other, i.e. there is a point $q \in W^s(p_n) \cap W^u(p_{n+1})$ whose neighbourhood is separated by $W^s(p_n)$ in two parts and $W^u(p_{n+1})$ locally visits both.

Notice that if $W^s(p_{n-1})$ crosses $W^u(p_n)$ and $W^s(p_n)$ crosses $W^u(p_{n+1})$, then $W^s(p_{n-1})$ crosses $W^u(p_{n+1})$. This proves that by induction that for $k, n \in \mathbb{Z}$ we have $W^s(p_k)$ crosses $W^u(p_n)$. In other words, p_n's form a transition chain. Now choose a sequence of positive numbers ε_n which tends to 0 as $n \to \infty$. Choose a sequence of points $q_n \in W^s(p_{-n}) \cap W^u(p_n)$ so that q_n is ε_n-close to p_0, but $\varepsilon / \|\mathcal{P}\|_{C^1}$-away from p_0. By definition q_n its α-limit is the trajectory of p_{-n} and ω-limit — of p_n. Now recall that Aubry-Mather sets depend continiously on rotation number (Thm. 11.3 [MF]). So $\Sigma_{\omega_n} \to \Sigma_\omega$ as $\omega_n \to \omega$ in Hausdorff distance, in particular, $p_n \to p$ as $n \to -\infty$. Similarly, $p_n \to p'$ as $n \to +\infty$. Therefore, one can choose a subsequence p_{n_k} such that $q_{n_k} \to q$ and q is different from p_0 and ε-close to it. Moreover, α-limit (resp. ω-limit) set of q contains $p \in \Sigma_\omega$ (resp. $p' \in \Sigma_{\omega'}$). This proves Mather's Connecting Theorem for EAPTs with KS property. Q.E.D.

4. THE PROOF OF MATHER'S ACCELERATING THEOREM

Let's make several preliminary remarks in order to show connection of this problem with EAPTs of the cylinder discussed above.

The Hamiltonian phase space of the geodesic flow is the cotangent bundle of the torus $\mathbf{T}\mathbb{T}^2$ which is isomorphic to $\mathbb{R}^2 \times \mathbb{T}^2$. Recall that we denote by q coordinates on $\mathbb{T}^2$ and by p coordinates in the cotangent space $\mathbf{T}_q^*\mathbb{T}^2$ which is isomorphic to $\mathbb{R}^2$. The geodesic flow is Hamiltonian with respect to the Darboux form $\omega = dp_1 \wedge dq_1 + dp_2 \wedge dq_2$ with the Hamiltonian function

$$H_0(p, q) = T_q(p) = \rho_q(p, p)/2, \ p \in \mathbf{T}_q^*\mathbb{T}^2.$$

Recall that ρ_q is the metric in $\mathbf{T}^*\mathbb{T}^2$. Denote by Φ_t the time t map of the geodesic flow. For each E, we denote by $\mathcal{L}_E^0 = \{(p, q) : H_0(p, q) = E\}$ the corresponding energy level. In particular, $S^*\mathbb{T}^2 = \mathcal{L}_{1/2}^0$ is the unit energy level. Denote also $\hat{\mathcal{L}}_{E_0}^0 = \cup_{E \geq E_0} \mathcal{L}_E^0$. We use energy as one of coordinates. It is not difficult to see that $\mathcal{L}_E^0$ is a 3-dimensional manifold invariant under the geodesic flow Φ_t and diffeomorphic to $\mathbb{T}^2 \times \mathbb{T}^1$. We can view any geodesic as the map "γ": $\mathbb{R} \to \mathbb{T}^2$ and denote by $\gamma_E(t) = (\gamma_E^q(t), \gamma_E^p(t))$ the orbit of the geodesic flow which belongs to the energy level $\mathcal{L}_E^0$ with a fixed origin $\gamma_E(0) \in \mathcal{L}_E^0$. These conditions determine the orbit $\{\gamma_E(t)\}_{t \in \mathbb{R}}$ uniquely.

Notice that an appropriate rescaling of time of an orbit $\{\gamma_E(t)\}_{t \in \mathbb{R}}$ on the energy level $\mathcal{L}_E^0$ leads to an orbit on the energy level $\mathcal{L}_{1/2}^0$.

$$(9) \qquad (\gamma_E^q(t), \gamma_E^p(t)) = \left(\sqrt{2E} \gamma_{1/2}^q(\sqrt{2E}t), \sqrt{2E} \gamma_{1/2}^p(\sqrt{2E}t) \right).$$

Recall that $\Gamma \subset \mathcal{L}_{1/2}^0$ is the shortest hyperbolic periodic geodesic in the homology class $h \in H_1(\mathbb{T}^2, \mathbb{Z})$ from non-degeneracy Hypothesis 1. Then if $\Gamma = \Gamma_{1/2}$ has period T the period of the same geodesic with rescaled time $\Gamma_E(\sqrt{2E}t) \equiv \Gamma_{1/2}(t)$ is $T/\sqrt{2E}$.

4.1. Hyperbolic Persistent Cylinder. Let's give a rigorous statement of the fact stated after Non-degeneracy hypothesis in Section 1.2.

Theorem 7. (Morse, Anosov, Mather) *For a C^r generic[5] Riemannian metric ρ on $\mathbb{T}^2$ with $r \geq 2$, for a homology class $h \in H_1(\mathbb{T}^2, \mathbb{Z})$, and for any energy $H_0(p, q) = E > 0$ there is a periodic geodesic $\Gamma^0_{1/2} = \cup_{t \in \mathbb{R}} \Gamma^0_{1/2}(t)$ in the homology class h, whose time rescaled copies $\mathbb{A}^0_E = \cup_{E' \geq E} \Gamma^0_{E'}$, as in (9), form an invariant normally hyperbolic manifold (cylinder) in $\mathbb{A}^0_E \subset T^* \mathbb{T}^2$. Its stable and unstable manifolds $W^s_{\mathbb{A}^0_E}$ and $W^u_{\mathbb{A}^0_E}$ respectively are 2-dimensional and on each energy level Γ^0_E there is a homoclinic orbit $\Lambda^0_E(t)$ given by*

$$\Lambda^0_E \subset (W^s_{\Gamma^0_E} \setminus \Gamma^0_E) \cap (W^u_{\Gamma^0_E} \setminus \Gamma^0_E), \tag{10}$$

whose projection into $\mathbb{T}^2$ corresponds to the Morse geodesic Λ. Moreover, this intersection of invariant manifolds restricted to any energy level $\mathcal{L}^0_E$ is transversal. Also on the energy level $E = 1/2$ for some a and b we have

$$\begin{aligned}
\rho(\Lambda^0_{1/2}(s), \Gamma^0_{1/2}(s + a)) &\to 0 \quad \text{as} \quad s \to -\infty \\
\rho(\Lambda^0_{1/2}(s), \Gamma^0_{1/2}(s + b)) &\to 0 \quad \text{as} \quad s \to +\infty.
\end{aligned} \tag{11}$$

Transversal intersection of two 2-dimensional manifolds $W^s_{\Gamma^0_E}$ and $W^u_{\Gamma^0_E}$ in 3-dimensional energy level $\mathcal{L}_E$ is 1-dimensional and by the theorem on implicit function is locally isolated curve.

Now formalize meaning of normal hyperbolicity of the cylinder $\mathbb{A}^0_E$:

Lemma 4. *(e.g. [DLS1]) Let $\mathbb{A}^0_E = \cup_{E' \geq E} \Gamma^0_{E'}$ be the invariant cylinder diffeomorphic to $[E, +\infty) \times \mathbb{T}^1$ and the canonical symplectic form ω restricted to $\mathbb{A}^0_E$ is non-degenerate and invariant under the geodesic flow Φ_t. Moreover, for some $C, \alpha > 0$ and for all $x \in \mathbb{A}^0_E$ we have*

$$T_x \mathcal{L}^0_E = E^s_x \oplus E^u_x \oplus T_x \hat{\Gamma}^0_E \tag{12}$$

with $\|d\Phi_t(x)|_{E^s_x}\| \leq C \exp(-\alpha t)$ for $t \geq 0$, $\|d\Phi_t(x)|_{E^u_x}\| \leq C \exp(\alpha t)$ for $t \leq 0$, and $\|d\Phi_t(x)|_{T_x \Gamma^0_E}\| \leq C$ for all $t \in \mathbb{R}$.

4.1.1. High energy motion. Recall that the original Hamiltonian has the form $H(p, q, t) = \rho_q(p, p)/2 + U(q, t)$. So if energy is of order ε^{-2} for a sufficiently small $\varepsilon > 0$, then it is convenient to scale the Hamiltonian $\varepsilon^2 H(p/\varepsilon, q, t) = \rho_q(p, p)/2 + \varepsilon^2 U(q, t)$. Introduce new: time $\bar{t} = t/\varepsilon$, impulse $\bar{p} = \varepsilon p$, and symplectic form $\bar{\omega} = d\bar{p}_1 \wedge dq_1 + d\bar{p}_2 \wedge dq_2 = \varepsilon \omega$. Then the rescaled Hamiltonian can be written as $H_\varepsilon(\bar{p}, q, \varepsilon \bar{t}) = \rho_q(\bar{p}, \bar{p})/2 + \varepsilon^2 U(q, \varepsilon \bar{t})$.

Fix a sufficiently small ε. Notice that $H_\varepsilon(\bar{p}, q, \varepsilon \bar{t})$ is a small perturbation of the geodesic Hamiltonian system $H_0(\bar{p}, q)$. By Sacker-Fenichel theorem [Sa], [Fe], [HPS] the hyperbolic invariant cylinder $\mathbb{A}^0_E$ with a sufficiently large $E > \varepsilon^{-2}$ for the geodesic Hamiltonian system $H_0(\bar{p}, q)$ persists under a small

---

[5]Here generic means open dense set in the space of C^r metrices with the uniform C^r topology

perturbation and as smooth as H_0 is. Therefore, the rescaled Hamiltonian system $H_\varepsilon(\bar p, q, \varepsilon \bar t)$ has a hyperbolic invariant cylinder $\mathbb{A}^0_{\bar E}$ which is close to the hyperbolic invariant cylinder $\hat\Gamma^0_{\bar E}$. Rescaling back to the initial Hamiltonian $H(p, q, t) = \rho_q(p, p)/2 + U(q, t)$ we see that $H(p, q, t)$ also has a hyperbolic invariant cylinder $\hat\Gamma_E$ close to $\hat\Gamma^0_E$ with $E = \bar E \varepsilon^{-2}$. This cylinder has to belong not to the phase space $\mathbf{T}^*\mathbb{T}^2$ of the geodesic Hamiltonian as $\mathbb{A}^0_E$ does, but the time *extended* phase space $\mathbf{T}^*\mathbb{T}^2 \times \mathbb{T}$. However, to avoid considering the extended phase space we take the time 1 map for the Hamiltonian system $H_0(p, q)$, denoted $\Phi^0 = \Phi^0_1 : \mathbf{T}^*\mathbb{T}^2 \to \mathbf{T}^*\mathbb{T}^2$ and for the initial Hamiltonian $H(p, q, t)$, denoted $\Phi = \Phi_1 : \mathbf{T}^*\mathbb{T}^2 \to \mathbf{T}^*\mathbb{T}^2$. Then Φ^0 has $\mathbb{A}^0_E$ as a hyperbolic invariant cylinder and for a sufficiently large E the map Φ also has a hyperbolic invariant cylinder, denoted by $\mathbb{A}_E \subset \mathbf{T}^*\mathbb{T}^2$. This cylinder has 3-dimensional stable and unstable manifolds $W^s(\mathbb{A}_E)$ and $W^u(\mathbb{A}_E)$ respectively which in a neighborhood of the cylinder Γ_E are small perturbation of stable and unstable manifolds $W^s(\mathbb{A}^0_E)$ and $W^u(\mathbb{A}^0_E)$ for the cylinder $\mathbb{A}^0_E$. By Hypothesis 2' $W^s(\mathbb{A}^0_E)$ and $W^u(\mathbb{A}^0_E)$ have transversal intersection. Therefore, $W^s(\mathbb{A}_E)$ and $W^u(\mathbb{A}_E)$ also have transversal intersection which is the family of homoclinic trajectories $\hat\Lambda_E$ to the cylinder $\mathbb{A}_E$[6].

4.2. **The inner and outer maps.** Consider $\Phi^0 : \mathbf{T}^*\mathbb{T}^2 \to \mathbf{T}^*\mathbb{T}^2$ the time 1 map for $H_0(p, q)$, restricted to the invariant cylinder $\hat\Gamma_E$. Wlog assume that period of the minimal periodic geodesic $\Gamma^0_{1/2}$ is 1, otherwise, rescaled the metric. Denote by $(\theta, r) \in \mathbb{S}^1 \times \mathbb{R} = \mathbb{A} \supset \mathbb{A}^0_E$ the natural coordinates on the invariant cylinder $\hat\Gamma_E$ chosen so that the time 1 map for $H_0(p, q)$ has the form $\mathcal{P}^0(\theta, r) = (\theta + r, r)$. It is possible by scaling arguments. It is easy to see that $\mathcal{P}_0$ is an EAPT. Denote by ω_Γ the restriction of the standard symplectic 2-form $\omega = dp_1 \wedge dq_1 + dp_2 \wedge dq_2$ to the invariant cylinder $\mathbb{A}^0_E$.

Similarly, consider $\Phi : \mathbf{T}^*\mathbb{T}^2 \to \mathbf{T}^*\mathbb{T}^2$ the time 1 map for $H(p, q, t)$. We know it has a hyperbolic invariant cylinder $\mathbb{A}_E$ for a sufficiently large E. Restrict $\Phi|_{\mathbb{A}_E}$ to this cylinder and denote it by

$$(13) \qquad \mathcal{P}_E = \Phi|_{\mathbb{A}_E} : \mathbb{A}_E \to \mathbb{A}_E.$$

Using standard arguments one can check that restriction of the standard symplectic form ω to $\mathbb{A}_E$ gives an area form ω_E on $\mathbb{A}_E$. Moreover, $\mathcal{P}_E$ is an EAPT ω_Γ-preserving. Following terminology proposed in [DLS1] call $\mathcal{P}_E$ *the inner map*, because it acts within the invariant cylinder $\mathbb{A}_E$. Implicitly the inner map is used in [Ma6], but it is encoded into Mather's variational principle.

To define the *outer* map recall that the cylinder $\mathbb{A}_E$ has 3-dimensional stable and unstable manifolds $W^s(\mathbb{A}_E)$ and $W^u(\mathbb{A}_E)$ which intersect transversally in 4-dimensional space $\mathbf{T}^*\mathbb{T}^2$. Since contraction and expansion along

---

[6]If $W^u(\mathbb{A}^0_E)$ and $W^s(\mathbb{A}^0_E)$ are not transversal we need intersaction to be an isolated curve. The fact that this intersation is non-empty follows from existence of Morse Class A geodesics defined in sect. 1.1

transversal to $\mathbb{A}_E$ invariant directions dominate the inner dynamics on $\mathbb{A}_E$ for each point $x \in \mathbb{A}_E$ there is a unique trajectory containing $z_+(x)$ (resp. $z_-(x)$) in $W^s(\mathbb{A}_E) \cap W^u(\mathbb{A}_E)$ such that for some $\beta > 0$, a sufficiently large E, and all sufficiently large $n \in \mathbb{Z}_+$

(14)
$$\text{dist}\left(\Phi^{-n}x, \Phi^{-n}z_-(x)\right) \leq \exp(\beta n)$$
$$\text{dist}\left(\Phi^{n}x, \Phi^{n}z_+(x)\right) \leq \exp(\beta n)$$

This defines the map

(15) $\mathcal{S}_E : \mathbb{A}_E \to \mathbb{A}_E \quad \mathcal{S}_E(x_-) = x_+,$

where trajectories of $z_-(x_-)$ and $z_+(x_+)$ being the same. In other words, $\mathcal{S}_E$ sends x_- to x_+ if there is a heteroclinic trajectory from x_- to x_+. It is important for definition of the outer map that transversal to the cylinder $\mathbb{A}_E$ dynamics (expansion/contraction) dominates dynamics on the cylinder itself. It is certainly true in our case, because the inner map $\mathcal{P}_E$ is a perturbation of an integrable map with no expansion/contraction. Following terminology proposed in [DLS1] call $\mathcal{S}_E : \mathbb{A}_E \to \mathbb{A}_E$ *the outer map*. Again implicitly the outer map is used in [Ma6], even though it is encoded in action-minimizing trajectories. It is not difficult to see that $\mathcal{S}_E$ is continuous and even can be shown to be C^r-smooth area-preserving (see [DLS1]), but we don't use this fact.

5. DIFFUSION STRATEGY

The idea of mixed diffusion following inner and outer dynamics goes back to the original work of Mather [Ma6]. Provided the Hamiltonian system (3) is sufficiently smooth Arnold's approach [Ar] of whiskered tori is also applicable (as shown in [BT] and [DLS1]). Our goal is to show existence of trajectories traveling indefinitely far along the invariant cylinder $\mathbb{A}_E$. Since the inner dynamics on $\mathbb{A}_E$, defined by the inner map $\mathcal{P}_E : \mathbb{A}_E \to \mathbb{A}_E$, is described by an EAPT, we can apply arguments and results from the first part of the paper. Namely, between any two adjacent invariant curves on $\mathbb{A}_E$ we have an EAPT acting inside of a BRI (Birkhoff region of instability). Therefore, there is a trajectory which goes from an arbitrary small neighborhood of the "bottom" invariant curve to an arbitrary small neighborhood of the "top" one. The problem which arises is that the inner map $\mathcal{P}_E$ is a small perturbation of the completely integrable map $\mathcal{P}^0(\theta, r) = (\theta + r, r)$. Indeed, every horizontal circle on $\mathbb{A}^0$ is $\mathcal{P}^0$-invariant, therefore, by KAM theory [He] after a small perturbation most of these invariant curves will survive. Thus, $\mathcal{P}_E$ has a lot of invariant curves. So, using the inner map $\mathcal{P}_E$ it is impossible to "jump" over an invariant curve Γ for $\mathcal{P}_E$ to increase energy. To overcome the problem we shall use the outer map $\mathcal{S}_E$. Mather [Ma6] shows that $\mathcal{S}_E(\Gamma)$ has a part strictly above Γ which provides the "jump" over Γ. Our strategy is to pick a neighborhood in $\mathbf{T}^*\mathbb{T}^2$ of a properly chosen point in $\mathbb{A}_E$ and iterate this neighborhood using alternating series of inner and outer maps.

To describe the strategy with more details introduce a bit of terminology: Let $\Gamma \subset \mathbb{A}_E$ be an invariant curve, which always means invariant for the inner map $\mathcal{P}_E$. Denote by $\mathbb{A}_\Gamma^+ \subset \mathbb{A}_E$ an open topological annulus infinite on one side and bounded by Γ on the other. We say that an invariant curve $\Gamma \subset \mathbb{A}_E$ is top (resp. bottom) isolated if it has a neighborhood in $\mathbb{A}_\Gamma^+$ (resp. $\mathbb{A}_E \setminus \mathbb{A}_\Gamma^+$) free from invariant curves.

Notice also that for a large enough energy any invariant curve Γ is a small perturbation of the curve $\Gamma^0 = \{H_0 = \omega_\Gamma^2/2\} \cap \mathbb{A}_0$, where ω_Γ is the rotation number induced on Γ by $\mathcal{P}_E$. Thus, $H(\cdot, t)$ is almost constant on Γ.

It turns out that under non-degeneracy Hypothesis 3 for any $d > 1$ and any invariant curve $\Gamma \subset \mathbb{A}_E$ the outer image $\mathcal{S}_E(\Gamma) \subset \mathbb{A}_E$ deviates up and down from Γ so that there is a point $x_+ \in \mathcal{S}_E(\Gamma) \cap \mathbb{A}_\Gamma^+$ (resp. $x_- \in \mathcal{S}_E(\Gamma) \setminus \mathbb{A}_\Gamma^+$) such that we have

$$(16) \qquad\qquad \mathrm{dist}(x_\pm, \Gamma) > \omega_\Gamma^{-d}.$$

Notice that for our method it does not matter an exact value of d in contract to the standard geometric method from [BT] or [DLS1]. Moreover, ω_Γ^{-d} can be replaced by a flat function, e.g. $\exp(-\omega_\Gamma)$. In [BT] and [DLS1] the value of d is crucial to overcome the so-called gap-problem about gaps between nearby invariant curves/tori. In [DLS3] modification of Arnold's whiskered tori approach is proposed to prove existence of diffusion in Arnold's example [Ar]. The authors use resonant tori in $\mathbb{A}_E$ as additional elements of transition chains there. In [X] extension of Mather's variational approach is given to show diffusion in this example. In future publicatons we shall extend mixed geometric/variational approach presented in this paper to include Arnold's example.

Let's distinguish two ways of diffusing: inner and outer.

Inner diffusion (or Birkhoff diffusion): Let x belong to a top isolated curve $\Gamma \subset \mathbb{A}_E$. Suppose $\Gamma' \subset \mathbb{A}_\Gamma^+$ is an adjacent invariant curve, i.e. $\mathcal{P}_E(\Gamma') = \Gamma'$ and the annulus between Γ and Γ' inside $\mathbb{A}_\Gamma^+ \setminus \mathbb{A}_{\Gamma'}^- \subset \mathbb{A}_E$ is a BRI, i.e $\mathbb{A}_\Gamma^+ \setminus \mathbb{A}_{\Gamma'}^-$ is free from other invariant curves (sect. 1.3). Then by Birkhoff invariant set theorem for any neighborhood U_x which has x in the closure and any neighborhood $U_{\Gamma'}$ of Γ' both in $\mathbf{T}^*\mathbb{T}^2$ there is $n_0 \in \mathbb{Z}_+$ depending on all the above quantities such that $\mathcal{P}_E^n(U_x \cap \mathbb{A}_E) \cap U_{\Gamma'} \neq \emptyset$. Moreover, $\cup_{n \in \mathbb{Z}_+} \mathcal{P}_E^n(U_x \cap \mathbb{A}_E)$ contains Γ' in the closure.

Outer diffusion: Let $\Gamma \subset \mathbb{A}_E$ be an invariant curve which is bottom isolated and $U \subset \mathbf{T}^*\mathbb{T}^2$ is an open set containing Γ in the closure. Consider the outer image $\mathcal{S}_E(\Gamma)$. There are two case:
· the first — $\mathbb{A}_\Gamma^+$ has a topological annulus C^{inv} close to Γ foliated by invariant curves and $\mathcal{S}_E(\Gamma)$ does not intersect a top isolated invariant curve Γ above C^{inv} in $\mathbb{A}_E$ and
· the second — $\mathcal{S}_E(\Gamma)$ intersects a top isolated invariant curve Γ' and does not intersect any topological annulus $C^{inv} \subset \mathbb{A}_{\Gamma'}^+$ separating $\mathbb{A}_E$ and foliated by invariant curves.

In the first case of outer diffusion we need the following

Lemma 5. *Let* $\mathcal{P}_E : \mathbb{A} \to \mathbb{A}$ *be a* C^1-*EAPT of the annulus* $\mathbb{A}$. *Suppose there is a topological annulus* $C \subset \mathbb{A}$ *separating* $\mathbb{A}$, *and consisting of invariant curves, i.e. every* $x \in C$ *belongs to a rotational curve* $\Gamma = \mathcal{P}(\Gamma)$. *Then for any neighborhood* $U \subset C$ *there is* $n \in \mathbb{Z}_+$ *so that* $\cup_{k=0}^{n} \mathcal{P}^k U$ *contains an invariant curve* Γ^*.

We shall prove this lemma at the end of this section for completeness. Clearly such a situation is easily destroyable by a pertubation, but for us it is easier to prove this lemma. In the outer case under consideration there is an open set $V \subset \mathbb{A}_E$ sufficiently close to Γ so that $U = \mathcal{S}_E(V)$ intersects only those invariant curves $\tilde{\Gamma}$ in $\mathbb{A}_\Gamma^+$ that for the point $x_+ \in \mathcal{S}_E(\Gamma)$ satisfy (16) we have

$$(17) \qquad \text{dist}(x_+, \tilde{\Gamma}) > \omega_\Gamma^{-2}/2.$$

Then application of the above lemma shows that for some $n \in \mathbb{Z}_+$ we get $\cup_{k=0}^{n} \mathcal{P}_E U$ contains an open neighborhood of some invariant curves $\tilde{\Gamma} \subset \mathbb{A}_\Gamma^+$ satisfying (17). In other words, first we send a neighborhood $V \subset \mathbb{A}_E$ below Γ by the outer map $\mathcal{S}_E(V)$ first and then by iterate by the inner map so that its images contain a neighborhood of an invariant $\tilde{\Gamma}$ which is "higher" than Γ by at least $\omega_\Gamma^{-d}/2$. If $\mathbb{A}_{\tilde{\Gamma}}^+$ is foliated by invariant circles, then iterating this procedure of applying the outer map and than the inner map a finite number of times we can increase energy indefinitely. If $\mathbb{A}_{\tilde{\Gamma}}^+$ is not foliated by invariant curves, then a number of iterations if this procedure will brings as to the second case above.

In the second outer case we know that $\mathcal{S}_E(\Gamma)$ intersects Γ' and has a part in $\mathbb{A}_{\Gamma'}^+$. Thus, if an open subset $U \subset \mathbb{A}_E$ has Γ in the closure its outer image $\mathcal{S}_E(U)$ closure has nonempty intersection with Γ' and there is a neighborhood U' in $\mathcal{S}_E(U) \cap \mathbb{A}_{\Gamma'}^+$ which has Γ' in the closure and fits to apply the inner diffusion. We also either require Γ' to satisfy (17) with $\tilde{\Gamma} = \Gamma'$ or pick the lowest invariant curve Γ^* in $\mathbb{A}_{\tilde{\Gamma}}^+$ to satisfy (17) with $\tilde{\Gamma} = \Gamma^*$. This completes the proof of Mather's acceleration theorem based on lemma 5 and estimate (16) about oscillations of the outer map. Now we proof lemma 5 and in the next section describe Mather's variational approaches to prove estimate (16) which is *the key.*

Proof of lemma 5: By Birkhoff invariant curve theorem ([MF] Thm.15.1) every rotation invariant curve Γ of an EAPT $\mathcal{P}$ is a Lipschitz graph over the base $\mathbb{S}^1$. This naturally induces a coordinate system on each invariant curve and gives that $\mathcal{P}|_\Gamma$ is a homeomorphism on each invariant curve Γ.

Notice that by area-preservation of $\mathcal{P}$ and invariance of C for some $n \in \mathbb{Z}_+$ we have $U \cap \mathcal{P}^k U \neq \emptyset$. U consists of intersections with invariant Lipschitz curves $U_\Gamma = U \cap \Gamma$. So for an open set of them $U_\Gamma \cap \mathcal{P}^k U_\Gamma \neq \emptyset$ and by area-preservation and twist conditions $\mathcal{P}^k U_\Gamma \setminus U_\Gamma \neq \emptyset$. Pick a curve Γ with irrational rotation number. Since $\mathcal{P}^k U_\Gamma \cap U \neq \emptyset$, the same is true for $\mathcal{P}^{rk} U_\Gamma$

and $\mathcal{P}^{(r+1)k}U_\Gamma$ with $r \in \mathbb{Z}$. Therefore, $\cup_{r=0}^s \mathcal{P}^{rk}U_\Gamma$ is always connected and by irrationality of rotation number of Γ should cover Γ for some s. This completes the proof of the lemma. Q.E.D.

6. The Jump Lemma

In this section we outline Mather's variational approach to prove the oscillation property (16) of the outer map $\mathcal{S}_E : \mathbb{A}_E \to \mathbb{A}_E$, defined in (15). The general idea of Mather's method is to construct trajectories of a Hamiltonian system as solutions to a variational problem. In our case time 1 map of the Hamiltonian system (3), denoted by $\Phi : \mathbf{T}^*\mathbb{T}^2 \to \mathbf{T}^*\mathbb{T}^2$, possesses an invariant cylinder $\mathbb{A}_E \subset \mathbf{T}^*\mathbb{T}^2$ and restriction of $\Phi|_{\mathbb{A}_E} = \mathcal{P}_E$ is a C^2-EAPT. Thus, by Aubry-Mather theory [AL], [Ma1], [MF], or [Ba1] for every rotation number ω there is an action-minimizing (Aubry-Mather) invariant set $\Sigma_\omega \subset \mathbb{A}_E$ (Theorem 3). By twist condition if $\Gamma = \mathbb{A}_E$ is an invariant curve with rotation number $\omega = \omega_\Gamma$, then $\Sigma_\omega \subseteq \Gamma$ and for any $\omega' > \omega$ the corresponding Aubry-Mather $\Sigma_{\omega'}$ belongs to $\mathbb{A}_\Gamma^+$, i.e. is above Σ_ω. Similarly, if $\omega' < \omega$, then $\Sigma_{\omega'} \subset \mathbb{A}_E \setminus \mathbb{A}_\Gamma^+$. Moreover, the map from Aubry-Mather sets to $\mathbb{R}$ according to their rotation numbers can be extended to a Lipschitz map (see e.g. [Do]). Since the EAPT $\mathcal{P}_E$ induced on the invariant cylinder $\mathbb{A}_E$ is a small perturbation of the completely integrable map $\mathcal{P}(\theta, r) = (\theta + r, r)$ for large rotation numbers, to prove existence of orbits with *arbitrarily increasing energy* it suffices to prove existence of orbits which consequently visit neighborhoods of invariant sets with *arbitrarily increasing rotation numbers* on $\mathbb{A}_E$. Therefore, *if for some $\omega' < \omega < \omega''$ with $|\omega' - \omega''| > \omega^{-d}$ there is a trajectory whose α-limit set is $\Sigma_{\omega'}$ and ω-limit set is $\Sigma_{\omega''}$*[7] or, almost equivalently, $\mathcal{S}_E(\Sigma_{\omega'}) \cap \Sigma_{\omega''} \neq \emptyset$, then this along with the arguments from the previous section proves Mather's acceleration theorem. The rest of the paper is devoted to a formal statement of the Jump lemma and outline of Mather's variational approach to prove it.

6.1. Duality between Hamiltonian and Lagrangian systems. Recall that $U : \mathbb{T}^2 \times \mathbb{T} \to \mathbb{R}$ be a C^2-smooth time periodic function on $\mathbb{T}^2$ which we use as the potential energy. We associate the kinetic energy to the metric ρ

$$(18) \qquad T^q(v) = \rho^q(v, v)/2, \quad v \in \mathbf{T}_q\mathbb{T}^2,$$

where $\mathbf{T}_q\mathbb{T}^2$ denotes the tangent bundle of $\mathbb{T}^2$ at q.

Definition 8. The *Lagrangian* L is a real valued function, defined on the phase space $\mathbf{T}\mathbb{T}^2 \times \mathbb{T}$ by

$$(19) \qquad L(q, v, t) = T^q(v) - U(q, t).$$

---

[7]we shall call this statement "the Jump Lemma"

Define the *Euler-Lagrange* flow associated to L. This is a flow on the phase space $T\mathbb{T}^2 \times \mathbb{T}$ such that trajectories are associated to the variational condition

$$(20) \qquad \delta \int_a^b L(d\gamma(t), t)dt = 0,$$

where $\gamma : [a, b] \to \mathbb{T}^2$ is a C^1 curve, $d\gamma(t)$ stands for the 1-jet $(\gamma(t), \dot\gamma(t))$, and the variation is taken relative to fixed endpoints. If γ satisfies this variational condition, then

$$(21) \qquad \xi(t) = (d\gamma(t), t \bmod 1)$$

is a trajectory of the Euler-Lagrange flow. All trajectories, defined on $[a, b]$ are of this form. Equivalently, trajectories of the Euler-Lagrange flow are solutions of the Euler-Lagrange equation

$$(22) \qquad \frac{d}{dt}(L_{\dot\theta}) = L_\theta,$$

where $\theta = (\theta_1, \theta_2)$ is the standard angular coordinate system on $\mathbb{T}^2$.

Lemma 6. *(see e.g. [Fa] sect. 2.3) In the setting above trajectories of the Hamiltonian system with the Hamiltonian H, defined by (3), are in one-to-one correspondence with trajectories of the Euler-Lagrange flow for the Lagrangian L, given by (19). Moreover, a trajectory for H is mapped into a trajectory for (22) by the Legendre transform*

$$(23) \qquad H(q, p, t) = \langle p, v\rangle_q - L(q, v, t), \quad \text{where} \quad p = \frac{\partial L}{\partial v}(q, v)$$

$$\text{and} \quad \langle \, , \, \rangle_q : T_q\mathbb{T}^2 \times T_q^*\mathbb{T}^2 \to \mathbb{R} \quad \text{is a dual pairing.}$$

6.2. Action-Minimizing Probabilities and Mather sets.

In this section we discuss Mather's theory of minimal or action minimizing measures. This theory can be considered as an extension of KAM theory. Namely, it provides a large class of invariant sets for an Euler-Lagrange flow (or the dual Hamiltonian flow). KAM invariant tori is an example of *Mather sets*. Mather sets are also generalization of Aubry-Mather sets from two degrees of freedom to arbitrary number of degrees of freedom.

Following Mather [Ma4] we say that μ is a probability if it is a Borel measure and of total mass one. A probability on $T\mathbb{T}^2 \times \mathbb{T}$ is *invariant* if it is invariant under Euler-Lagrange flow. If η is a one form on $\mathbb{T}^2 \times \mathbb{T}$, we may associate to it a real valued function $\hat\eta$ on $T\mathbb{T}^2 \times \mathbb{T}$, as follows: express η in the form

$$(24) \qquad \eta = \eta_1 d\theta_1 + \eta_2 d\theta_2 + \eta_\tau d\tau$$

with respect to the standard angular coordinates (θ_1, θ_2) on $\mathbb{T}^2$ and τ on $\mathbb{T}^1$ and set

$$(25) \qquad \hat\eta = \eta_{\mathbb{T}^2} + \eta_\tau \circ \pi,$$

where $\pi : \mathbf{TT}^2 \times \mathbf{T} \to \mathbb{T}^2 \times \mathbf{T}$ denotes the natural projection. This function has the property

$$(26) \qquad \int_a^b \hat{\eta}(d\gamma(t), t)dt = \int_{\gamma, \tau} \eta,$$

for every C^1 curve $\gamma : [a, b] \to \mathbb{T}^2$ with the right side being the usual integral and $(\gamma, \tau) : [a, b] \to \mathbb{T}^2 \times \mathbf{T}$ defined by $(\gamma, \tau)(t) = (\gamma(t), t \bmod 1)$.

If μ is an invariant probability on $\mathbf{TT}^2 \times \mathbf{T}$, its *average action* is defined as

$$(27) \qquad A(\mu) = \int L d\mu.$$

Since L is bounded below, this integral is defined, although it may be $+\infty$. If $A(\mu) < +\infty$, one can define the *rotation vector* $\rho(\mu) \in H_1(\mathbb{T}^2, \mathbb{R})$ of μ by

$$(28) \qquad \langle \rho(\mu), [\eta]_{\mathbb{T}^2} \rangle + [\eta]_{\mathbf{T}} = \int \hat{\eta} d\mu$$

for every C^1 one form η on $\mathbb{T}^2 \times \mathbf{T}$, where

$$(29) \qquad [\eta] = ([\eta]_{\mathbb{T}^2}, [\eta]_{\mathbf{T}}) \in H^1(\mathbb{T}^2 \times \mathbf{T}, \mathbb{R}) = H^1(\mathbb{T}^2, \mathbb{R}) \times \mathbb{R}$$

denotes de Rham cohomology class and $\langle \, , \, \rangle$ denotes dual pairing $H_1(\mathbb{T}^2, \mathbb{R}) \times H^1(\mathbb{T}^2, \mathbb{R}) \to \mathbb{R}$. Mather introduced this concept in [Ma4] in the case of time independent one forms, but in time dependent case arguments are the same. In [Ma4] using Krylov-Bogoliuboff arguments Mather proved that

Lemma 7. *For every homology class $h \in H_1(\mathbb{T}^2, \mathbb{R})$ there exists an invariant probability μ such that $A(\mu) < +\infty$ and $\rho(\mu) = h$.*

Such a probability is called *minimal* or *action-minimizing* if

$$(30) \qquad A(\mu) = \min\{A(\nu) : \rho(\nu) = \rho(\mu)\},$$

where ν ranges over invariant probabilities such that $A(\nu) < +\infty$. If $\rho(\mu) = h$, we also say that μ is *h-minimal*.

The rotation vector has a natural geometric interpretation as an asymptotic direction of motion. More exactly, for a μ-generic trajectory of the Euler-Lagrange flow $\gamma : \mathbb{R} \to \mathbb{T}^2$ for $T > 0$ let z_T be the closed curve consisting of two parts: $\gamma|_{[-T,T]}$ and the shortest geodesic connecting $\gamma(-T)$ and $\gamma(T)$ on $\mathbb{T}^2$. Then

$$(31) \qquad \rho(\mu) = \lim_{T \to +\infty} \frac{1}{2T}[z_T].$$

We say that an invariant probability is *c-minimal* (for $c \in H^1(\mathbb{T}^2, \mathbb{R})$) if it minimizes $A(\nu) - \langle \rho(\nu), c \rangle$ over all invariant probabilities. Mather [Ma4] also proved

Lemma 7'. *For every cohomology class $c \in H^1(\mathbb{T}^2, \mathbb{R})$ there exists an invariant c-minimal probability μ such that $A(\mu) < +\infty$.*

We say that an invariant probability μ is minimal if and only if there is a one form η on $\mathbb{T}^2 \times \mathbb{T}$ such that μ minimizes $\int (L - \hat{\eta}) d\nu = A(\nu) - \langle \rho(\nu), [\eta]_{\mathbb{T}^2} \rangle - [\eta]_{\mathbb{T}}$ over invariant probabilities ν (see [Ma6]).

Definition 9. We call the function

(32) $\beta : H_1(\mathbb{T}^2, \mathbb{R}) \to \mathbb{R}$, $\beta(h) = A(\mu)$,   where μ is h-minimal.

Mather's β-function and we call

(33) $\alpha : H^1(\mathbb{T}^2, \mathbb{R}) \to \mathbb{R}$ $\alpha(c) = \inf\limits_{h \in H_1(\mathbb{T}^2, \mathbb{R})} \{\beta(h) - \langle h, c \rangle\}$.

Mather's α-function. It is well defined by lemma 7.

Thus, the α-function is conjugate to the β-function by the Legendre transform. It follows that both functions are convex. By definition

(34) $\beta(h) + \alpha(c) \geq \langle h, c \rangle$, $h \in H_1(\mathbb{T}^2, \mathbb{R})$, $c \in H^1(\mathbb{T}^2, \mathbb{R})$.

The β-Legendre transform

$\mathcal{L}_\beta : H_1(\mathbb{T}^2, \mathbb{R}) \to \{\text{compact, convex, non-empty subsets of } H^1(\mathbb{T}^2, \mathbb{R})\}$

is defined by putting $\mathcal{L}_\beta(h)$ as the set of $c \in H^1(\mathbb{T}^2, \mathbb{R})$ for which the inequality above becomes equality. The α-Legendre transform

(35) $\mathcal{L}_\alpha : H^1(\mathbb{T}^2, \mathbb{R}) \to H_1(\mathbb{T}^2, \mathbb{R})$

as the inverse of $\mathcal{L}_\beta$. In what follows, we shall identify a h-minimal invariant probability with a c-minimal invariant probability, provided that $c \in \mathcal{L}_\beta(h)$. We call a *Mather set* $\mathcal{M}_h = supp \; \mu \subset \mathbf{T}\mathbb{T}^2$ with a rotation vector $h \in H_1(\mathbb{T}^2, \mathbb{R})$ support of an h-minimal probability μ.

We say that an absolutely continuous curve $\gamma : [a, b] \to \mathbb{T}^2$ is *c-minimal* if for some one form $[\eta] = c$ the curve minimizes

(36) $\int_a^b (L - \hat{\eta})(d\gamma(t), t) dt$

subject to a fixed endpoint condition. In other words, we require that

(37) $\int_a^b (L - \hat{\eta})(d\gamma(t), t) dt \leq \int_a^b (L - \hat{\eta})(d\gamma_1(t), t) dt$

for any absolutely continuous curve $\gamma_1 : [a, b] \to \mathbb{T}^2$ such that $\gamma(a) = \gamma_1(a)$ and $\gamma(b) = \gamma_1(b)$. Note that γ and γ_1 does not have to be homologous and c-minimality is independent of the choose of the closed one form. We say that an absolutely continuous curve $\gamma : \mathbb{R} \to \mathbb{T}^2$ is *c-minimal* if every segment $[a, b] \subset \mathbb{R}$ gives a c-minimal curve subject to a fixed point condition $\gamma(a)$ and $\gamma(b)$.

6.3. **Mather sets in the phase space and Aubry-Mather sets on the invariant cylinder** $\mathbb{A}_E$. By duality between the Hamiltonian flow (3) and the Euler-Lagrange flow (22) we have that the time one map of the latter flow has an invariant cylinder in $\mathbf{T}\mathbb{T}^2$ which for simplicity we also denote by $\mathbb{A}_E$ as the dual one for the former flow. Denote by $\tilde{\mathbb{A}}_E \subset \mathbf{T}\mathbb{T}^2 \times \mathbb{T}$ the suspension of $\mathbb{A}_E$ by the flow (22). It is easy to see that $\tilde{\mathbb{A}}_E$ is locally diffeomorphic to $\mathbb{A}_E \times \mathbb{R}$.

Recall that $h_0 \in H_1(\mathbb{T}^2, \mathbb{R})$ denotes the homology class of the unique hyperbolic minimal geodesic $\Gamma \subset \mathbb{T}^2$ for the metric ρ chosen in Hypothesis 1. Let $\mathcal{N} \subset \mathbb{T}^2$ be a neighborhood of Γ. Denote by $l_{h_0} \subset H_1(\mathbb{T}^2, \mathbb{R})$ the ray in the space of homologies given by $\{h = Eh_0 : E \in \mathbb{R}_+\}$. Denote by $r_{h_0} = \mathcal{L}_\beta(l_{h_0})$ the image of this ray under the Legendre transform. Mather [Ma6] proved that

Lemma 8. *The Legendre image $r_{h_0} \subset H^1(\mathbb{T}^2, \mathbb{R})$ of the ray l_{h_0} contains a truncated cone, i.e. a cone centered at the origin with non-empty interior and without a large ball around the origin.*

If $h = E'h_0$ for some sufficiently large $E' \in \mathbb{R}_+$ and $c \in \mathcal{L}_\beta(h)$, then for any $c \in \mathcal{L}_\beta(h)$ any c-minimal curve $\gamma : \mathbb{R} \to \mathbb{T}^2$ has its image in $\mathcal{N}$.

This lemma can also be deduced from Bangert's result [Ba2] on (non-)differentiability on β-function or so-called stable norm. This lemma implies the following

Corollary 10. *For a cohomology class $c \in \mathcal{L}_\beta(l_{h_0})$ as above any c-minimal curve belongs to the suspended cylinder $\tilde{\mathbb{A}}_E \subset \mathbf{T}\mathbb{T}^2 \times \mathbb{T}$. The natural projection $\pi : \mathbf{T}\mathbb{T}^2 \times \mathbb{T} \to \mathbb{T}^2$ of support of a c-minimal probability μ is contained in $\tilde{\mathbb{A}}_E$.*

Proof: If we consider the time one map $\Phi : \mathbf{T}\mathbb{T}^2 \to \mathbf{T}\mathbb{T}^2$ of the Euler-Lagrange flow, then $\mathbb{A}_E$ is a hyperbolic invariant cylinder. Projection of this cylinder into $\mathbb{T}^2$ by scaling arguments (sect. 4.1.1) has to belong to a neighborhood $\mathcal{N}$ of Γ. Therefore, a c-minimal trajectory belongs to a neighborhood of the cylinder $\mathbb{A}_E$ in $\mathbf{T}\mathbb{T}^2$. But $\mathbb{A}_E$ is hyperbolic with 1-dimensional stable and unstable direction at every point of $\mathbb{A}_E$, so the only trajectories which stay in a neighborhood $\pi^{-1}(\mathcal{N})$ of $\mathbb{A}_E$ for all time are those which belong to $\mathbb{A}_E$. This proves the first part of the Corollary. To prove the second part recall support of a c-minimal probability is contained in a union of all c-minimal trajectories and, therefore, has to belong to $\tilde{\mathbb{A}}_E$. Q.E.D.

Corollary 11. *If $c, c' \in r_{h_0} \subset H^1(\mathbb{T}^2, \mathbb{R})$ and $\langle h_0, c \rangle = \langle h_0, c' \rangle$, then c-minimal orbits defined on all $\mathbb{R}$ are the same as c'-minimal orbits.*

Proof: This immediately follows from the fact that c and c'-minimal curves belong to $\mathcal{N}$. Q.E.D.

6.4. **A Variational Principle.** Choose a smooth curve S embedded in $\mathbb{T}^2$ that

· S does not intersect the periodic geodesic Γ for the metric ρ, i.e. S is topologically parallel to Γ, and
· S intersects homoclinic (Morse) geodesic Λ transversally at one point.

Denote $G = \{t \in \mathbb{T}^1 : G(t) = \min_{s \in \mathbb{T}^1} G(s)\}$ the set of moments of time when Melnikov integral (5) takes its minimal value.

We say that η (or $[\eta] \in H^1(\mathbb{T}^2, \mathbb{R}) \times \mathbb{R}$) is *subcritical, critical,* or *super-critical* according as $\int (L - \hat{\eta})d\mu$ is positive, zero, or negative respectively. In other words, η is subcritical when $[\eta]_{\mathbb{T}} < (A(\mu) - [\eta]_{\mathbb{T}^2})$, critical when $[\eta]_{\mathbb{T}} = (A(\mu) - [\eta]_{\mathbb{T}^2})$, and supercritical when $[\eta]_{\mathbb{T}} > (A(\mu) - [\eta]_{\mathbb{T}^2})$ holds.

Given a closed one form η on $\mathbb{T}^2 \times \mathbb{T}$, which is critical and its homology class on $\mathbb{T}^2$ satisfies $\langle h_0, [\eta]_{\mathbb{T}^2}\rangle$, we define the variational principles $h_\eta^\pm : (S \times \mathbb{T}) \to \mathbb{R} \cup \{-\infty\}$ as follows:

$$(38) \qquad h_\eta^+(\sigma, \tau) = \inf_\gamma \left\{ \liminf_{T \to +\infty} \int_{t_0}^T (L - \hat{\eta})(d\gamma(t), t)dt \right\},$$

where the infimum is taken over the set of all absolutely continuous curves $\gamma : [t_0, +\infty) \to \mathbb{T}^2$ such that $t_0 \equiv \tau(\mathrm{mod}\ 1)$, $\gamma(t_0) = \sigma$, and $\gamma(t) \in \mathcal{N}$ for all large t. Similarly,

$$(39) \qquad h_\eta^-(\sigma, \tau) = \inf_\gamma \left\{ \liminf_{T \to +\infty} \int_{-T}^{t_0} (L - \hat{\eta})(d\gamma(t), t)dt \right\},$$

where the infimum is taken over the set of all absolutely continuous curves $\gamma : (-\infty, t_0] \to \mathbb{T}^2$ such that $t_0 \equiv \tau(\mathrm{mod}\ 1)$, $\gamma(t_0) = \sigma$, and $\gamma(-t) \in \mathcal{N}$ for all large t.

Mather's Fundamental Lemma. [Ma6] *The functions $h_\eta^\pm$ are finite and continuous and minima are achieved by minimizing trajectories $\gamma^+ : [0, +\infty) \to \mathbb{T}^2$ and $\gamma^- : (-\infty, 0] \to \mathbb{T}^2$ respectively.*

Let η and η' be critical closed one forms such that $[\eta], [\eta'] \in \mathcal{L}_\beta(l_{h_0})$, $\|[\eta]\|, \|[\eta']\|$ are sufficiently large, $\|[\eta] - [\eta]\| < \|[\eta]\|^{-2},$[8] and η, η' coincide in a small neighborhood of $\{S \cap \Lambda\} \times G \subset \mathbb{T}^2 \times \mathbb{T}$. Consider the following variational problem

$$(40) \qquad h_{\eta,\eta'} = \inf_{(\sigma,\tau) \in S \times \mathbb{T}^1} \left(h_\eta^+(\sigma, \tau) + h_{\eta'}^-(\sigma, \tau) \right).$$

Then there is a trajectory $\gamma : \mathbb{R} \to \mathbb{T}^2$ of the Euler-Lagrange flow (22) such that it realizes the minimum for $h_{\eta,\eta'}$, $[\eta]$-minimal for any segment in $t \geq 0$ and $[\eta']$-minimal for $t \leq 0$ and

$$(41) \qquad \begin{aligned} \mathrm{dist}(d\gamma(t), \mathcal{M}_{[\eta]}) &\to 0 \quad \text{as } t \to +\infty \\ \mathrm{dist}(d\gamma(t), \mathcal{M}_{[\eta']}) &\to 0 \quad \text{as } t \to -\infty. \end{aligned}$$

Moreover, γ intersects the Poincare section S only once and $\gamma(t)$ belongs to the neighbourhood $\mathcal{N}$ of Γ for all $|t|$ sufficiently large.

---

[8] 2 can be replaced by any $d \geq 1$

Consider the time one map $\Phi : \mathbf{TT}^2 \to \mathbf{TT}^2$ of the Euler-Lagrange flow (22). Recall that by definition if a closed one form $\eta^* \in r_{h_0} \subset H^1(\mathbb{T}^2, \mathbb{R})$ is critical and $c^* = [\eta^*]_{\mathbb{T}^2}$, then the corresponding Mather set $\mathcal{M}_{c^*}$ belongs to the suspended invariant cylinder $\tilde{\mathbb{A}}_E \subset \mathbf{TT}^2 \times \mathbb{T}$. If we consider the time one map Φ, i.e. discretize (22), then the invariant set for Φ is the cylinder $\mathbb{A}_E \subset \mathbf{TT}^2$. Moreover, if $h = \mathcal{L}_\alpha(c^*)$ is a dual homology class to c^*, then the restriction of the Mather set $\mathcal{M}_{c^*}$ to the time one map Φ gives an invariant set for the inner map $\mathcal{P}_E : \mathbb{A}_E \to \mathbb{A}_E$ which we denote by $\Sigma_\omega \subset \mathbb{A}_E$ with $h = \omega h_0$. So for each Mather set $\mathcal{M}_{c^*}$ there is an invariant set $\Sigma_{\omega(c^*)}$, which can be proved to be an Aubry-Mather for $\mathcal{P}_E$ (see [Ma4]), but we are not going to use this fact in general. We shall use it only in the case $\Sigma_{\omega(c^*)}$ is an invariant curve and $\mathcal{M}_{c^*}$ is an invariant two-torus, when this is straightforward.

Corollary 12. (Jump Lemma) *With the notations of Mather's Fundamental lemma and above we suppose that for some critical closed one forms η and η' with $c = [\eta]_{\mathbb{T}^2}$ and $c' = [\eta']_{\mathbb{T}^2}$ from r_{h_0} and $|c - c'| < |c|^{-2}$. Suppose also that η and η' coincide in a small neighborhood of $S \cap \Lambda \times G \subset \mathbb{T}^2 \times \mathbb{T}^1$ and there is $c^* \in r_{h_0}$ such that $\mathcal{M}_{c^*}$ is an invariant two-torus and*

$$(42) \qquad\qquad \langle h_0, c \rangle < \langle h_0, c^* \rangle < \langle h_0, c' \rangle.$$

Then there is a trajectory $\gamma : \mathbb{R} \to \mathbb{T}^2$ of the Euler-Lagrange flow (22) which
· *crosses a small neighborhood of $\{S \cap \Lambda\} \times G$;*
· *is c-minimal for any segment $t \geq 0$ and c'-minimal for any segment $t \leq 0$;*
· *$\gamma(t)$ belongs to $\mathcal{N}$ for any $|t|$ sufficiently large.*

Moreover, either the invariant set $\Sigma_{\omega(c')} \subset \mathbb{A}_E$ corresponding to $\mathcal{M}_{c'}$ belongs to a nonempty BRI $C \subset \mathbb{A}_E$ with the "top" frontier $\Sigma_{c''}$ satisfying $|c'' - c^*| > |c|^{-2}/2$ or $\Sigma_{\omega(c')}$ is an invariant curve itself.

Acknowledgment: I would like to thank John Mather for numerous fruitful conversations and remarks, Bill Cowieson, Dima Dolgopyat, Basam Fayad, Yakov Sinai, and Jeff Xia for useful discussions, and Corinna Ulcigrai for reading the manuscript and pointing out its defects.

References

[Ar]   Arnold, V. Instabilities of dynamical systems with several degrees of freedom, Sov. Math. Dokl., 5 (1964) 581–585;

[An]   Anosov, D. Generic properties of closed geodesics. (Russian) Izv. Akad. Nauk SSSR Ser. Mat. 46 (1982), no. 4, 675–709, 896;

[AL]   Aubry, S. LeDaeron, P. The discrete Frenkel-Kontorova model and its extension I: exact results for the ground states, Physica, 8D (1983), 381–422;

[Ba1]   Bangert, V. Mather Sets for Twist Maps and geodesics on Tori. Dynamics Reported, v.1, 1988, John Wiley & Sons, 1–56;

[Ba2]   Bangert, V. Geodesic rays, Busseman functions and monotone twist maps, Calc. Var. Part. Diff. Eqns, 2 (1994), 59–63;

[B1]   Birkhoff, G. Surface transformation and their dynamical applications, Acta Math., v. 43, 1920, 1–119, aussi dans Collected Math. Papers, v. II, 111–229;

[B2]   Birkhoff, G. Nouvelles recherches sur les systmes dynamiques, Memoriae Pont.
          Acad. Sci. Novi Lyncaei, v.1, 1935, 85–216 et Collected Math. Papers, v. II,
          530–661

[BT]   Bolotin, S. Treschev, D. Unbounded growth of energy in non-autonomous Hamil-
          tonian systems, Nonlinearity 12 (1999), no. 2, 365–388;

[DLS1]   Delshams, A. de la Llave, R. Seara, T. A geometric approach to the existence of
          orbits with unbounded energy in generic periodic perturbations by a potential of
          generic geodesic flows of T^2, Comm. Math. Phys. 209 (2000), no. 2, 353–392;

[DLS2]   Delshams, A. de la Llave, R. Seara, T. Orbits of unbounded energy in generic
          quasiperiodic perturbations of geodesic flows of certain manifolds, preprint;

[DLS3]   Delshams, A. de la Llave, R. Seara, T. A geometric mechanism for diffusion in
          Hamiltonian systems overcoming the large gap problem: heuristics and rigorous
          verification of the model, preprint;

[Do]   Douady, R. Regular dependence of invariant curves and Aubry-Mather sets of
          twist maps of an annulus. Erg. Th.& Dynam. Syst. 8 (1988), no. 4, 555–584;

[Fa]   Fathi, A. weak KAM theorem in lagrangian dynamics, preprint of a book, pp.
          148;

[Fe]   Fenichel, N. Persistence and smoothness of invariant manifolds for flows, Indiana
          Univ. Math. J. 21 (1971), 193–226

[Ha]   Hall, G. A topological version of a theorem of Mather on twist maps, Ergod. Th.
          & Dynam. Syst. (1984), no. 4, 585–603;

[He]   Herman, M. Sur les courbes invariantes parles difféomorphismes de L'anneau,
          Astérisque, Soc. Math. Fr., vol. 103–104, 1983;

[HPS]   Hirsch, M. Pugh, C. Shub, M. Invariant manifolds, LNM 583, Springer-Verlag,
          1977;

[Ka]   Kaloshin, V. Some prevalent properties of smooth dynamical systems. Proc.
          Steklov Inst. Math. 1996, no. 2 (213), 115–140;

[KH]   Katok, A. Hasselblatt, B. Introduction to the modern theory of dynamical sys-
          tems. Encyc. of Math. and its Apps., 54. Cambridge University Press, Cambridge,
          1995;

[L1]   Le Calvez, P. Propriété dynamique des regions d'instabilité, Ann. Sci. École
          Norm. Sup. (4), 20 (1987), no. 3, 443–464;

[L2]   Le Clavez, P. Dynamical properties of Diffeomorphisms of the Annulus and of the
          Torus, transl from the 1991 French original by Ph. Mazaud. SMF/AMS Texts &
          Monographs, 4. AMS, Providence, RI; SMF, Paris, 2000. SMF/AMS Texts &
          Monographs, 2000.

[Ma1]   Mather, J. Existence of quasiperiodic orbits for twist homeomorphisms of the
          annulus, Topology 21 (1982), no. 4, 457–467;

[Ma2]   Mather, J. Differentiability of the minimal average action as a functon of the
          rotaton number, Bo. Soc. Bras. Mat., 21, (1990), 59–70;

[Ma3]   Mather, J. Variational construction of orbits of twist diffeomorphisms. J. Amer.
          Math. Soc. 4 (1991), no. 2, 207–263;

[Ma4]   Mather, J. Action minimizing invariant measures for positive definite Lagrangian
          systems. Math. Z. 207 (1991), no. 2, 169–207;

[Ma5]   Mather, J. Variational construction of connecting orbits. Ann. Inst. Fourier
          (Grenoble) 43 (1993), no. 5, 1349–1386;

[Ma6]   Mather, J. Variational construction of trajectories for time periodic Lagrangian
          systems on the two torus, preprint, unpublished;

[Ma7]   Mather, J. Arnold's diffusion, lecture course, Princeton, fall 2002;

[MF]   Mather, J.; Forni, G. Action minimizing orbits in Hamiltonian systems. Transition to chaos in classical and quantum mechanics (Montecatini Terme, 1991), 92–186, Lecture Notes in Math., 1589, Springer, Berlin, 1994;

[Mo]   Moser, J. Stable and random motion in dynamical systems, Ann. of Math. Studies 77, Princeton NJ, Princeton Univ. Press, 1973;

[Ro]   Robinson, C. Generic properties of conservative systems. Amer. J. Math. 92 (1970), 562–603;

[Sa]   Sacker, R. A perturbation theorem for invariant manifolds and Holder continuity. J. Math. Mech. 18, (1969) 705–762;

[Sm]   Smale, S. Bull. Amer. Math. Soc. 73 (1967), 747–817;

[X]   Xia, J. Arnold Diffusion and Instabilities in Hamiltonian dynamics, preprint, 34pp. www.math.neu.edu/ xia/preprint/arndiff.ps.

CURRENT: INSTITUTE FOR ADVANCED STUDY, EINSTEIN DRIVE, PRINCETON, NJ, 08540

ON LEAVE: DEPARTMENT OF MATHEMATICS 253-37, CALTECH, PASADENA, CA 91125

E-mail address: kaloshin@math.mit.edu, kaloshin@math.princeton.edu

STRUCTURAL STABILITY IN 1D DYNAMICS

OLEG KOZLOVSKI

ABSTRACT. This is a set of notes of a series of lectures given by the author on the International conference on Dynamical Systems and Ergodic Theory, Katseveli 2000 where we will demonstrate a various techniques and methods used in the study of interval maps. In particular, we will focus on the negative Schwarzian derivative condition and on the density of Axiom A maps.

CONTENTS

1. INTRODUCTION

The aim of this paper is to introduce the Structural stability problem and to demonstrate different techniques and methods of the theory of Dynamical systems on examples arising from this problem. Though these methods usually are quite technical, we will try to avoid giving all the details and instead we will try to demonstrate the underlying ideas.

The Structural stability problem deals with the following question: What dynamical systems do not change their topological behaviour under small perturbations? Such dynamical systems are called structurally stable:

Definition 1. *Let M be a manifold. The C^k map $f : M \to M$ is called C^k structurally stable if there is $\epsilon > 0$ such that any C^k map $g : M \to M$, $\|f - g\|_{C^k} < \epsilon$, is topologically conjugate to f (i.e. there exists a homeomorphism h of M such that $g \circ h = h \circ f$).*

107

This definition was first explicitly given by Andronov and Pontrjagin, though we should also mention Poincar and Fatou who had worked on this problem before. For a long time it was believed that the structurally stable maps should have only finitely many periodic orbits. However in 1960' Smale constructed his famous horseshoe example showing that a structurally stable diffeomorphism can have infinitely many periodic points. This discovery led to definitions of hyperbolic sets and Axiom A maps (we define Axiom A maps below). In [Rob76] Robinson proved that Axiom A and the transversality condition implies C^1 (and, therefore, C^k) structural stability and the converse in the C^1 case was shown by Mañè in [Mañ98]. We should also mention other people who heavily contributed in the solution of the C^1 structural stability problem: Anosov, Franks, Havashi, Mather, Palis, Peixoto, Robbin.

However, it is not known if C^k structural stability implies Axiom A for $k > 1$. This question is in Smail's list of the most challenging math problems of the 21'st century.

Much more is known about unimodal maps, *i.e.* C^k maps of an interval which have one critical point. First, it was proven that Axiom A maps are dense in the quadratic family $Q_c : x \mapsto x^2 + c$. It was noticed by Sullivan that if one can prove that if quadratic maps Q_{c_1} and Q_{c_2} are topologically conjugate, then these maps are quasiconformally conjugate, then this would imply that Axiom A maps are dense in the family Q. Now this conjecture is completely proven in the case of real c and many people made contributions to its solution: Yoccoz proved it in the case of the finitely renormalizable quadratic maps, [Yoc90]; Sullivan — the case of the infinity renormalizable unimodal maps of "bounded combinatorial type", [Sul91], [Sul92]. Finally, it was proved in the general case by Światek, Graczyk [GŚ97] and Lyubich [Lyu97].

Based on this result for the quadratic family Kozlovski has proved density of Axiom A in the space of C^k unimodal maps for $k > 1$. The similar result is not proven in the multimodal case or even in the case of the family $x \mapsto x^4 + c$.

2. C^1 STABILITY VERSUS C^k STABILITY

So, one can see that the C^1 Structural stability problem was solved 30 years ago, however its C^k counterpart, $k > 1$, is done only in the simplest case. Why is the C^k Structural stability problem qualitatively more complicated than the C^1 one? It seems the problem here is that it is almost impossible to produce *local* C^k perturbations. That is to say that to solve the C^1 Structural stability problem one can use just local perturbation, but in the C^k case one has to construct *global* perturbations. We will illustrate this by a simple example below.

First we introduce a definition of Axiom A maps in the one dimensional case. In fact, the standard definition of Axiom A maps is different, however due to [Mañ85] they coincide for C^2 maps.

Definition 2. *Let f be a C^2 map of an interval such that*

- *the forward iterates of all critical points of f converge to the periodic attracting points;*
- *all periodic points of f are hyperbolic.*

Then the map f satisfies Axiom A.

Suppose that we have a unimodal map f and assume that the iterates of its critical point c do not converge to a periodic attractor. Thus, this map cannot satisfy Axiom A. Moreover, suppose that the critical point is recurrent and that f is even (so $c = 0$ in this case). Let us see if we can construct a perturbation of f which is localized in a neighborhood of a preimage of c.

Denote $c_m := f^m(c)$. Let m_i be a sequence of returns of the critical point to itself, *i.e.* $m_1 = 1$, m_{i+1} is the smallest integer larger than m_i such that $c_{m_{i+1}} \in I_{m_i}$, where $I_{m_i} := [-c_{m_i}, c_{m_i}]$. Denote a connected component of $f^{-1}(I_{m_i})$ containing $c_{m_{i+1}-1}$ by $\hat{I}_{m_i}$. Construct a sequence of functions $\phi_i \in C^\infty$ such that $\phi_i(x) = 0$ for $x \notin \hat{I}_{m_i}$ and $\phi_i(c_{m_{i+1}-1}) = -c_{m_{i+1}}$. The maps $g_i := \phi_i + f$ have a periodic critical point and, hence, cannot be conjugate to f.

A rough estimate gives us

$$\|g_i - f\|_{C^k} \sim K \frac{|c_{m_{i+1}}|}{|c_{m_i}|^k}$$

for some constant K. Thus, if $k = 1$ and the ratio $\frac{|c_{m_{i+1}}|}{|c_{m_i}|}$ decays to zero (and it does in many cases), then the maps g_i give required perturbations. Now consider the case $k = 2$. If the sequence $\frac{|c_{m_{i+1}}|}{|c_{m_i}|^2}$ decays to zero, then it is easy to check that $\frac{|c_{m_{i+1}}|}{|c_{m_i}|} = O(\lambda^{2^i})$ for some $\lambda \in (0,1)$, so this ratio decays superexponentially fast. However, it was shown in [NvS88] that this is impossible.

However one can construct smarter perturbations. Let us suppose that the critical point is non degenerate, so our map is $f(x) = c_1 + ax^2 + o(x^2)$. Now instead of taking forward images of the critical point we take backward images and instead of producing a perturbation near a preimage of c we are going to do it at the critical point itself. Take a sequence $\{I_n\}$ of symmetric intervals around the critical point such that the forward images of the boundary points of I_n never belong to the interior of I_n (such intervals will be called *nice*) and the sizes of I_n go to zero. Consider the first return maps R_n of f to I_n and let d_n be a preimage of the critical point 0 closest to 0. Construct a sequence of C^∞ functions ϕ_n such that $\phi_n(0) = f(d_n) - f(0)$ and $\phi_n(x) = 0$ for $x \notin I_n$. Again, the maps $g_n = f + \phi_n$ have a periodic critical point and are not conjugate to f.

The norm of the perturbations is

$$\|g_n - f\|_{C^k} \sim K \frac{|f(d_n) - f(0)|}{|I_n|^k} \sim K' \frac{d_n^2}{|I_n|^k}$$

Now one can see that if the ratio $d_n/|I_n|$ is not bounded away from zero (and in many cases it is not), g_n will give the required C^2 perturbations of f. These ideas are implemented in [BM98], [BM97] and [She].

So, it is fairly easy to construct a C^1 local perturbation and this method works for C^2 perturbations as well but nobody has yet managed to use it for constructing C^k perturbations for $k > 2$. We do not claim that it is impossible to make a C^k perturbation of the map which is localized in an arbitrarily small neighborhood of the critical point. In fact, such perturbations always exist. More precisely, let f be a C^ω unimodal map with a recurrent periodic point and suppose that f has no neutral periodic points; then for any $\epsilon > 0$ there is a unimodal map g such that $f(x) = g(x)$ for $x \in [c - \epsilon, c + \epsilon]$, $\|f - g\|_{C^k} < \epsilon$ and the maps f and g are not conjugate ([Koz98], [Tod]). However the proof of this result requires quite a different technique (compared to the C^1 case).

3. REAL ESTIMATES

The proof of the Structural stability theorem is based on both real and complex estimates. In this section we demonstrate some real methods to get such estimates.

Our unimodal map has a critical point, so one cannot hope to get a good bound for its non–linearity. There are several ways to overcome this problem, the most useful ones are Koebe lemma and estimates for the cross–ratios.

Since we cannot control the distortion of a map with critical points, the ratio of lengths of 2 adjacent intervals can change dramatically under iterates of the map. So, instead of considering 3 consecutive points, we consider 4 points and we measure their position by their cross-ratio. We will use 2 types of the cross–ratios:

$$\mathbf{a}(M, J) = \frac{|J||M|}{|M^- \cup J||J \cup M^+|}$$

$$\mathbf{b}(M, J) = \frac{|J||M|}{|M^-||M^+|}$$

where $J \subset M$ are intervals and M^-, M^+ are connected components of $M \setminus J$.

If $f : M \to f(M)$ is a diffeomorphism of an interval, we will measure how this map distorts the cross-ratios and introduce the following notation:

$$\mathbf{A}(f, M, J) = \frac{\mathbf{a}(f(M), f(J))}{\mathbf{a}(M, J)}$$

$$\mathbf{B}(f, M, J) = \frac{\mathbf{b}(f(M), f(J))}{\mathbf{b}(M, J)}$$

The following result is well-known and can be found in [dMvS88], [dMvS89] and [vS90]. It is based on the very simple idea: nearby the non-flat critical point the map has negative Schwarzian derivative (and, as we will see later, such maps increase the cross-ratios) and outside of a fixed neighborhood of the critical point the distortion of the map is bounded.

Lemma 3.1 ([dMvS89]). *Let X be an interval, $f : X \to X$ be a C^{2+1} map whose critical points are non-flat. Then there exists a constant C with the following property. If $M \supset J$ are intervals such that f^m is a diffeomorphism on M and $M \setminus J$ consists of two components M^- and M^+ then:*

$$\mathbf{A}(f^m, M, J) \geq \exp\left\{ -C \sum_{i=0}^{m-1} |f^i(M^-)| \, |f^i(M^+)| \right\},$$

$$\mathbf{B}(f^m, M, J) \geq \exp\left\{ -C \sum_{i=0}^{m-1} |f^i(M)|^2 \right\}.$$

The maps which the next lemma can be applied to, should have *a non-flat* critical point. If the map is smooth and one of its higher derivatives does not vanish at the critical point, this map automatically has a non-flat critical point. If the map f is only C^{2+1}, then we say that a C^{2+1} map f has a *non-flat* critical point if there is a local C^{2+1} diffeomorphism ϕ with $\phi(c) = 0$ such that $f(x) = |\phi(x)|^\alpha + f(c)$ for some real $\alpha \geq 2$. Thus, if the map is smooth and one of its higher derivatives does not vanish at the critical point, this map automatically has a non-flat critical point.

Maps which do have a flat critical point, may have completely different properties. For example, such maps can have wandering intervals.

Once we have the control of the cross-ratios, we can also control the distortion of the map using Koebe lemma:

Lemma 3.2 (Koebe lemma). *Let $J \subset M$ be intervals, $f : M \to \mathbf{R}$ be a C^1 diffeomorphism, C be a constant such that $0 < C < 1$. Assume that for any interval J^* and M^* with $J^* \subset M^* \subset M$ one has*

$$\mathbf{B}(f, M^*, J^*) \geq C.$$

If $f(M)$ contains a τ-scaled neighborhood of $f(J)$, then

$$\frac{1}{K(C,\tau)} \leq \frac{Df(x)}{Df(y)} \leq K(C,\tau)$$

where $x, y \in J$ and $K(C,\tau) = \frac{(1+\tau)^2}{C^6 \tau^2}$.

Here we say that an interval M is *τ-scaled* neighborhood of the interval J, if M contains J and if each component of $M \setminus J$ has at least length $\tau|J|$.

The proof of Lemma 3.2 can be found in [dMvS93].

This lemma remains true if we replace the inequality $\mathbf{B}(f, M^*, J^*) \geq C$ by $\mathbf{A}(f, M^*, J^*) \geq C$.

Notice that these two lemmas give us the control of the distortion of the branches of iterates of the map f if $f^n : M \to f(M)$ is a diffeomorphism and we have a good bound for $\sum_{i=0}^{m-1} |f^i(M^-)| \, |f^i(M^+)|$ or $\sum_{i=0}^{m-1} |f^i(M)|^2$. Often it is quite hard to estimates these sums. However, in the next section we introduce a tool which allow us to use Koebe lemma without their estimations.

3.1. Schwarzian derivative.

The Schwarzian derivative plays a fundamental role in the theory of one–dimensional maps. First the negative Schwarzian derivative condition in the content of maps of interval was introduced by Herman and Singer, see [Her79] and [Sin78]. Guckenheimer, Misiurewicz, van Strien showed the importance of the Schwarzian derivative for the study of several dynamical properties, [Guc79], [Mis81], [vS81]. Later, a large number of papers appeared where an extensive use of the negative Schwarzian derivative condition was made. For a comprehensive list of these papers see [dMvS93].

Let f be a C^3 map of an interval. The Schwarzian derivative Sf of the map f is defined for non-critical points of f by the formula:

$$Sf(x) = \frac{D^3 f(x)}{Df(x)} - \frac{3}{2} \left(\frac{D^2 f(x)}{Df(x)} \right)^2 .$$

One can easily check the following expression of the Schwarzian derivative of a composition of two maps:

$$S(fg)(x) = Sf(g(x)) \, (Dg(x))^2 + Sg(x).$$

¿From this formula we can deduce an important property of maps having negative Schwarzian derivative (*i.e.* $Sf(x) < 0$ where $Df(x) \neq 0$): all iterates of such maps also have negative Schwarzian derivative.

One of the main property of maps with negative Schwarzian derivative in given in the following well known lemma:

Lemma 3.3. *Let f be a C^3 map with negative Schwarzian derivative and M be an interval such that $f|_M$ is a diffeomorphism. Then for any subinterval $J \subset M$ we have*

$$A(f, M, J) \geq 1$$
$$B(f, M, J) \geq 1$$

Now we can see that we do not need to control the sums to use Koebe lemma: if the map f have negative Schwarzian and $f^n : M \to f(M)$ is a diffeomorphism, then we can apply Koebe lemma to this branch of f^n.

Obviously, not all maps have negative Schwarzian derivative. Moreover, a smooth change of the coordinate can spoil this property. However on small scales maps do have this negative Schwarzian:

Theorem 1. *Let $f : X \hookleftarrow$ be a C^3 map of an interval to itself. Let all its critical points are non-flat. Then for any critical point c there exists an* interval Z around the critical value $f(c)$ such that if $f^n(x) \in Z$ for $x \in X$ and $n > 0$, then $Sf^n(x) < 0$.

This theorem was first proved in [Koz00] in the case of unimodal maps (*i.e.* maps having only one critical point) and then generalized in [vSV].

We will outline the proof of this theorem in the unimodal case.

3.2. **Real bounds.** In this section we study an extremely powerful tool used now in virtually all papers devoted real one-dimensional dynamics.

Let f be a smooth unimodal map. A symmetric interval I around the critical point is called *nice* or *properly returning* if the forward orbit of the boundary of I do not intersect the interior of I. It is easy to see that there are nice intervals of arbitrarily small length.

The first entry map R of f to a nice interval I has particularly nice properties: there is one central branch with a critical point and all other branches are monotone and the range of these branches is I. Moreover, the branches of R can be decomposed as a composition of f and of a maps with a bounded distortion as we will see later.

Suppose that $g : X \hookleftarrow$ is a C^1 map and suppose that $g|_V : V \to J$ is a diffeomorphism of the interval V onto the interval J. If there is a larger interval $V' \supset V$ such that $g|_{V'}$ is a diffeomorphism, then we will say that the range of the map $g|_V$ can be *extended* to the interval $g(V')$.

Lemma 3.4. *Let f be a unimodal map, T be a nice interval, J be its central domain and V be a domain of the first entry map R_J to J which is disjoint from J, i.e. $V \cap J = \emptyset$. Then the range of the map $R_J : V \to J$ can be extended to T.*

Notice, that this lemma gives also some information in the case when $V \subset J$. In this case if $R_J|_V = f^m$, then it can be decomposed as $R_J|_V = f^{m-1}|_{f(V)} \circ f$, where the range of $f^{m-1}|_{f(V)}$ can be extended to T.

This lemma would be useless if we did not know the amount of space around J, *i.e.* if we did not know a bound for the ratio $|T|/|J|$. Fortunately such a bound exists.

Lemma 3.5 ([Koz00]). *Let f be a C^3 unimodal map with a non-flat non-periodic critical point. There is a constant $\tau > 0$ and a sequence $\{T_i, i = 1, \dots\}$ of nice intervals whose sizes shrink to 0 such that the range of any branch $R_{T_i} : V \to T_i$ of the first entry map can be extended to an interval which contains a τ-scaled neighborhood of T_i provided that the domain V is disjoint from T_i.*

Here we say that an interval M is τ-*scaled* neighborhood of the interval T, if M contains J and if each component of $M \setminus T$ has at least length $\tau|T|$.

The next lemma is similar to the Koebe lemma. It gives the estimate of the derivative of branches of iterates of the map from below. (In fact, one can use the Koebe lemma more or less straightforward here)

Lemma 3.6 ([Koz00]). *Let $f : X \hookleftarrow$ be a C^3 map with non-flat critical points and let $J \subset T$ be intervals such that $f^n|_T$ is monotone, the interval $f^n(T)$ contains a δ-scaled neighborhood of the interval $f^n(J)$ and the orbit $\{f^i(J), i = 0, \ldots, n - 1\}$ is disjoint. Then there exists a constant $C > 0$ depending only on the map f such that*

$$|Df^n(x)| > C \frac{\delta}{1+\delta} \frac{|f^n(J)|}{|J|}$$

where $x \in J$.

The proof of this lemma is a straightforward application of estimates 3.1.

3.3. **Proof of the Theorem.**

Now we have all tools needed to proof Theorem 1. We will prove here just the most interesting case when the critical point is recurrent.

Take T to be one of nice intervals around c from the sequence given by Lemma 3.5 and let T be so small that T is disjoint from the immediate basins of attractors. Let $f^n : V \to T$ be a branch of the first entry map to T if $V \not\subset T$ and let $n = 0$ if $V \subset T$. As we know the map $f^n : V \to T$ is a diffeomorphism and its range can be diffeomorphically extended to W where the interval W contains τ-scaled neighborhood of T where τ is given by Lemma 3.5. Due to Lemma 3.6 we can estimate the derivative of $f^{n-i} : f^i(V) \to T$ by the ratio of intervals: $|Df^{n-i}(x)| > C_1 \frac{|T|}{|f^i(V)|}$, $x \in f^i(V)$, $i = 0, \ldots, n$, where C_1 depends only on τ and the constant given by Lemma 3.6 and is independent of T.

The map f has a non-flat critical point and nearby the critical point has the form $f(x) = |\phi(x)|^\alpha + f(c)$, where ϕ is some local C^3 diffeomorphism with $\phi(c) = 0$, $\alpha \geq 2$. Applying the Schwarzian derivative to f it is easy to see that in a sufficiently small neighborhood of c we have $Sf(x) < -\frac{C_2}{(x-c)^2}$ for some $C_2 > 0$. We can assume that T is contained in this neighborhood. Since outside this neighborhood the Schwarzian derivative is bounded there is a constant $C_3 > 0$ such that $Sf(x) < C_3$ for all x.

Now let us estimate the Schwarzian derivative of the map $f^{n+1} : V \to f(T)$. Notice that this is a branch of the first entry map to $f(T)$.

$$\begin{aligned} S(f^{n+1})(x) &= Sf(f^n(x))|Df^n(x)|^2 + \sum_{i=0}^{n-1} Sf(f^i(x))|Df^i(x)|^2 \\ &= |Df^n(x)|^2 \left(Sf(f^n(x)) + \sum_{i=0}^{n-1} Sf(f^i(x))|Df^{n-i}(f^i(x))|^{-2} \right) \\ &\leq \left(\frac{|Df^n(x)|}{|T|} \right)^2 \left(-C_2 \left(\frac{|T|}{f^n(x) - c} \right)^2 + C_1 C_3 \sum_{i=0}^{n-1} |f^i(V)|^2 \right) \end{aligned}$$

$\frac{|T|}{|f^n(x)-c|}$ is always greater than 1 because $f^n(x) \in T$. The intervals from the orbit of V are disjoint, thus $\sum_{i=0}^{n-1} |f^i(V)|^2 < |X| \max_{0 \leq i < n} |f^i(V)|$. The map f does not have wandering intervals (see [MdMvS92]), this implies that the sizes of domains of the first entry map uniformly tend to zero as $|T| \to 0$. Every interval $f^i(V)$ is a domain of the first entry map to T (for $0 \le i < n$) therefore $\max_{0 \le i < n} |f^i(V)|$ tends to 0 uniformly with respect to domains of the first entry map if the size of T tends to 0. Then $Sf^{n+1}(x) < 0$ if T is small enough.

3.4. **Recent developments.** One can notice that though we have a Koebe lemma in real 1D Dynamics it is not quite similar to the Koebe lemma in the theory of functions of one complex variable. Indeed, the Complex Koebe lemma gives us some information on the higher derivatives of the map and the real Koebe lemma provides an estimate only for the distortion. However, often bounds for higher derivatives are required (specially if one wants to work on the C^k Structural stability conjecture).

Recently D. Sands and O. Kozlovski have found a generalization of the negative Schwarzian derivative condition. They have proved that if a map is a monotone matrix function of order n, then one has bounds on the higher derivatives up to order n similar to the Complex Koebe lemma. In this terminology, if a map is a monotone matrix function of the second order, then it has positive Schwarzian derivative (and thus its inverse has negative Schwarzian). More generally, there are n differential conditions for a map to be monotone matrix function of order n.

Thus there exists the Real Koebe lemma. Moreover, the situation here is similar to Theorem 1. If a map is enough differentiable, has non flat critical points, then for given n there is an interval Z around a critical value such that the inverse branches of the first entry map to Z are monotone matrix function of order n. For details see [KS].

4. COMPLEX TOOLS

Theorem 2 ([Koz98],[Koz02]). *Axiom A maps are dense in the space of C^k unimodal maps.*

k here can be $1, 2, \dots, \infty, \omega$.

This theorem completely describes C^k structurally stable maps in the space of unimodal maps: a unimodal map f is C^k structurally stable if and only if f satisfies Axiom A and its critical point is not periodic and not degenerate.

We are not going to give a complete proof of this theorem here. Instead we will prove one case (the case of infinitely renormalizable maps) of the following theorem which will imply the main one:

Theorem 3 ([Koz98],[Koz02]). *Let $f_\lambda : X \to X$ be a one parameter analytic family of unimodal real-analytic maps of an interval. Suppose that f_λ has negative Schwarzian derivative (w.r.t. the phase variable) for any λ. Moreover, suppose that this family is non-trivial in the sense that there is a parameter λ_0*

such that f_{λ_0} satisfies Axiom A. Then if the map f_{λ_1} does not satisfy Axiom A, then there is an interval Λ around λ_1 such that for any $\lambda \in \Lambda$ the maps f_{λ_1} and f_λ are not combinatorially equivalent. In particular, Axiom A maps are dense in this family.

Here we say that two unimodal maps f and $\hat{f}$ are *combinatorially equivalent* if the order of their forward critical orbit is the same. Obviously, if two maps are topologically conjugate, then they are combinatorially equivalent.

The condition $Sf_\lambda < 0$ in this theorem can be generalized, however it is not clear if it can be completely removed.

4.1. Quadratic-like maps and complex bounds.

Recall that we are going to derive Theorem 3 from the Rigidity theorem for the quadratic maps. The notion of the quadratic-like maps introduced in [DH85] provides a convenient tool which allows to transform some properties of the quadratic maps to the real-analytic maps.

Definition 3. *A holomorphic map $F : B \to A$ is called* quadratic-like *if the domain B is relatively compact in the domain A (i.e. $\bar{B} \subset A$) and F is a double covering map.*

The quadratic maps are obviously quadratic-like. Moreover, any infinitely renormalizable unimodal real-analytic map with a quadratic critical point can be "renormalized" to obtain a quadratic-like map. Before giving the precise statement let us define infinitely renormalizable maps.

Definition 4. *Let $f : X \to X$ be a unimodal map. Suppose there exists an interval T and an integer n such that $f^n : T \to T$ is unimodal. Then f is called* renormalizable. *If there are infinitely many such intervals and integers (different), then such a map is called* infinitely renormalizable.

Lemma 4.1 ([LvS98]). *Let $f : X \to X$ be a real-analytic unimodal infinitely renormalizable map of an interval with a quadratic critical point. Then there is a quadratic-like map $F : B \to A$ and an integer n such that $F|_T = f^n|_T$ where $T = B \cup \mathbf{R}$. Moreover, the modulus of the annulus $A \setminus B$ is universally bounded away from zero and the iterates of the critical point of F stay in V.*

In fact, first this theorem was proved in the case of negative Schwarzian derivative, however the theorems discussed in the previous section make this condition superficial.

Another important result ensures that the quadratic-like maps are "almost" quadratic:

Theorem 4 (Straightening theorem). *For any quadratic-like $F : B \to A$ there exists a quadratic map $Q : x \mapsto x^2 + c$ and a quasi conformal homeomorphism $H : A \to h(A)$ such that the maps F and Q are topologically conjugate by H.*

For the definition of quasi conformal homeomorphisms and other related theorems, see the Appendix.

4.2. Case of an infinitely renormalizable map.

If the reader is unfamiliar with the theory of quasiconformal maps, we suggest that he/she first read Appendix 6.

Suppose f_{λ_0} is infinitely renormalizable map and suppose that the conclusion of Theorem 3 does not hold, *i.e.* there is no interval around λ_0 which does not contain parameters with the corresponding maps not combinatorially equivalent to f_{λ_0}. Consider the set of parameters $\Upsilon \subset \Lambda$ such that for any $\lambda \in \Upsilon$ the maps f_{λ_0} and f_λ are combinatorially equivalent. We know that the family f_λ is not trivial so all maps in it cannot be combinatorially equivalent, hence $\Upsilon \neq \Lambda$.

Lemma 4.2. *The set Υ is closed.*

$\lhd$ Let $\lambda' \notin \Upsilon$ and let c_λ denote the critical point of f_λ. Consider 2 cases. First, suppose there is an integer n such that $f_{\lambda'}^n(c_{\lambda'}) = c_{\lambda'}$. Thus we have a superattractive periodic point and for parameters close to λ' the corresponding maps will also have a periodic attractor (which attracts iterates of the critical point). Such maps cannot be infinitely renormalizable and $\lambda' \notin \tilde{\Upsilon}$.

The other case: there exists an integer n such that the points $f_{\lambda'}^n(c_{\lambda'})$ and $f_{\lambda_0}^n(c_{\lambda_0})$ are on different sides from the corresponding critical points. Then, if parameter λ is close enough to λ', the order of these points will be the same. Therefore, there is an interval around λ' which does not contain points from Υ. This implies that Υ is a closed set. $\rhd$

So, Υ is closed and not equal to Λ, hence there is a point on the boundary of Υ such that there are points of Υ arbitrarily closed to this point. We can assume that this point is 0.

The map f_0 is real-analytic and infinitely renormalizable. From Theorem 4.1 we know that there is an induced quadratic-like map $F_0 : B \to A$, where $B \subset A \subset \mathbf{C}$ are simply connected domains and the modulus of the annulus $A \setminus B$ is not zero and $F_0|_B = f_0^n$ for some n. If we take a small neighborhood $D \subset \mathbf{C}$ of 0 in the parameter space, then the map f_λ^n will have the extension to some domain which contains B for any $\lambda \in D$. Fix the domain A and let B_λ be a preimage of the domain A under the map F_λ where $F_\lambda = f_\lambda^n$, $\lambda \in D$. Taking smaller disk D if necessary we can get $\bar{B}_\lambda \subset A$.

Define the map $\phi_\lambda : \partial B_0 \cup \partial A \to \partial B_\lambda \cup \partial A$ by the following rule: ϕ_λ is identity on ∂A and $\phi_\lambda(z) = F_\lambda^{-1} \circ F_0(z)$ for $\lambda \in D$, $z \in \partial B_0$. The map F_λ is not invertible, but ϕ is defined uniquely if we make it continuous with respect to λ and $\phi_0 = \mathrm{id}$.

For fixed z the map $\phi_\lambda(z)$ is holomorphic with respect to λ. Shrinking the neighborhood D if necessary, we can suppose that the map $z \mapsto \phi_\lambda(z)$ is injective for fixed $\lambda \in D$. Due to λ–lemma (theorem 6.3) the map ϕ_λ can be extended to the annulus $A \setminus B_0$ in the quasi conformal way. Denote this extension by $h_\lambda^0 : A \setminus B_0 \to A \setminus B_\lambda$. Thus, h_λ^0 is a q.c. homeomorphism and its Beltrami coefficient ν_λ^0 is a holomorphic function with respect to $\lambda \in D$.

Denote the pullback of the Beltrami coefficient ν_λ^0 by the map F_0 as ν_λ, *i.e.* if $F_0^k(z) \in A \setminus B$, then $\nu_\lambda(z) = F_0^{k*}\nu_\lambda^0(F_0^k(z))$. On the filled Julia set of F_0 and outside of the domain A we put ν_λ to be equal to 0. It is easy to see that since $\lambda \mapsto \nu_\lambda^0(z)$ is analytic the map $\lambda \mapsto \nu_\lambda(z)$ is analytic as well.

According to the Measurable Riemann mapping Theorem 6.1 there is a family a q.c. homeomorphisms $h_\lambda : \mathbf{C} \to \mathbf{C}$ whose Beltrami coefficient is ν_λ and which is normalized in such a way $h_\lambda(\infty) = \infty$, $h_\lambda(a^-) = a^-$, $h_\lambda(a^+) = a^+$ where a are 2 points of the intersection of ∂A with the real line.

Since the map F_0 conserves the Beltrami coefficient ν_λ the map

$$G_\lambda = h_\lambda \circ F_0 \circ h_\lambda^{-1} : B_\lambda \to A$$

is holomorphic. Due to the Ahlfors-Bers Theorem 6.2 the map $\lambda \mapsto G_\lambda(z)$ is analytic for the fixed point z. Thus G is an analytic family of holomorphic quadratic-like maps.

The map G_λ is topologically conjugate to F_0 for any $\lambda \in D$. Moreover, we have tried to construct this family in such a way that G_λ is equal to F_λ if their dynamics are the same. This idea is formalized in the following lemma.

Lemma 4.3. *The maps f_0 and f_λ are combinatorially equivalent if and only if $F_\lambda = G_\lambda$.*

◁ If $F_\lambda = G_\lambda$, then F_λ and F_0 are topologically conjugate, hence f_λ and f_0 are combinatorially equivalent.

If f_0 and f_λ are combinatorially equivalent, then the maps F_0 and F_λ are combinatorially equivalent as well. Due to the Rigidity theorem and Straightening theorem we know that there is a q.c. homeomorphism $\tilde{H} : \mathbf{C} \to \mathbf{C}$ which is a conjugacy between F_0 and F_λ on their Julia sets, *i.e.* $\tilde{H} \circ F_0|_J = F_\lambda \circ \tilde{H}|_J$ where J is the Julia set of the map F_0.

Define a new q.c. homeomorphism H^0 in the following way:

$$H^0(z) = \begin{cases} z & \text{if } z \notin A \\ h_\lambda^0(z) & \text{if } z \in A \setminus B \\ \tilde{H}(z) & \text{if } z \in U \end{cases}$$

where U is a neighborhood of the Julia set J such that $\bar{U} \subset B$. In the annulus $B \setminus U$ the q.c. homeomorphism H^0 is defined in an arbitrary way.

Consider the sequence of q.c. homeomorphisms H^i which are defined by the formula $H^{i+1} = F_\lambda^{-1} \circ H^i \circ F_0$. The map F_λ is not invertible, but H^{i+1} is defined correctly because the homeomorphism $\tilde{H}$ and, as a consequence, the homeomorphism H^i maps the orbit of the critical point of F_0 onto the orbit of the critical point of F_λ. Since the maps F_0 and F_λ are holomorphic the distortion of H^i does not increase with i. So the sequence $\{H^i\}$ is normal and we can take a subsequence convergent to some limit $\hat{H}$ which is also a q.c. homeomorphism. In fact, the whole sequence $\{H^j\}$ converges to $\hat{H}$ because from the definition of the sequence it follows that for any point $z \in \mathbf{C}$ there is an integer N such that $H^j(z) = H^N(z)$ for all $j > N$. Taking a limit in the equality $H^{i+1} = F_\lambda^{-1} \circ H^i \circ F_0$ we obtain that the homeomorphism $\hat{H}$ is a conjugacy between F_0 and F_λ, i.e. $F_\lambda \circ \hat{H} = \hat{H} \circ F_0$. On the other hand, it is easy to see that the Beltrami coefficient of $\hat{H}$ coincides with the Beltrami coefficient ν_λ. Indeed, outside of A both coefficients are zero. In the domain $A \setminus J$ both coefficients are obtained by pulling back the Beltrami coefficient ν_λ^0. On the Julia set the Beltrami coefficient of $\hat{H}$ is equal to the Beltrami coefficient of $\tilde{H}$ which is 0 because of the Rigidity theorem. The homeomorphism $\hat{H}$ is normalized in the same way as h_λ, so by the measurable Riemann mapping theorem these homeomorphisms coincide. From the very definition of the map G_λ we obtain that $F_\lambda = G_\lambda$. $\triangleright$

The point 0 is a limit point of the set Υ. We have $F_\lambda = G_\lambda$ for all $\lambda \in D \cap \Upsilon$. The both families are analytic, hence $F_\lambda = G_\lambda$ for all $\lambda \in D$. This implies $D \cap \mathbf{R} \subset \Upsilon$ and the point 0 is an interior point of Υ. This is a contradiction.

4.3. Remarks on the non renormalizable case.

If the map f_{λ_0} is non renormalizable (or finitely renormalizable) the strategy of the proof remains the same, however some extra difficulties appear. In the rest of this section we will assume that the critical point is recurrent (if it is not, the proof is much simpler).

First of all, we do not have an induced quadratic-like map. If the $\omega-$limit set of the critical point is minimal, we can generalize the definition of the quadratic-like maps and allow the domain of definition B to be a finite union of simply connected domains. Then the corresponding map maps every but one such a domain onto the range A univalently and one domain (called *central*) is mapped in the 2-to-1 way onto A. In this case one can proceed with the proof exactly as in the case of infinitely renormalizable maps.

If the $\omega-$limit set of the critical point is non minimal, we have to generalize the definition of the quadratic-like maps even further and allow the range A consists of finitely many connected components (as well as the domain of definition B). Moreover, we have also allow intersections of the boundaries of the domains A and B. (Later another approach was introduced in [LvS00]; the range A is simply connected but the domain B is a union of infinitely many connected domains.)

Notice that in the previous section we never used the negative Schwarzian derivative condition. And indeed, in the case of infinitely renormalizable maps it is not needed. However, in the case of the non minimal $\omega-$limit set of the critical point we have to be sure that the iterates of the critical point do not accumulate at a neutral periodic point. The negative Schwarzian condition ensures that this does not happen.

5. Further remarks and open questions

So, the Structural stability conjecture is completely proven in the case of the unimodal maps. The main remaining open question is if it holds for the multimodal maps as well. Now, let us discuss the difficulties arising in this case. Here we need to introduce the following construction.

Suppose we have a non renormalizable unimodal map f with a critical point c. Let T_0 be some nice interval. Consider a first return map R_0 to T_0 and let T_1 be a connected component of the domain of definition of R_0 containing the critical point (T_1 will be called *a central domain* of R_0). Now we can continue this procedure and obtain the first return map R_1 to T_1. We get a sequence of maps R_i and of intervals T_i. Since we have assumed that f is non-renormalizable map the sizes of the intervals T_i go to zero.

Let us take a subsequence T_{i_k} such that $R_{i_k}(c) \notin T_{i_k+1}$. Such the sequence is called a *sequence of non-central returns*.

The decay of sizes of $|T_{i_k}|$ plays a crucial role in the study of properties of unimodal maps. If the critical point is not degenerate, this decay is always superexponential (in fact, the ratio $r_k = |T_{i_k+1}|/|T_{i_k}|$ decays exponentially fast). However, if the critical point is degenerate (or the map f is multimodal), the ratio r_k can stay bounded away from zero or it can even oscillate.

In order to apply a method used in the previous section to prove the Structural stability conjecture one should show that the ratios r_k for two different topologically conjugate maps behave in a similar way. So far it is an open problem.

Recently W. Shen proved the C^2 case for multimodal maps using the following idea: if the ratio r_k is not bounded away from zero, one can use local perturbations as shown in Section 2, and if the ratio r_k is bounded away from zero, one can use a technique developed for the quadratic maps similar to what we have used in Section 4.1.

6. Appendix. Quasiconformal homeomorphisms

In this section we will give a short overview of definitions and results connected with quasiconformal maps. For the details the reader can consult books [Ahl79], [LV73].

There are many different equivalent definitions of the quasiconformal (q.c.) homeomorphism. We will use the following:

Definition 5. *Let $U \subseteq \bar{\mathbb{C}}$ be a domain in the complex plane. The map $h : U \to h(U)$ is called* quasiconformal *homeomorphism if*

- *h is an orientation preserving homeomorphism between the domains U and $h(U)$;*
- *the real part $\Re(h)$ and the imaginary part $\Im(h)$ of h are absolutely continuous on almost all verticals and almost all horizontals in the sense of Lebesgue;*

- *there exists a constant $k < 1$ such that for*

$$\mu_h(z) = \frac{d_{\bar{z}}f(z)}{d_z f(z)}$$

one has

$$|\mu_h(z)| < k$$

for almost all $z \in U$ where $d_{\bar{z}}h = \frac{dh}{d\bar{z}}$ and $d_z h = \frac{dh}{dz}$.

The function μ_h is called the *Beltrami coefficient* of q.c. homeomorphism h.

To the Beltrami coefficient μ one can associate a field of infinitesimal ellipses. The eccentricities of these ellipses are given by $\frac{1+|\mu(z)|}{1-|\mu(z)|}$ and their directions of the major axes are given by $\sqrt{\mu(z)}$.

If f is a holomorphic map, we can pull back this field of ellipses even if f is not injective. This pullback we will denote as $f^*\mu$ and it is equal to

$$(f^*\mu)(z) = \mu(f(z))\frac{\overline{d_z f(z)}}{d_z f(z)}.$$

This is a list of theorems we are going to use later on.

Theorem 6.1 (Measurable Riemann Mapping Theorem). *Let $\mu : \mathbf{C} \to \mathbf{C}$ be a measurable function such that $|\mu| < k < 1$ almost everywhere. Then there exists a unique q.c. homeomorphism $h : \bar{\mathbf{C}} \to \bar{\mathbf{C}}$ whose Beltrami coefficient is μ and which is normalized such that $h(0) = 0$, $h(1) = 1$ and $h(\infty) = \infty$.*

Theorem 6.2 (Ahlfors-Bers Theorem). *Let $\Lambda \subset \mathbf{C}^n$ be an open set and $\mu : \mathbf{C}\Lambda \to \mathbf{C}$ be a measurable function satisfying:*

- *$|\mu(z, \lambda)| < k < 1$ for all $\lambda \in \Lambda$ and for almost all $z \in \mathbf{C}$;*
- *the map $\lambda \mapsto \mu(z, \lambda)$ is holomorphic in λ for almost all $z \in \mathbf{C}$.*

Then there exists a unique function $H : \mathbf{C}\Lambda \to \mathbf{C}$ such that

- *$H(0, \lambda) = 0$, $H(1, \lambda) = 1$, $H(\infty, \lambda) = \infty$;*
- *for fixed $\lambda \in \Lambda$ the map $z \mapsto F(z, \lambda)$ is a q.c. homeomorphism whose Beltrami coefficient is $\mu(\cdot, \lambda)$;*
- *the map $\lambda \mapsto F(z, \lambda)$ is holomorphic for almost every z.*

The first version of the next theorem appeared in [MSS83] and after it was generalized several times: [BR86], [Slo91].

Theorem 6.3 (λ–lemma). *Let $Z \subset \bar{\mathbf{C}}$ be a set, D be an open unit disk in the complex plane and let $h : ZD \to \bar{\mathbf{C}}$ satisfy the following conditions:*

- *$h(z, 0) = z$ for any $z \in Z$;*
- *for fixed $z \in Z$ the function $\lambda \mapsto h(z, \lambda)$ is holomorphic for $\lambda \in D$;*
- *for fixed $\lambda \in D$ the map $z \mapsto h(z, \lambda)$ is injective for all $z \in Z$.*

Then there exists $H : \bar{\mathbf{C}}D \to \bar{\mathbf{C}}$ such that

- *$H(z, \lambda) = h(z, \lambda)$ for $\lambda \in D$ and $z \in Z$;*

- $H(z,0) = z$ for $z \in \bar{\mathbf{C}}$;
- for fixed $z \in \bar{\mathbf{C}}$ the function $\lambda \mapsto H(z,\lambda)$ is holomorphic for $\lambda \in D$;
- for fixed $\lambda \in D$ the map $z \mapsto H(z,\lambda)$ is a q.c. homeomorphism;
- for almost every $z \in \bar{\mathbf{C}}$ the Beltrami coefficient of H depends holomorphically on λ.

Since the Beltrami coefficient of a q.c. homeomorphism is not defined everywhere we have to clarify the last item in the previous theorem. We say that the Beltrami coefficient depends holomorphically on λ for almost every z if there is a function $\mu(z,\lambda)$ such that for almost every z the function $\lambda \mapsto \mu(z,\lambda)$ is holomorphic and for fixed λ the equality $\mu(z,\lambda) = \mu_{H(\lambda,\cdot)}(z)$ holds almost everywhere.

Theorem 6.4 (Compactness of the set of q.c. homeomorphisms). *If H is a family of q.c. homeomorphisms of $\bar{\mathbf{C}}$ whose Beltrami coefficients are uniformly bounded by a constant $k < 1$, then any sequence in H has a subsequence which converges uniformly and the limit either a constant or a q.c. homeomorphism whose Beltrami coefficient is bounded by k.*

Theorem 6.5. *If f is holomorphic, then $\mu_{f \circ h} = \mu_h$ and $\mu_{h \circ f}(z) = \mu_h(f(z))\overline{\frac{d_z f(z)}{d_z f(z)}}$.*

The real counterpart of q.c. homeomorphisms are quasisymmetric homeomorphisms of the real line.

Definition 6. *The homeomorphism $h : \mathbf{R} \to \mathbf{R}$ is called* quasisymmetric *if there is a constant $C > 0$ such that for any three points $x_{-1} < x_0 < x_1$ such that $x_0 - x_{-1} = x_1 - x_0$ the following inequality holds:*

$$C^{-1} < \frac{|h(x_1) - h(x_0)|}{|h(x_0) - h(x_{-1})|} < C.$$

The following theorem describes relations between quasiconformal and quasisymmetric homeomorphisms:

Theorem 6.6. *Let h^c be a quasiconformal homeomorphism of the complex plane such that its restriction h^r to the real line is a real function. Then this restriction is a quasisymmetric homeomorphism.*

If h^r is a quasisymmetric homeomorphism of the real line, then there is a quasiconformal homeomorphism $h^c : \mathbf{C} \to \mathbf{C}$ such that the restriction of h^c to the real line is h^r.

REFERENCES

[Ahl79]   L.V. Ahlfors, *Complex analysis*, McGraw-Hill, New York, 1979.
[BM97]   A. Blokh and M. Misiurewicz, *Dense set of negative Schwarzian maps whose critical points have minimal limit sets*, Discrete Contin. Dynam. Sys **4** (1997), 141–158.
[BM98]   A. Blokh and M. Misiurewicz, *Collet-Eckmann maps are unstable*, Comm. Math. Phys. **191** (1998), 61–70.

[BR86]   L. Bers and H.L. Royden, *Holomorphic families of injections*, Acta MAth. **157** (1986), 259–286.

[DH85]   A. Douady and J.H. Hubbard, *On the dynamics of polynomial-like mappings*, Ann. Sc. E.N.S. 4e série **18** (1985), 287–343.

[dMvS88]   W. de Melo and S. van Strien, *One-dimensional dynamics: the Schwarzian derivative and beyond*, Bull. Amer. Math. Soc. (N.S.) **18** (1988), no. 2, 159–162.

[dMvS89]   W. de Melo and S. van Strien, *A structure theorem in one-dimensional dynamics*, Ann. of Math. (2) **129** (1989), no. 3, 519–546.

[dMvS93]   W. de Melo and S. van Strien, *One-dimensional Dynamics*, Springer, 1993.

[GŚ97]   J. Graczyk and G. Światek, *Generic hyperbolicity in the logistic family*, Ann. of Math. (2) **146** (1997), no. 1, 1–52.

[Guc79]   John Guckenheimer, *Sensitive dependence to initial conditions for one-dimensional maps*, Comm. Math. Phys. **70** (1979), no. 2, 133–160.

[Her79]   Michael-Robert Herman, *Sur la conjugaison différentiable des difféomorphismes du cercle à des rotations*, Inst. Hautes Études Sci. Publ. Math. (1979), no. 49, 5–233.

[Koz98]   O.S. Kozlovski, *Structural stability in one dimensional dynamics*, Ph.D. thesis, Amsterdam University, 1998.

[Koz00]   O. S. Kozlovski, *Getting rid of the negative Schwarzian derivative condition*, Ann. of Math. (2) **152** (2000), no. 3, 743–762.

[Koz02]   O.S. Kozlovski, *Axiom A maps are dense in the space of unimodal maps in the C^k topology*, to appear in Ann. of Math., 2002.

[KS]   O.S. Kozlovski and D. Sands, *Real Koebe lemma*, in preparation, 2002.

[LV73]   O. Lehto and K.I. Virtanen, *Quasiconformal mappings in the plane*, Springer, Berlin New York, 1973.

[LvS98]   G. Levin and S. van Strien, *Local connectivity of the Julia set of real polynomials*, Ann. of Math. (2) **147** (1998), no. 3, 471–541.

[LvS00]   Genadi Levin and Sebastian van Strien, *Bounds for maps of an interval with one critical point of inflection type. II*, Invent. Math. **141** (2000), no. 2, 399–465.

[Lyu97]   M. Lyubich, *Dynamics of quadratic polynomials I-II*, Acta Math. **178** (1997), 185–297.

[Mañ85]   R. Mañè, *Hyperbolicity, sinks and measure in one-dimensional dynamics*, Commum. Math. Phys. **100** (1985), 495–524.

[Mañ98]   R. Mañè, *A proof of the C^1 stability conjecture.*, Publ. Math., Inst. Hautes Etud. Sci. 66, 161-210 (1998) (English).

[MdMvS92]   M. Martens, W. de Melo, and S. van Strien, *Julia-Fatou-Sullivan theory for real one-dimensional dynamics*, Acta Math. **168** (1992), no. 3-4, 273–318.

[Mis81]   Michał Misiurewicz, *Structure of mappings of an interval with zero entropy*, Inst. Hautes Études Sci. Publ. Math. (1981), no. 53, 5–16.

[MSS83]   R. Mañè, P. Sad, and D. Sullivan, *On the dynamics of rational maps*, Ann. Sci. Ec. Norm. Sup. **16** (1983), 193–217.

[NvS88]   T. Nowicki and S. van Strien, *Absolutely continuous invariant measures for C^2 unimodal maps satisfying the Collet-Eckmann conditions*, Invent. Math. **93** (1988), no. 3, 619–635.

[Rob76]   Clark Robinson, *Structural stability of C^1 diffeomorphisms*, J. Differential Equations **22** (1976), no. 1, 28–73.

[She]   W. Shen, *On the C^2 density of Axiom A maps*, preprint.

[Sin78]   D. Singer, *Stable orbits and bifurcation of maps of the interval*, SIAM J. Appl. Math. **35** (1978), no. 2, 260–267.

[Slo91]   Z. Slodkowski, *Holomorphic motions and polinomial hulls*, Proc. AMS **111** (1991), 347–255.

[Sul91]   D. Sullivan, *The universalities of Milnor, Feigenbaum and Bers*, Topological methods in modern mathematics, SUNY at Stony Brook, 1991, Proceedings of the Symposium held in honor of John Milnor's 60th birthday, pp. 14–21.

[Sul92]   D. Sullivan, *Bounds, quadratic differentials, and renormalization conjectire*, AMS Centennial Publications, vol. 2, 1992.

[Tod]   M. Todd, *Local perturbation of unimodal maps*, personal communication.

[vS81]   Sebastian J. van Strien, *On the bifurcations creating horseshoes*, Dynamical systems and turbulence, Warwick 1980 (Coventry, 1979/1980), vol. 898, Springer, Berlin, 1981, pp. 316–351.

[vS90]   Sebastian van Strien, *Hyperbolicity and invariant measures for general C^2 interval maps satisfying the Misiurewicz condition*, Comm. Math. Phys. **128** (1990), no. 3, 437–495.

[vSV]   S. van Strien and E. Vargas, *Real bounds, ergodicity and negative schwarzian for multimodal maps*, preprint.

[Yoc90]   J.-C. Yoccoz, *On the local connectivity of the Mandelbrot set*, Preprint, 1990.

MATHEMATICS INSTITUTE, UNIVERSITY OF WARWICK, COVENTRY CV4 7AL, UK
E-mail address: oleg@maths.warwick.ac.uk

PERIODIC POINTS OF NONEXPANSIVE MAPS: A SURVEY

BAS LEMMENS

ABSTRACT. In this paper we survey the research on periodic points of nonexpansive maps. Since the pioneering paper [3] by Akcoglu and Krengel in the nineteen-eighties remarkable progress has been made in this field. This paper brings together the main results and it discusses some of the open problems. At the same time we hope that it will be an invitation for others to become acquainted with the subject.

CONTENTS

1. INTRODUCTION

Let P be an $n \times n$ column stochastic matrix and let $f : \mathbb{R}^n \to \mathbb{R}^n$ be given by $f(x) = Px$ for $x \in \mathbb{R}^n$. It is then well-known that one can use the theory of Perron and Frobenius, concerning the eigenvalues of nonnegative matrices, to predict the asymptotic behaviour of the sequence of iterates $(f^k(x))_k$ for $x \in \mathbb{R}^n$. Indeed one can show (see [36, Section 9]) that there exists an integer $p \geq 1$ such that the sequence $(f^{kp}(x))_k$ converges for each $x \in \mathbb{R}^n$ to a periodic point of f, and moreover p is the order of a permutation on n letters. There are two properties of the map f that cause this behaviour. To begin the map f is *nonnegative*, that is to say it leaves the positive cone in $\mathbb{R}^n$ invariant. Secondly, the map f is *nonexpansive* in the 1-norm, i.e.

$$\|f(x) - f(y)\|_1 \leq \|x - y\|_1 \quad \text{for all } x, y \in \mathbb{R}^n.$$

Here $\|z\|_1 = \sum_i |z_i|$ denotes the *1-norm* for $z = (z_1, \dots, z_n)$.

Surprisingly often the nonexpansiveness property is sufficient to give this type of asymptotic behaviour of the iterates. This is illustrated by the following remarkable theorem.

Theorem 1.1. *Let $\|\cdot\|$ be a norm on $\mathbb{R}^n$ for which the unit ball is a polyhedron and let $X \subset \mathbb{R}^n$ be closed. If $f : X \to X$ is nonexpansive with respect to this norm and there exists $\eta \in X$ such that $(\|f^k(\eta)\|)_k$ is bounded, then the following assertions are true.*

- *For each $x \in X$ there exists $p_x = p \geq 1$ such that $(f^{kp}(x))_k$ converges to a periodic point $\xi \in X$ of f of minimal period p, that is $f^p(\xi) = \xi$ and $f^j(\xi) \neq \xi$ for $0 < j < p$.*
- *For each polyhedral norm there exists an integer $\rho(n)$, which only depends on the dimension, such that the minimal period of each periodic point of f is at most $\rho(n)$.*

The main point of this theorem is that the nonexpansiveness property causes the limit behaviour of the iterates of certain nonlinear maps to be periodic. Important examples of polyhedral norms on $\mathbb{R}^n$ are the 1-norm and the *sup-norm*: $\|z\|_\infty = \max_i |z_i|$ for $z = (z_1, \ldots, z_n)$.

Theorem 1.1 raises a number of questions.

Question 1.1. *For which nonexpansive maps $f : X \to X$, with $X \subset \mathbb{R}^n$, is the asymptotic behaviour of the sequence of iterates $(f^k(x))_k$ periodic for each $x \in X$?*

Question 1.2. *Given a polyhedral norm and a domain $X \subset \mathbb{R}^n$, can one determine the finite set of integers $p \geq 1$ for which there exist a nonexpansive map $f : X \to X$ and a periodic point of f of minimal period p?*

Of course, not every nonexpansive map exhibits this type of behaviour. One can think for instance of rotations under an irrational angle in the plane. Such maps are nonexpansive with respect to the Euclidean norm. In connection with the second question Nussbaum [26] made the following conjecture.

Conjecture 1.1 (Nussbaum). *The minimal period of each periodic point of a sup-norm nonexpansive map $f : X \to X$, with $X \subset \mathbb{R}^n$, is at most 2^n.*

At present the conjecture is known to be true for the dimensions $n = 1, 2$, and 3 (see [23]).

The remainder of the paper has the following outline. In Section 2 we give a brief history of Theorem 1.1 and provide some motivation to study the iterative behaviour of nonexpansive maps. Section 3 is used to explain the main ideas behind the proof of Theorem 1.1. It moreover discusses some results concerning Question 1.2. In Section 4 we review the results on Question 1.2 in case the polyhedral norm is the 1-norm. Subsequently we discuss in Section 5 the connection between lattice homomorphisms and nonnegative nonexpansive maps. We conclude with Section 6 in which some numerical data is given.

2. Historical Remarks and Motivation

Pioneering research on the behaviour of nonlinear nonexpansive maps was done by Akcoglu and Krengel in the nineteen-eighties. In [3] they proved Theorem 1.1 in case the polyhedral norm is the 1-norm. Their results however did not provide an upper bound $\rho(n)$. An upper bound was obtained by Misiurewicz [25]. In fact, he showed (in case of the 1-norm) that the minimal period of periodic points of 1-norm nonexpansive maps $f : X \to X$, with $X \subset \mathbb{R}^n$, is at most $n!2^m$, where $m = 2^n$. In his thesis [44] Weller generalised the result of Akcoglu and Krengel to polyhedral norms on $\mathbb{R}^n$. Thereafter various upper bounds for $\rho(n)$ for different polyhedral norms were derived by: Blokhuis and Wilbrink [9], Lemmens, Nussbaum and Verduyn Lunel [21], Lo [22], Martus [24], Nussbaum [26], and Sine [42].

Further investigations on the periodic points of 1-norm nonexpansive maps were made by Scheutzow. In [39] and [40] he showed that if $f : \mathbb{R}^n \to \mathbb{R}^n$, with $f(0) = 0$, is a 1-norm nonexpansive map, then there exists an integer $p \geq 1$ such that the sequence $(f^{kp}(x))_k$ is convergent for each $x \in \mathbb{R}^n$, and moreover p is a divisor of the least common multiple of the integers $1, 2, \ldots, 2n$. If in addition f leaves the positive cone in $\mathbb{R}^n$ invariant, then p divides the least common multiple of the integers $1, 2, \ldots, n$. Later his ideas were further developed by Nussbaum and Scheutzow in [27] and [34]. The results from these papers eventually allowed Nussbaum, Scheutzow, and Verduyn Lunel to give a complete characterization (in arithmetical and combinatorial constraints) of the possible minimal periods of periodic points of 1-norm nonexpansive maps $f : \mathbb{R}^n \to \mathbb{R}^n$ that leave the positive cone in $\mathbb{R}^n$ invariant and have zero as a fixed point (see [35, Theorem 3.1]). Further improvements for the possible minimal periods of periodic points of general 1-norm nonexpansive maps $f : \mathbb{R}^n \to \mathbb{R}^n$ were obtained by Lemmens in [19] (see also [20]). We will give a detailed overview of these result in Section 4.

Let us now provide some motivation to study nonexpansive maps. One of the reasons to study nonexpansive maps is that they arise in applications. For instance, 1-norm nonexpansive maps can be used as models for diffusion processes on a finite state space. (see [3] and [32]). A simple example of such a process is the following. Suppose there are n containers $C_1, C_2, \ldots, C_n$ each having an infinite volume. Let x_i denote the amount of sand in container C_i and define $x = (x_1, \ldots, x_n) \in \mathbb{R}^n$ to be the distribution vector. With each container C_i, where $1 \leq i \leq n$, a sequence of buckets $(b_{ij})_{j \geq 1}$ is associated. For each bucket b_{ij} the volume is denoted by a_{ij}, and it is assumed that

$$\sum_{j=1}^{\infty} a_{ij} = \infty \quad \text{for } 1 \leq i \leq n.$$

We start the following procedure to pour sand from the containers into the buckets. For each container C_i pour sand into bucket b_{i1} until either b_{i1} is full or C_i is empty. If b_{i1} is full, then pour the remaining sand in C_i into

b_{i2} until either b_{i2} is full or C_i is empty. Continue in the same manner until C_i is empty. If we let $M_{ik}(x)$ denote the amount of sand in bucket b_{ik} after the procedure we find that

$$M_{ik}(x) = \min\{a_{ik}, \max\{x_i - \sum_{j=0}^{k-1} a_{ij}, 0\}\}, \quad \text{where } a_{i0} = 0.$$

Now let $\gamma : \{1, \ldots, n\} \times \mathbb{N} \to \{1, \ldots, n\}$ be a map. This map will serve as a rule to pour sand from the buckets back into the containers. For each bucket b_{ik} pour sand in container $C_{\gamma(i,k)}$. The new distribution $y = (y_1, \ldots, y_n)$ of sand in the containers is given by

$$y_j = \sum_{\gamma(i,k)=j} M_{ik}(x) \quad \text{for } 1 \le j \le n.$$

More formally, we can define a map $f : \mathbb{K}^n \to \mathbb{K}^n$ by

$$f(x)_j = \sum_{\gamma(i,k)=j} M_{ik}(x) \quad \text{for } 1 \le j \le n \text{ and } x \in \mathbb{K}^n,$$

where $\mathbb{K}^n = \{x \in \mathbb{R}^n : x_i \ge 0 \text{ for } 1 \le i \le n\}$ is the *positive cone* in $\mathbb{R}^n$. The map f is usually called a *sand-shift map* and it was introduced by Nussbaum in [32].

To see that sand-shift maps are 1-norm nonexpansive one can use a result of Crandall and Tartar [10], which says: If $f : X \to X$, where $X = \mathbb{R}^n$ or $\mathbb{K}^n$, is *integral-preserving*, i.e. $\sum_i f(x)_i = \sum_i x_i$ for all $x \in X$, then f is 1-norm nonexpansive if and only if f is *order-preserving*, i.e. $f(x) \le f(y)$ for all $x, y \in X$ with $x \le y$. Here one should read the inequalities coordinate-wise.

An interesting class of sup-norm nonexpansive maps is provided by another observation of Crandall and Tartar [10]. If $f : \mathbb{R}^n \to \mathbb{R}^n$ is *additive homogeneous*, i.e. $f(x + h\mathbf{1}) = f(x) + h\mathbf{1}$ for all $h \in \mathbb{R}$ and $x \in \mathbb{R}^n$, then f is sup-norm nonexpansive if and only if f is order-preserving. Here $\mathbf{1}$ denotes the vector in $\mathbb{R}^n$ with all coordinates unity. Examples are so called *max-plus functions*, which are defined as follows. Let $A = (a_{ij})$ be a real $n \times n$ matrix and let $f : \mathbb{R}^n \to \mathbb{R}^n$ be given by

$$f(x)_i = \max_j \{a_{ij} + x_j\} \quad \text{for } 1 \le i \le n \text{ and } x \in \mathbb{R}^n.$$

Max-plus functions appear in various applications such as statistical mechanics (see [29]) and the analysis of discrete event systems (see [6], [11], and [14]).

Other, more general, examples of homogeneous and order-preserving maps are maps $g : \mathbb{R}^n \to \mathbb{R}^n$, where each component $g(x)_i$ consists of finitely many expressions of the form $x_j + c$, where $1 \le j \le n$ and $c \in \mathbb{R}$, which are joined by $\wedge$ or $\vee$ operations that are defined by $a \wedge b = \min\{a, b\}$ and $a \vee b = \max\{a, b\}$.

An example is the map $g : \mathbb{R}^3 \to \mathbb{R}^3$ given by

$$g(x)_1 = (x_2 + 2) \vee (x_3 \wedge (x_1 - 3)),$$
$$g(x)_2 = (x_1 + 1) \wedge (x_2 + 5) \wedge (x_3 - 6),$$
$$g(x)_3 = (x_1 \vee (x_3 - 3)) \wedge ((x_2 - 2) \vee x_3) \quad \text{for } x \in \mathbb{R}^3.$$

These maps are often called *min-max functions*. From Theorem 1.1 it follows that the minimal period of periodic points of min-max functions $g : \mathbb{R}^n \to \mathbb{R}^n$ is bounded above by a number that only depends on the dimension n. It is believed that for these functions the optimal upper bound is n choose $[n/2]$ (see [14, page 25]), but at the present time no proof is known. A variety of other applications of nonexpansive maps can be found in: [26], [30], [31], and [37]

Besides the applications there are theoretical reasons to be interested in nonexpansive maps. One such reason is that there exists a connection between the iterative behaviour of nonexpansive maps and the geometry of the underlying normed space (see [2] and [13]). Another reason is that Question 1.2, concerning the possible minimal periods of periodic points of nonexpansive maps, is related to nice problems in combinatorial geometry.

3. LIMIT SETS OF NONEXPANSIVE MAPS

To understand the iterative behaviour of a map $f : X \to X$ one has to study the structure of the ω-*limit* sets:

$$\omega(x) = \{y \in X : y = \lim_{i \to \infty} f^{k_i}(x) \text{ for some integer sequence } k_i \to \infty\}.$$

Indeed to prove Theorem 1.1 it is sufficient to show the following assertion. If $X \subset \mathbb{R}^n$ is closed, $f : X \to X$ is nonexpansive with respect to a polyhedral norm $\| \cdot \|$, and $(\|f^k(\eta)\|)_k$ remains bounded for some $\eta \in X$, then there exists an integer $\rho(n)$ such that the cardinality of $\omega(x)$ is at most $\rho(n)$ for each $x \in X$.

As $X \subset \mathbb{R}^n$ is closed, the ω-limits of f are closed. Moreover, since f has a bounded orbit the nonexpansiveness of f implies that the ω-limit sets of f are bounded, and hence compact. Furthermore it is shown in [12] that for each $x \in X$ the restriction of f to $\omega(x)$ is an isometry that maps $\omega(x)$ onto itself. With these observations in mind Misiurewicz [25] formulated the following property.

Definition 3.1. A set S in $(\mathbb{R}^n, \| \cdot \|)$ has a *transitive and commutative family of isometries* if there exists a commutative family Γ of isometries (with respect to $\| \cdot \|$) of S onto itself, such that for each $x, y \in S$ there exists $F_{x,y} \in \Gamma$ with $F_{x,y}(x) = y$.

He then showed the following proposition (see [25, Lemma 1]).

Proposition 3.1. *Let $\| \cdot \|$ be a norm on $\mathbb{R}^n$ and let $X \subset \mathbb{R}^n$ be closed. If $f : X \to X$ is nonexpansive with respect to this norm and there exists $\eta \in X$*

such that $(\|f^k(\eta)\|)_k$ *remains bounded, then for each* $x \in X$ *the limit set* $\omega(x)$ *has a transitive and commutative family of isometries.*

Thus, in order to prove Theorem 1.1 it is sufficient to give for each polyhedral norm on $\mathbb{R}^n$ an upper bound for the cardinality of compact sets in $\mathbb{R}^n$ that have a transitive and commutative family of isometries.

In the next subsection we will see how an upper bound for the cardinality of such sets is derived in case the polyhedral norm is the sup-norm. By using the fact that for any polyhedral norm $\| \cdot \|$ on $\mathbb{R}^n$ the space $(\mathbb{R}^n, \| \cdot \|)$ can be isometrically embedded into $(\mathbb{R}^m, \| \cdot \|_\infty)$, where m is sufficiently large, an upper bound can be derived for any polyhedral norm.

3.1. Upper bounds for the cardinality of limit sets.
We begin by recalling several definitions. A sequence $x^1, x^2, \dots, x^m$ in $\mathbb{R}^n$ is called an *additive chain* (with respect to the sup-norm) if

$$\|x^1 - x^m\|_\infty = \sum_{i=1}^{m-1} \|x^i - x^{i+1}\|_\infty.$$

The *length* of a sequence is the number of distinct points in it.

For each $i = 1, \dots, n$ a partial ordering $\leq_i$ on $\mathbb{R}^n$ is defined by $x \leq_i y$ if $\|x - y\|_\infty = y_i - x_i$. A sequence $x^1, x^2, \dots, x^m$ is called an *i-chain* if

$$x^1 \leq_i x^2 \leq_i \dots \leq_i x^m \text{ or } x^m \leq_i x^{m-1} \leq_i \dots \leq_i x^1.$$

The set of all i for which the sequence $x^1, x^2, \dots, x^m$ is an i-chain is denoted by $I(x^1, x^2, \dots, x^m)$.

By using the definition of the sup-norm one can verify that $x^1, x^2, \dots, x^m$ is an additive chain if and only if $I(x^1, x^2, \dots, x^m)$ is not empty. Furthermore, if $x^{k_1}, \dots, x^{k_r}$ is a subsequence of an additive chain $x^1, x^2, \dots, x^m$ with $x^{k_1} = x^1$ and $x^{k_r} = x^m$, then it is easy to show that

$$I(x^{k_1}, \dots, x^{k_r}) = I(x^1, \dots, x^m).$$

For each $x, y \in \mathbb{R}^n$ we define the set

$$W(x,y) = \{z \in \mathbb{R}^n : x, y, z \text{ is an additive chain}\}.$$

Further let $W^\circ(x,y)$ denote the interior of $W(x,y)$ (with respect to the Euclidean norm). The definition of $W(x,y)$ is illustrated in Figure 1. It is not difficult to show that

$$W^\circ(x,y) = \{z \in W(x,y) : I(x,y,z) = I(y,z) \text{ and } z \neq y\}.$$

By using these definitions the following result can be stated.

Proposition 3.2. *If S is a compact subset of $\mathbb{R}^n$ and S has a transitive and commutative family of sup-norm isometries, then $W^\circ(x,y) \cap S = \emptyset$ for all $x, y \in S$ with $x \neq y$.*

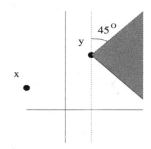

FIGURE 1. The set $W(x,y)$

Proof. The argument goes by contradiction. So, suppose there exist $x, y, z \in S$ with $x \neq y$ and $z \in W^o(x,y)$. Since $x \neq y$ and $y \neq z$ we can find $\varepsilon > 0$ such that $\|x - y\|_\infty \geq \varepsilon$ and $\|y - z\|_\infty \geq \varepsilon$.

Define $\mathcal{F}$ to be the collection of additive chains in S that start with the sequence x, y, z and that are such that the distance between any two consecutive points in the sequence is at least ε. Since S is a compact subset of $\mathbb{R}^n$ there exists an upper bound on the length of sequences in $\mathcal{F}$, say r.

Now let $x^1 = x, x^2 = y, x^3 = z, \dots, x^r$ be a sequence of maximal length in $\mathcal{F}$. For integers $1 \leq k, l \leq r$ let $F_{k,l} : S \to S$ be a sup-norm isometry of the commutative family that maps x^k to x^l, and put $x^{r+1} = F_{1,2}(x^r)$.

We claim that $x^2, x^3, \dots, x^r, x^{r+1}$ is again an additive chain and the sup-norm distance between two consecutive elements in this sequence is at least ε. To prove this claim, we first show that the distance between consecutive elements is at least ε. It suffices to verify that $\|x^r - x^{r+1}\|_\infty \geq \varepsilon$. Since $x^r = F_{1,r}(x^1)$ we have that

$$
\begin{aligned}
\|x^r - x^{r+1}\|_\infty &= \|F_{1,r}(x^1) - F_{1,2}(F_{1,r}(x^1))\|_\infty \\
&= \|F_{1,r}(x^1) - F_{1,r}(F_{1,2}(x^1))\|_\infty \\
&= \|x^1 - x^2\|_\infty,
\end{aligned}
$$

so that

$$
(1) \qquad \|x^r - x^{r+1}\|_\infty = \|x^1 - x^2\|_\infty,
$$

and this shows $\|x^r - x^{r+1}\|_\infty \geq \varepsilon$. ¿From (1) it follows that

$$
\begin{aligned}
\|x^2 - x^{r+1}\|_\infty &= \|F_{1,2}(x^1) - F_{1,2}(x^r)\|_\infty \\
&= \|x^1 - x^r\|_\infty \\
&= \sum_{i=1}^{r-1} \|x^i - x^{i+1}\|_\infty \\
&= \|x^r - x^{r+1}\|_\infty + \sum_{i=2}^{r-1} \|x^i - x^{i+1}\|_\infty \\
&= \sum_{i=2}^{r} \|x^i - x^{i+1}\|_\infty,
\end{aligned}
$$

and hence $x^2, x^3, \ldots, x^{r+1}$ is an additive chain.

¿From the claim it follows that $I(x^2, x^3, \ldots, x^{r+1})$ is nonempty. Now let $i \in I(x^2, x^3, \ldots, x^{r+1})$. As $z \in W^\circ(x, y)$ we know that

$$
I(x^2, x^3) = I(y, z) = I(x, y, z) = I(x^1, x^2, x^3).
$$

Combining this with $I(x^2, x^3, \ldots, x^{r+1}) \subset I(x^2, x^3)$ gives $i \in I(x^1, x^2, x^3)$. Therefore the extended sequence $x^1, x^2, \ldots, x^r, x^{r+1}$ is an i-chain and hence an additive chain in $\mathcal{F}$. This however, contradicts the fact that r is maximal. $\square$

This result motivates the following definition. A set S in $\mathbb{R}^n$ is called ∞-separated if $W^\circ(x, y) \cap S$ is empty for all $x, y \in S$ with $x \neq y$. By using this geometric property one can now easily obtain cardinality estimates.

Proposition 3.3. *If S is an ∞-separated set in $\mathbb{R}^n$, then $|S| \leq (n+1)^n$.*

Proof. Let $x^1, x^2, \ldots, x^m$ be an additive chain S of length m. Then clearly

$$
(2) \qquad I(x^1, x^2) \supset I(x^1, x^2, x^3) \supset \ldots \supset I(x^1, x^2, \ldots, x^m).
$$

Observe that each inclusion in (2) is strict. Because if $I(x^1, x^2, \ldots, x^k) = I(x^1, x^2, \ldots, x^{k+1})$ for some $1 < k < m$, then

$$
\begin{aligned}
I(x^{k+1}, x^k, x^1) &= I(x^1, x^k, x^{k+1}) = I(x^1, x^2, \ldots, x^{k+1}) \\
&= I(x^1, x^2, \ldots, x^k) = I(x^1, x^k) = I(x^k, x^1).
\end{aligned}
$$

Hence $x^1 \in W^\circ(x^{k+1}, x^k)$, which contradicts the fact that S is ∞-separated. Since $I(x^1, x^2) \subset \{1, 2, \ldots, n\}$ we conclude that the length of every additive chain in S is at most $n + 1$.

Now for $x \in S$ and $1 \leq i \leq n$ let $h_i(x)$ be the length of the longest decreasing i-chain starting in x, and consider $h : S \to \{1, \ldots, n+1\}^n$ given by $h(x) = (h_1(x), \ldots, h_n(x))$ for $x \in S$. Then for each $x, y \in S$ with $x \neq y$ we have that $\|x - y\|_\infty = |x_i - y_i| > 0$ for some $1 \leq i \leq n$, so that $h_i(x) \neq h_i(y)$. Therefore h is injective and hence $|S| \leq (n+1)^n$. $\square$

A combination of the Propositions 3.2 and 3.3 now shows that every compact set in $\mathbb{R}^n$ with a commutative family of sup-norm isometries has at most $(n+1)^n$ elements. This together with the observations at the beginning of this section gives a proof of Theorem 1.1.

There exist several proofs of Theorem 1.1 in the literature. The proof presented here is based on ideas that can be found in [9], [21], and [25]. However, very different ideas were used by Martus in [24]. Another nice proof of Theorem 1.1 is given by Nussbaum in [26].

At this point one might wonder what the optimal upper bound for $\rho(n)$ in Theorem 1.1 is, for a given polyhedral norm. This however appears to be a very difficult combinatorial geometric question. Of course Nussbaum's conjecture says that in case of the sup-norm the optimal bound for $\rho_\infty(n)$ is 2^n. But his conjecture is proved solely for $n = 1, 2$, and 3 (see [23]). The best known general upper bound for $\rho_\infty(n)$ is $n!2^n$, which was obtained by Martus in [24].

In case of the 1-norm even less is known. Misiurewicz [25] proved for the 1-norm that $\rho_1(n) \leq n!2^m$, where $m = 2^n$. Improvements on this bound for $n = 2, 3, 4$ and 5 were given in [21]. Indeed it is shown there that $\rho_1(n) \leq n^m$, where $m = 2^{n-1}$, for $n \geq 2$. But even in dimension 3 this last bound is expected to be far from sharp. It is generally believed that $\rho_1(n) = \mathcal{O}(c^n)$ for some $c \geq 2$.

The difficulty in case of the 1-norm is related to the fact that for a 1-norm nonexpansive map $f : X \to \mathbb{R}^n$, with $X \subset \mathbb{R}^n$, there may not exist a 1-norm nonexpansive map $F : \mathbb{R}^n \to \mathbb{R}^n$ that extends f (see [16] or [45]). For the sup-norm however there always exists, by the Aronszajn-Panitchpakdi Theorem [5], a sup-norm nonexpansive extension to the whole of $\mathbb{R}^n$.

3.2. Lower bounds for the maximum cardinality of limit sets. To get some feeling for the optimal upper bound for $\rho(n)$ one can try to find examples of periodic points which have a large minimal period. To generate such examples one can look for so called regular polygons.

Definition 3.2. A finite sequence of p distinct points $x^0, x^1, \ldots, x^{p-1}$ in $(\mathbb{R}^n, \| \cdot \|)$ is called a *regular polygon of size p* or simply a *regular p-gon* if

$$\|x^{k+l} - x^k\| = \|x^l - x^0\| \quad \text{for all } k, l \geq 0,$$

where the indices are counted modulo p.

Of course, for every regular p-gon $x^0, x^1, \ldots, x^{p-1}$ in $(\mathbb{R}^n, \| \cdot \|)$ the map f given by $f(x^i) = x^{i+1 \bmod p}$ for $0 \leq i < p$ is an isometry with respect to $\| \cdot \|$, and moreover f has a periodic point of minimal period p. On the other hand, each periodic orbit of a nonexpansive map is a regular polygon. Therefore the optimal upper bound for the integer $\rho(n)$ in Theorem 1.1 for a given polyhedral norm is precisely the maximum size of a regular polygon in $\mathbb{R}^n$ under this norm.

An example of a regular 2^n-gon in $\mathbb{R}^n$ under the sup-norm is formed by the set of vertices of the n-dimensional cube $\{x \in \mathbb{R}^n : x_i = \pm 1 \text{ for } 1 \le i \le n\}$, as the distance between any two distinct points in this set is 2. This example shows that $\rho_\infty(n)$ is at least 2^n for each $n \ge 1$, and hence the upper bound suggested in Nussbaum's conjecture is sharp.

Regular polygons with an exponential size under the 1-norm are harder to obtain. However, several constructions for such regular polygons were given by Lemmens, Nussbaum, and Verduyn Lunel in [21]. These constructions yield the following result.

Theorem 3.1. *For each $n \ge 3$ there exists a regular polygon of size $3 \cdot 2^{n-1}$ in $\mathbb{R}^n$ under the 1-norm.*

An example of a regular 12-gon in dimension 3 under the 1-norm is given by the sequence $x^0, -x^0, x^1, -x^1, \ldots, x^5, -x^5$, where

$$
\begin{aligned}
x^0 &= (0,1,2), & x^1 &= (0,2,1), \\
x^2 &= (1,2,0), & x^3 &= (2,1,0), \\
x^4 &= (2,0,1), & x^5 &= (1,0,2).
\end{aligned}
$$

Theorem 3.1 shows that $\rho_1(n) \ge 3 \cdot 2^{n-1}$ for each $n \ge 3$. At the present time no regular p-gons in $\mathbb{R}^n$, with $p > 3 \cdot 2^{n-1}$ are known under the 1-norm. Furthermore for $n = 3$ it can be shown that there exist regular polygons under the 1-norm of size 1, 2, 3, 4, 5, 6, 7, 8, and 12 (see [19]), but it is an open problem to decide whether there exist regular polygons of size 9, 10, and 11 in this space. As an aside we like to mention that the maximum size of a "trivial" 1-norm regular polygon in $\mathbb{R}^n$, that is a regular polygon in which any two distinct points are at the same 1-norm distance, is not known. It is generally believed that the maximum size of such regular polygons is $2n$, but at present this conjecture is proved only for $n = 1, 2, 3$, and 4 (see [8], [15], and [17]).

4. PERIODS OF 1-NORM NONEXPANSIVE MAPS

In the previous section we have seen that not much is known for the possible minimal periods of periodic points of 1-norm nonexpansive maps $f : X \to X$ when X can be an arbitrary subset of $\mathbb{R}^n$. On the other hand, if X is the whole of $\mathbb{R}^n$ or X is the positive cone in $\mathbb{R}^n$, then there are many detailed results. In this section we will give an overview of these results. We begin by discussing nonnegative 1-norm nonexpansive maps.

4.1. **Nonnegative 1-norm nonexpansive maps.** Motivated by the models for diffusion processes on a finite state space one has studied in [27], [34], [35], [36], and [39] the set $P^*(n)$ which consists of integers $p \ge 1$ for which there exist a 1-norm nonexpansive map $f : \mathbb{K}^n \to \mathbb{K}^n$, with $f(0) = 0$, and a periodic point of f of minimal period p. Surprisingly the set $P^*(n)$ admits a complete characterization in terms of arithmetical and combinatorial constraints. Indeed it is shown in [34] and [35] together that $P^*(n)$ is precisely the set of possible periods of an admissible array on n symbols. Here an admissible array is defined as follows.

Definition 4.1. Let $(L, <)$ be a finite totally ordered set and let Σ be a set with n elements. For each $i \in L$ let $\vartheta_i : \mathbb{Z} \to \Sigma$ be a map. The sequence $\vartheta = (\vartheta_i : \mathbb{Z} \to \Sigma \mid i \in L)$ is called an *admissible array on n symbols* if the maps ϑ_i satisfy the following properties:

   (i) For each $i \in L$ there exists an integer p_i with $1 \le p_i \le n$ such that the map $\vartheta_i : \mathbb{Z} \to \Sigma$ is periodic with period p_i and moreover $\vartheta_i(s) \neq \vartheta_i(t)$ for each $1 \le s < t \le p_i$.
   (ii) If $m_1 < m_2 < \ldots < m_{r+1}$ is an increasing sequence of distinct points in L and

$$\vartheta_{m_i}(s_i) = \vartheta_{m_{i+1}}(t_i) \quad \text{for } 1 \le i \le r,$$

   then

$$\sum_{i=1}^{r}(t_i - s_i) \not\equiv 0 \bmod \rho, \text{ where } \rho = \gcd(\{p_{m_i} : 1 \le i \le r+1\}).$$

Here $\gcd(S)$ denotes the greatest common divisor of the elements of the set S. The *period* of an admissible array is said to be the least common multiple of p_i, where $i \in L$. Thus, if one defines for each $n \in \mathbb{N}$ the set

(3) $Q(n) = \{p \in \mathbb{N} : p \text{ is the period of an admissible array on } n \text{ symbols}\}$,

then the characterization of the set $P^*(n)$ reads as follows.

Theorem 4.1 ([35], Theorem 3.1). $P^*(n) = Q(n)$ *for each* $n \in \mathbb{N}$.

The main idea behind the proof of the inclusion $P^*(n) \subset Q(n)$ is to relate to each periodic point of a 1-norm nonexpansive map $f : \mathbb{K}^n \to \mathbb{K}^n$, with $f(0) = 0$, a so called lower semi-lattice homomorphism. Via this lower semi-lattice homomorphism an admissible array can be constructed such that its period corresponds to the minimal period of the original periodic point. Let us explain this procedure in more detail.

4.1.1. *Lower semi-lattice homomorphisms.* We begin by collecting several definitions. On $\mathbb{R}^n$ a partial ordering $\le$ is defined by $x \le y$ if $x_i \le y_i$ for each $1 \le i \le n$. In particular, we write $x < y$ if $x \le y$ and $x \neq y$. Of course, $x \le y$ if and only if $y - x \in \mathbb{K}^n$. Further for $x, y \in \mathbb{R}^n$ we let $x \wedge y$ be the vector in $\mathbb{R}^n$ given by $(x \wedge y)_i = \min\{x_i, y_i\}$ for $1 \le i \le n$. Similarly, $x \vee y$ denotes the vector with coordinates $(x \vee y)_i = \max\{x_i, y_i\}$ for $1 \le i \le n$.

A set $V \subset \mathbb{R}^n$ is called a *lower semi-lattice* if $x \wedge y \in V$ for all $x, y \in V$. If in addition, $x \vee y \in V$ for all $x, y \in V$, then V is called a *lattice*. If S is a subset of $\mathbb{R}^n$, then V_S denotes the smallest (in the sense of inclusion) lower semi-lattice that contains S. The set V_S is called the *lower semi-lattice generated by* S. A map $g : V \to V$, where V is a lower semi-lattice is said to be a *lower semi-lattice homomorphism* if $g(x) \wedge g(y) = g(x \wedge y)$ for all $x, y \in V$. In a similar way *lattice homomorphisms* can be defined. The relation of these notions with nonnegative 1-norm nonexpansive maps is given by the following observation of Scheutzow [39].

Theorem 4.2. *Let* $f : \mathbb{K}^n \to \mathbb{K}^n$ *be a 1-nonexpansive map, with* $f(0) = 0$, *and let* $\xi \in \mathbb{K}^n$ *be a periodic point of* f *of minimal period* p. *If* $V \subset \mathbb{K}^n$ *is the lower semi-lattice generated by* $\{f^j(\xi) : 0 \le j < p\}$, *then the restriction of* f *to* V *is a lower semi-lattice homomorphism that maps* V *onto itself.*

This results motivates a further study of periodic points of lower semi-lattice homomorphisms. To do so several more notions have to be introduced. If A is a subset of a lower semi-lattice V in $\mathbb{R}^n$ and there exists $\beta \in V$ such that $\alpha \le \beta$ for each $\alpha \in A$, we say that A is *bounded above* in V, and β is called an *upper bound* of A in V. By replacing $\le$ with $\ge$ *lower bounds* can be defined in the same manner. If A is bounded above in V, then there exists a unique $\alpha \in V$ upper bound of A in V such that $\gamma < \alpha$ implies γ is not an upper bound of A in V, and α is called the *supremum* of A in V, which will be denoted by $\sup_V(A)$. Analogously the *infimum* of A in V, denoted $\inf_V(A)$, is said to be the unique lower bound $\alpha \in V$ of A such that no $\beta > \alpha$ is a lower bound of A in V.

For each x in a finite lower semi-lattive V the *height* of x, denoted $h_V(x)$, is defined by

(4) $h_V(x) = \sup\{k \ge 0 :$ there exist $y^0, \ldots, y^k \in V$ such that

$$y^k = x \text{ and } y^j < y^{j+1} \text{ for } 0 \le j < k\}.$$

If no $y \in V$ exists with $y < x$, then we say that $h_V(x) = 0$.

For every $x \in V$ put $S_x = \{y \in V : y < x\}$. An element $x \in V$ is called *irreducible* in V if either S_x is empty or

(5) $x > \sup_V(S_x)$.

If $x \in V$ is irreducible in V and S_x is nonempty, then $I_V(x)$ is said to be

(6) $I_V(x) = \{i : x_i > z_i\}$,   where $z = \sup_V(S_x)$.

In case S_x is empty, that is $x = \inf_V(V)$, then $I_V(x) = \{1, 2, \ldots, n\}$.

Using these notions we can now state the following result of Scheutzow [39]. A proof of this version of the lemma can be found in [34].

Lemma 4.1. *Let* $j \in \mathbb{Z}$, *let* V *be a finite lower semi-lattice in* $\mathbb{R}^n$, *and let* $f : V \to V$ *be a lower semi-lattice homomorphism of* V *onto itself. If* $y \in V$ *and* $f^j(y) \ne y$, *then* y *and* $f^j(y)$ *are incomparable, and* $h_V(y) = h_V(f^j(y))$, *where* $h_V(\cdot)$ *is the height function given in (4). If* y *is irreducible in* V, *then* $f^j(y)$ *is irreducible in* V. *If* $\eta \in V$ *and* $\zeta \in V$ *are not comparable, and* η *and*

ζ *are irreducible in* V, *then*

$$(7) \hspace{4cm} I_V(\eta) \cap I_V(\zeta) = \emptyset.$$

If $y \in V$ *is irreducible in* V *and* y *is a periodic point of* f *of minimal period* p, *then* $1 \leq p \leq n$.

The following technical definition forms the base from which the admissible arrays will be constructed.

Definition 4.2. Let W be a lower semi-lattice in $\mathbb{R}^n$, let $g : W \to W$ be a lower semi-lattice homomorphism, and let $\xi \in W$ be a periodic point of g of minimal period p. Let V denote the lower semi-lattice generated by $\{g^j(\xi) : j \geq 0\}$ and let f be the restriction of g to V. A finite sequence $(y^i)_{i=1}^m \subset V$ is called a *complete sequence for* ξ, if it satisfies:

   (i) For $1 \leq i \leq m$ we have $y^i \leq \xi$.
  (ii) For $1 \leq i \leq m$ the element y^i is irreducible in V.
 (iii) If p_i is the minimal period of y^i under f, then p is the least common multiple of $\{p_i : 1 \leq i \leq m\}$.
  (iv) For $1 \leq i < m$ we have $h_V(y^i) \leq h_V(y^{i+1})$, where $h_V(\cdot)$ is the height function given by equation (4).
   (v) For $1 \leq i < j \leq m$, the sets $\{f^k(y^i) : k \geq 0\}$ and $\{f^k(y^j) : k \geq 0\}$ are disjoint.
  (vi) For $1 \leq i < j \leq m$, the elements y^i and y^j are not comparable.

The next result says that every periodic point of a lower semi-lattice homomorphism has a complete sequence (see [34, Proposition 1.1]).

Proposition 4.1. *If* W *is a lower semi-lattice in* $\mathbb{R}^n$, $g : W \to W$ *is a lower semi-lattice homomorphism, and* $\xi \in W$ *is a periodic point of* g, *then there exists a complete sequence for* ξ.

Using the complete sequences one can now construct the admissible arrays.

4.1.2. *Admissible arrays.* Let W be a lower semi-lattice in $\mathbb{R}^n$ and let $g : W \to W$ be a lower semi-lattice homomorphism. Suppose that $\xi \in W$ is a periodic point of g of minimal period p. Let V denote the lower semi-lattice generated by $\{g^j(\xi) : j \geq 0\}$ and let f be the restriction of g to V. Now by Proposition 4.1 there exists a complete sequence $(y^i)_{i=1}^m \subset V$ for ξ. Moreover it follows from property (ii) in Definition 4.2 and Lemma 4.1 that $f^j(y^i)$ is irreducible in V for $1 \leq i \leq m$ and $j \in \mathbb{Z}$, so that the set $I_V(f^j(y^i))$ (as defined in (6)) is nonempty for $1 \leq i \leq m$ and $j \in \mathbb{Z}$. Let p_i denote the minimal period of y^i under f. Select for $1 \leq i \leq m$ and $0 \leq j < p_i$ an integer $a_{ij} \in I_V(f^j(y^i))$, and define for $1 \leq i \leq m$ and general $j \in \mathbb{Z}$ the integer a_{ij} by

$$a_{ij} = a_{ik}, \quad \text{where } 0 \leq k < p_i \text{ and } j \equiv k \mod p_i.$$

The semi-infinite matrix (a_{ij}), where $1 \leq i \leq m$ and $j \in \mathbb{Z}$, is called an *array of ξ*. Now there exists the following connection with the admissible arrays on n symbols (see [34, Propostion 1.2]).

Proposition 4.2. *Let W be a lower semi-lattice in $\mathbb{R}^n$, $g : W \to W$ be a lower semi-lattice homomorphism, and $\xi \in W$ be a periodic point of g of minimal period p. Let (a_{ij}), where $1 \leq i \leq m$ and $j \in \mathbb{Z}$, be an array of ξ. Further, let $L = \{1, \ldots, m\}$ be equipped with the usual ordering and let $\Sigma = \{1, 2, \ldots, n\}$. If $\vartheta = (\vartheta_i : \mathbb{Z} \to \Sigma \mid i \in L)$ is defined by*

$$\vartheta_i(j) = a_{ij} \quad \text{for } i \in L \text{ and } j \in \mathbb{Z},$$

then ϑ is an admissible array on n symbols with period p.

A combination of Theorem 4.2 and Propositions 4.1 and 4.2 yields the inclusion $P^*(n) \subset Q(n)$. The other inclusion $Q(n) \subset P^*(n)$ is shown in [35]. In fact, the following stronger result is proved there.

Theorem 4.3 ([35], Theorem 3.1). *For each $p \in Q(n)$ there exist a sand-shift map $f : \mathbb{K}^n \to \mathbb{K}^n$ and a periodic point of f of minimal period p.*

Although the set $Q(n)$ is described in terms of arithmetical and combinatorial constraints it is difficult to compute it. Despite this difficulty the set $Q(n)$ has been determined up to dimension 50 in [36] with the aid of a computer. From these computations it follows that the set $Q(n)$ has a highly irregular structure, and therefore we do not think that there exists a simple description of $Q(n)$. In Table 1 in Section 6 we only show the largest elements of $Q(n)$ for $1 \leq n \leq 20$. To conlude this subsection we like to mention that there exist several overviews of the results discussed here, see for instance [32], [33], and [43].

4.2. **1-Norm nonexpansive maps on the whole space.** Another interesting set of periods that has been studied is the set $R(n)$, which consists of integers $p \geq 1$ for which there exist a 1-norm nonexpansive map $f : \mathbb{R}^n \to \mathbb{R}^n$ and a periodic point of f of minimal period p. It turns out that the results for nonnegative 1-norm nonexpansive maps say something about the set $R(n)$. More precisely, it is shown in [40] and [32] that $R(n) \subset P^*(2n)$ for all $n \geq 1$, so that by Theorem 4.1 the inclusion $R(n) \subset Q(2n)$ holds for each $n \geq 1$. This upper bound is however not optimal. A sharper bound was obtained by Lemmens in [19] (see also [20]). We will discuss this upper bound in this subsection.

To obtain a sharper bound one uses the following consequence of the proof of the inclusion $R(n) \subset P^*(2n)$: *For each $p \in R(n)$ there exist a 1-norm nonexpansive map $f : \mathbb{K}^{2n} \to \mathbb{K}^{2n}$, with $f(0) = 0$, and a periodic point ξ of f of minimal period p, such that $f^j(\xi) \in \mathbb{E}^{2n}$ for each $j \geq 0$, where $\mathbb{E}^{2n} = \{(x, y) \in \mathbb{K}^n \times \mathbb{K}^n : x \wedge y = 0\}$. As $\mathbb{E}^{2n}$ is a lower semi-lattice in $\mathbb{K}^{2n}$ this observation and the results from the previous subsection suggest that one should study the arrays of periodic points of lower semi-lattice homomorphisms $g : W \to W$, where $W \subset \mathbb{E}^{2n}$.

It turns out that one can derive two additional properties for such arrays. These properties motivate the notion of a strongly admissible array on $2n$ symbols, which we will give now. To exhibit the definition of a strongly admissible array a final piece of notation is needed. If $a \in \{1, 2, \ldots, 2n\}$, then we write $a^+ = a + n$ if $1 \leq a \leq n$, and $a^+ = a - n$ if $n + 1 \leq a \leq 2n$.

Definition 4.3. Suppose that $(L, <)$ is a finite totally ordered set and let $\Sigma = \{1, 2, \ldots, 2n\}$. Assume that $\vartheta = (\vartheta_i : \mathbb{Z} \to \Sigma \mid i \in L)$ is an admissible array on $2n$ symbols, and let p_i denote the period of ϑ_i, for $i \in L$. We call ϑ a *strongly admissible array on $2n$ symbols* if the maps ϑ_i satisfy:

(i) If m_1 and m_2 are distinct elements of L and $\vartheta_{m_1}(s) = \vartheta_{m_2}(t)^+$, then

$$s - t \not\equiv 0 \mod \pi, \quad \text{where } \pi = \gcd(p_{m_1}, p_{m_2}).$$

(ii) If $m_1 < m_2 < \ldots < m_{r+1}$ is an increasing sequence of distinct elements in L such that

$$\vartheta_{m_i}(s_i) = \vartheta_{m_{i+1}}(t_i) \quad \text{for } 1 \leq i \leq r,$$

and if $\vartheta_{m_1}(u) = \vartheta_{m_{r+1}}(v)^+$ for some $u, v \in \mathbb{Z}$, then

$$\sum_{i=1}^{r} (t_i - s_i) \not\equiv (v - u) \mod \rho, \quad \text{where } \rho = \gcd(\{p_{m_i} : 1 \leq i \leq r + 1\}).$$

The connection between the strongly admissible arrays and the arrays of periodic points of lower semi-lattice homomorphisms $g : W \to W$, where $W \subset \mathbb{E}^{2n}$, is given in the following proposition.

Proposition 4.3. *Let W be a lower semi-lattice in $\mathbb{E}^{2n}$, $g : W \to W$ be a lower semi-lattice homomorphism, and $\xi \in W$ be a periodic point of g of minimal period p. Let (a_{ij}), where $1 \leq i \leq m$ and $j \in \mathbb{Z}$, be an array of ξ. Further let $L = \{1, \ldots, m\}$ be equipped with the usual ordering and let $\Sigma = \{1, \ldots, 2n\}$. If $\vartheta = (\vartheta_i : \mathbb{Z} \to \Sigma \mid i \in L)$ is defined by*

$$\vartheta_i(j) = a_{ij} \quad \text{for } i \in L \text{ and } j \in \mathbb{Z},$$

then ϑ is a strongly admissible array on $2n$ symbols with period p.

Now if one defines for each $n \geq 1$ the set

(8) $T(n) = \{p \in \mathbb{N} : p$ is the period of a strongly admissible array

on $2n$ symbols$\}$,

then by using Proposition 4.3 the following theorem can be obtained.

Theorem 4.4 ([20], Theorem 2.1). *$R(n) \subset T(n)$ for each $n \in \mathbb{N}$.*

The set $T(n)$ is computed for $1 \leq n \leq 10$ in [19]. A list of the largest elements of $T(n)$ for $1 \leq n \leq 10$ is given in Table 2 in Section 6. It turns out that $T(n)$ is much smaller than $Q(2n)$ and moreover that $R(n) = T(n)$ for $n = 1, 2, 3, 4, 6, 7$, and 10. However, it is unknown whether the sets $R(n)$ and $T(n)$ are equal for all $n \in \mathbb{N}$. To determine $R(n)$ up to $n = 10$ it only remains to be decided whether or not $18 \in R(5)$, $90 \in R(8)$, and $126 \in R(9)$ (see [19]).

5. LATTICE HOMOMORPHISMS AND NONEXPANSIVE MAPS

We saw in the previous section that there exists a strong connection between lower semi-lattice homomorphisms and nonnegative 1-norm nonexpansive maps. It has been observed by Nussbaum in [28] that this connection can be extended to other nonnegative nonexpansive maps. In particular to maps that are nonexpansive in a strictly monotone norm. A norm $\| \cdot \|$ on $\mathbb{R}^n$ is called *strictly monotone* if $\|x\| < \|y\|$ for each $0 \leq x < y$. More precisely Nussbaum's result can be stated as follows.

Theorem 5.1 ([28]). *If $f : \mathbb{K}^n \to \mathbb{K}^n$, with $f(0) = 0$, is nonexpansive in a strictly monotone norm and f is order-preserving, then the following assertions are true:*

- *For each $x \in \mathbb{K}^n$ there exists an integer $p_x = p \geq 1$ such that $(f^{kp}(x))_k$ converges to a periodic point of f of minimal period p.*
- *If $\xi \in \mathbb{K}^n$ is a periodic point of f of minimal period p, and V denotes the lattice generated by $\{f^j(\xi) : 0 \leq j < p\}$, then the restriction of f to V is a lattice homomorphism that maps V onto itself.*
- *If $\xi \in \mathbb{K}^n$ is periodic point of f of minimal period p, then $p \in Q(n)$, where $Q(n)$ is given in (3).*

It is known that in this theorem the assumption that f is order-preserving is necessary. Indeed one can find a map $f : \mathbb{K}^n \to \mathbb{K}^n$, with $f(0) = 0$, which is nonexpansive in the Euclidean norm, but not order-preserving, for which the first assertion in Theorem 5.1 does not hold (see [33, p. 224]).

Further we like to remark that Theorem 5.1 can not be applied in case the map $f : \mathbb{K}^n \to \mathbb{K}^n$ is nonexpansive in the sup-norm, since this norm is not strictly monotone. In fact, there seems to be no relation between sup-norm nonexpansive maps and lattice homomorphisms. Therefore we feel that different ideas are needed to obtain a good upper bound for the minimal period of periodic points of sup-norm nonexpansive maps $f : \mathbb{K}^n \to \mathbb{K}^n$, with $f(0) = 0$.

Further Theorem 5.1 raises the following problem. Decide for a given strictly monotone norm whether the set $Q(n)$ is the optimal upper bound in the third assertion in Theorem 5.1. At present there exist no results in this direction, except for the 1-norm.

Another remark we like to make is that the second statement in Theorem 5.1 is suggested by the structure of the fixed point set of these nonexpansive maps. More precisely, one can show that if $f : \mathbb{K}^n \to \mathbb{K}^n$, with $f(0) = 0$, is nonexpansive in a strictly monotone norm and f is order-preserving, then $\{x \in \mathbb{K}^n : f(x) = x\}$ is a lattice (see [35, Proposition 2.1]). This statement is well-known and easy to see if in addition f is assumed to be a linear map (compare [1] or [38]). Indeed, if $z \in \{x : f(x) = x\}$, then as f is order-preserving $z = f(z) \leq f(z \vee 0)$. Since $0 = f(0) \leq f(z \vee 0)$ it follows that $z \vee 0 \leq f(z \vee 0)$. On the other hand, as f is nonexpansive $\|f(z \vee 0)\| \leq \|z \vee 0\|$. Therefore by using the fact that $\| \cdot \|$ is strictly monotone we obtain $f(z \vee 0) = z \vee 0$. If we now apply the linearity of f, then it follows from the expressions $y \vee z = (y - z) \vee 0 + z$ and $y \wedge z = -((-y) \vee (-z))$ that $\{x : f(x) = x\}$ is a lattice.

To conclude we like to mention that it is interesting to understand under which additional assumptions a map $f : \mathbb{R}^n \to \mathbb{R}^n$, with $f(0) = 0$, which is nonexpansive in a strictly monotone norm, satisfies the first assertion in Theorem 5.1. It is remarkable that even for linear maps there are no results concerning this problem.

6. NUMERICAL DATA

In the first table we list the largest element of $Q(n)$ for $1 \leq n \leq 20$. This table is taken from [36]. The second table contains the largest element of $T(n)$ for $1 \leq n \leq 10$ and is taken from [19]. We like to emphasize that more detailed lists can be found in [36] and [19].

TABLE 1. The largest element of $Q(n)$ for $1 \leq n \leq 20$.

n	largest element $Q(n)$	n	largest element $Q(n)$
1	1	11	60
2	2	12	120
3	3	13	120
4	4	14	168
5	6	15	180
6	12	16	336
7	12	17	420
8	24	18	420
9	24	19	840
10	60	20	1680

REFERENCES

[1] Y.A. Abramovitch, C.D. Aliprantis and O. Burkinshaw, An elementary proof of Douglas' theorem on contractive projections on L_1-spaces, *J. Math. Analysis Appl.*, 177: 641–644, 1993.

[2] J. Aaronson, E. Glasner and M. Misiurewicz, Rigid sets in the plane, *Israel J. Math.*, 68: 307–326, 1989.

TABLE 2. The largest element of $T(n)$ for $1 \leq n \leq 10$.

n	largest element $T(n)$
1	2
2	4
3	6
4	12
5	20
6	30
7	60
8	90
9	140
10	210

[3] M.A. Akcoglu and U. Krengel, Nonlinear models of diffusion on a finite space, *Probab. Theory Related Fields*, 76: 411–420, 1987.

[4] T. Ando, Contractive projections in L_p spaces", *Pacific J. Math.*, 17: 391–405, 1966.

[5] N. Aronszajn and P. Panitchpakdi", Extension of uniformly continuous transformations and hyperconvex metric spaces, *Pacific J. Math.*, 6: 405–439, 1956.

[6] F.L. Baccelli, G. Cohen, G.J. Olsder and J.-P. Quadrat, *Synchronization and Linearity, an Algebra for Discrete Event Systems*, John Wiley and Sons, New York, 1992.

[7] S. Banach, *Théorie des Opérations Linéaires*, Chelsea Publishing Co, New York, 1955.

[8] H.J. Bandelt, V. Chepoi and M. Laurent, Embedding into rectilinear spaces, *Discrete Comput. Geom.*, 19: 595–604, 1998.

[9] A. Blokhuis and H.A. Wilbrink, Alternative proof of Sine's theorem on the size of a regular polygon in $\mathbb{R}^k$ with the ℓ_∞ metric, *Discrete Comput. Geom.*, 7: 433–434, 1992.

[10] M.G. Crandall and L. Tartar, Some relations between nonexpansive and order preserving mappings, *Proc. Amer. Math. Soc.*, 78(3): 385–390, 1980.

[11] R. Cuninghame-Green, *Minimax Algebra*, Lecture Notes in Economics and Math. Systems, 166, Springer Verlag, Berlin, 1979.

[12] C.M. Dafermos and M. Slemrod, Asymptotic behaviour of nonlinear contraction semigroups, *J. Funct. Anal.*, 13: 97–106, 1973.

[13] G. DiLena and B. Messano and D. Roux, Rigid sets and nonexpansive mappings, *Proc. Amer. Math. Soc.*, 125: 3575–3580, 1997.

[14] J. Gunawardena, An introduction to idempotency, *Idempotency*, Ed. J. Gunawardena, Publ. Newton Inst., Cambridge Univ. Press, Cambridge, 11: 1–49, 1998.

[15] R.K. Guy and R.B. Kusner, An olla podria of open problems, often odly posed, *Amer. Math. Monthly*, 90: 196–199, 1983.

[16] M. Kirszbraun, Über die zusammenziehende und Lipschitzsche Transformationen, *Fund. Math.*, 22: 77–108, 1943.

[17] , J. Koolen, M. Laurent and A. Schrijver, Equilateral dimension of the rectilinear space, *Des. Codes Cryptogr.*, 21: 149–164, 2000.

[18] B. Lemmens, Integral rigid sets and periods of nonexpansive maps, *Indag. Mathem. (N.S.)*, 10(3): 437–447, 1999.

[19] B. Lemmens, *Iteration of nonexpansive maps under the 1-norm*, Vrije Universiteit Amsterdam, Amsterdam, The Netherlands, 2001.

[20] B. Lemmens, *Periods of periodic points of 1-norm nonexpansive maps*, Eurandom, Eindhoven, The Netherlands, 2001-028, 2001.

[21] , B. Lemmens, R.D. Nussbaum and S.M. Verduyn Lunel, Lower and upper bounds for ω-limit sets of nonexpansive maps, *Indag. Mathem. (N.S.)*, 12(2), 2001

[22] Shih-Kung Lo, *Estimates for rigid sets in $\mathbb{R}^n$ with l_∞ or polyhedral norms*, Diplomarbeit, Inst. für Stochastik der Georg-August Univ. zu Göttingen (in German), 1989.

[23] R. Lyons and R.D. Nussbaum, On transitive and commutative finite groups of isometries, *Fixed Point Theory and Applications*, Ed. Kok-Keong Tan, World Scientific, Singapore, 189–228, 1992.

[24] P. Martus, *Asymptotic properties of nonstationary operator sequences in the nonlinear case* (in German), Friedrich-Alexander Univ. Erlangen-Nürnberg,1989.

[25] M. Misiurewicz, *Rigid sets in finite dimensional l_1-spaces*, Mathematica Göttingensis Schriftenreihe des Sonderforschungsbereichs Geometrie und Analysis, 45, 1987.

[26] R.D. Nussbaum, Omega limit sets of nonexpansive maps: finiteness and cardinality estimates, *Differential Integral Equations*, 3: 523–540, 1990.

[27] R.D. Nussbaum, Estimates of the periods of periodic points of nonexpansive operators, *Israel J. Math.*, 76: 345–380, 1991.

[28] R.D. Nussbaum, Lattice isomorphisms and iterates of nonexpansive maps, *Nonlinear Anal.*, 22: 945–970, 1994.

[29] R.D. Nussbaum, Convergence of iterates of a nonlinear operator arising in statistical mechanics, *Nonlinearity*, 4: 1223-1240, 1991.

[30] R.D. Nussbaum, Hilbert's projective metric and iterated nonlinear maps, *Mem. Amer. Math. Soc.*, 75 (391), 1988.

[31] R.D. Nussbaum, Hilbert's projective metric and iterated nonlinear maps,II, *Mem. Amer. Math. Soc.*, 79 (401), 1989.

[32] R.D. Nussbaum, A nonlinear generalization of Perron-Frobenius theory and periodic points of nonexpansive maps, *Recent Developments in Optimization and Nonlinear Analysis*, Eds. Y. Censor and S. Reich, Contemporary Mathematics, 204, American Mathematical Society, Providence, R.I. 183–198, 1997.

[33] R.D. Nussbaum, Periodic points of nonexpansive operators, *Optimization and Nonlinear Analysis*, Edt. A. Ioffe, M. Marcus and S. Reich, Pitman Research Notes in Mathematics, 244, Longman, London, 214–226, 1992.

[34] R.D. Nussbaum and M. Scheutzow, Admissible arrays and a generalization of Perron-Frobenius theory, *J. London Math. Soc.*, 58(2): 526–544, 1998.

[35] R.D. Nussbaum and M. Scheutzow and S.M. Verduyn Lunel, Periodic points of nonexpansive maps and nonlinear generalizations of the Perron-Frobenius theory, *Selecta Math. (N.S.)*, 4: 1–41, 1998.

[36] R.D. Nussbaum and S.M. Verduyn Lunel, Generalizations of the Perron-Frobenius theorem for nonlinear maps, *Mem. Amer. Math. Soc.*, 138(659): 1–98, 1999.

[37] R.D. Nussbaum and E. Shustin, Nonexpansive periodic operators in l_1 with applications to superhigh-frequency oscillations in a discontinuous dynamical system with time delay, *J. Dynam. Differential Equations*, 13(2): 381–424, 2001.

[38] B. Randrianantoanina, Norm one projections in Banach spaces, *Taiwaneese J. Math.*, 5: 35–95, 2001.

[39] M. Scheutzow, Periods of nonexpansive operators on finite l_1-spaces, *European J. Combin.*, 9: 73–78, 1988.

[40] M. Scheutzow, Corrections to periods of nonexpansive operators on finite l_1-spaces, *European J. Combin.*, 12: 183–183, 1991.

[41] E. Seneta, *Nonnegative matrices and Markov chains*, Springer Series in Statistics, Springer-Verlag, New York, 1981.

[42] R. Sine, A nonlinear Perron-Frobenius theorem, *Proc. Amer. Math. Soc.*, 109: 331–336, 1990.

[43] S.M. Verduyn Lunel, The iteration of nonexpansive maps, *Infinite Dimensional Stochastic Analysis*, Eds. Ph. Clement, F. den Hollander, J. van Neerven and B. de Pagter, Koninklijke Nederlandse Akademie van Wetenschappen Verhandelingen, Afd. Natuurkunde, Eerste reeks, deel 52, North-Holland, Amsterdam, 269–282, 2000.

[44] D. Weller, *Hilbert's metric, part metric and self mappings of a cone*, Universität Bremen, 1987

[45] J. H. Wells and L. Williams, *Embeddings and Extensions in Analysis*, Springer-Verlag, New York, 1975.

EURANDOM, P.O. BOX 513, 5600 MB EINDHOVEN, THE NETHERLANDS
E-mail address: lemmens@eurandom.tue.nl

ARITHMETIC DYNAMICS

NIKITA SIDOROV

To Paul Glendinning on the occasion of the birth of his twins

ABSTRACT. This survey paper is aimed to describe a relatively new branch of symbolic dynamics which we call Arithmetic Dynamics. It deals with explicit arithmetic expansions of reals and vectors that have a "dynamical" sense. This means precisely that they (semi-) conjugate a given continuous (or measure-preserving) dynamical system and a symbolic one. The classes of dynamical systems and their codings considered in the paper involve:
- Beta-expansions, i.e., the radix expansions in non-integer bases;
- "Rotational" expansions which arise in the problem of encoding of irrational rotations of the circle;
- Toral expansions which naturally appear in arithmetic symbolic codings of algebraic toral automorphisms (mostly hyperbolic).

We study ergodic-theoretic and probabilistic properties of these expansions and their applications. Besides, in some cases we create "redundant" representations (those whose space of "digits" is *a priori* larger than necessary) and study their combinatorics.

CONTENTS

Key words and phrases. β-expansion, beta-expansion, rotational expansion, toral automorphism, arithmetic coding.

Supported by the EPSRC grant no GR/R61451/01.

1. Introduction

The present survey paper is devoted to the recent progress in the new branch of dynamical systems theory which we call *Arithmetic Dynamics* (AD). It is worth pointing out that there is no conventional agreement regarding what the expression AD actually stands for – at the moment when I am writing these words (the year 2002), different people seem to see it differently (see, *e.g.*, [54, 21]). This is not particularly surprising: since the term is not fixed, anything that has something to do with arithmetics and dynamics may qualify.

Nonetheless, in the present paper the scope will be more narrow than that; namely, we define AD as a discipline which deals with symbolic codings of continuous (or measure-preserving) dynamical systems (invertible or not) expressed in terms of **explicit** arithmetic expansions of real numbers of vectors. If one accepts this (rather vague) definition, the classic ergodic theory of continued fractions, for instance, quite fits into the scope of AD. The term in question in this setting was suggested first by A. Vershik in the mid-1990's (implying that it would be suitable for a full-developed theory in the future).

The expansions in questions will be usually called *arithmetic codings*. Here are some characteristic features of arithmetic codings:

- They extensively use number-theoretic methods and techniques;
- The arithmetic structure possessed by dynamical systems we encode, is fully preserved;
- As long as there are no obstacles of number-theoretic nature, they can be generalized.

Apart from the "normal" (i.e., one-to-one a.e.) expansions, there exists another important topic that may be regarded as a part of AD, namely, the theory of *redundant* or *excessive* representations of real numbers and vectors. The model is as follows: assume we have a fixed "normal" expansion and the natural set of its "digits" is not a Cartesian product but has, for instance, pairwise (or even more sophisticated) restrictions. Our goal is to study the pattern, in which we lift all the restrictions between **distinct** digits and leave only the minimal Cartesian hull for the original set of digits. For example, in the case of β-expansions with a non-integer β (see below) this eventually leads to the well-known *Bernoulli convolutions*. Special attention will be paid to the combinatorics of **all** representations of a given x as well as to the set of those x which despite lifting the restrictions, will have a *unique representation* in the class in question.

One important point has to be made: AD is still in the cradle, so to speak, i.e., definitely not yet a "full-scale" subarea of Dynamical Systems. Here are its lacks:

- At present there is hardly any systematic approach that would cover a more or less substantial variety of measure-preserving (or continuous) transformations. On the contrary, as we will see, for each class of maps under consideration the model turns out to be state-of-the-art.

- Another serious issue regarding AD is that the constructions in question are not yet quite robust and strongly depend on the arithmetic structure of a dynamical system in question.

Despite all this, most constructions look rather nice and are closely related to number-theoretic problems as well (especially for arithmetic codings of toral automorphisms). Thus, we believe that even in this intermediate state AD is worth a detailed description, with the hope that some day it will become an area with a more systematic approach. Such a description is the aim of the present paper.

Our intention is mostly to summarize the progress in AD in the recent 10 years. The reason for this particular figure is A. Vershik's seminal paper [74] which appeared in the winter of 1991–92 and which has stimulated quite a number of new works and a great deal of ideas in AD. This paper presents an arithmetic coding of the Fibonacci automorphism of the 2-torus (see Section 4 for the definitions) based on the two-sided generalization of the corresponding adic transformation suggested by the same author in the late 1970's [72, 73] (see Section 3 and Appendix).

The structure of the paper is as follows: **Section 2** is devoted to the β-*expansions*, i.e., radix representations in (generally speaking, non-integer) bases β with $\beta > 1$. In particular, we will briefly outline some well-known results about the *greedy* and *lazy* beta-expansions, then describe in detail the recent progress in the theory of *unique* beta-representations in a fixed alphabet and finally, will deal with the space of *all* beta-representations of a given real number. Besides, we are going to devote a part of this section to the case of the one-parameter family of *intermediate* beta-expansions, i.e., those which lie in between greedy and lazy ones (in the sense of the lexicographic ordering).

Section 3 deals with, generally speaking, quite a different subject, namely, arithmetic codings of a dynamical system with a purely discrete cyclic spectrum. More precisely, we present two models for adic realization of an irrational rotation of the circle $S^1 = \mathbb{R}/\mathbb{Z}$. Both schemes deal with expansions of the elements of S^1 in bases involving the sequence of best approximation of the angle of rotation, which we call *rotational expansions*. In this case the compacta we obtain, are, generally speaking, non-stationary, and the map is not a traditional shift but the *adic transformation* (see Appendix for the definition).[1] This map is in a way transversal to the shift (if the latter is well defined). Putting it simply, the adic transformation is a generalization of the "adding machine" in the ring of p-adic integers to more general Markov compacta.

---

[1] There is nonetheless some intersection with Section 2 – the two models coincide if the angle of rotation is $\frac{1}{2}(-a + \sqrt{a^2 + 4})$ for some $a \in \mathbb{N}$.

We study the distribution of the "digits" in both cases and give sufficient conditions for the laws of large numbers and the central limit theorem to hold for them. In the end of Section 3 we – similarly to Section 2 – consider the set of unique rotational expansions, prove a number of claims that fully describe it and discuss the effect of non-stationarity, which leads to an essential difference with the beta-expansions.

Finally, in **Section 4** we consider arithmetic codings of hyperbolic automorphisms of a torus; as was mentioned above, it was a discovery in this area that initiated most of the research in AD in the last 10 years. After the model case of the *Fibonacci automorphism* of $\mathbb{T}^2$, i.e., the one given by the matrix $\begin{pmatrix} 1 & 1 \\ 1 & 0 \end{pmatrix}$, was studied in detail in [74], A. Vershik [75] asked the question whether it is possible to generalize this construction to all hyperbolic (or even ergodic) automorphisms. More precisely, is it possible to find a symbolic coding of a given automorphism T such that certain structures (the stable and unstable foliation, homoclinic points) have a clear expression in the corresponding symbolic compactum. As is well known, the classical models that deal with Markov partitions [70, 11, 36] do not have this property (apart from possibly the case of dimension two).

Thus, in Section 4 we describe the evolution of this area in the past 10 years. This includes the Fibonacci automorphism (as well as ergodic automorphisms of the 2-torus considered in [69, 78]), the Pisot automorphisms, i.e. those whose unstable (or stable) foliation is one-dimensional, and finally, we will try to summarize all attempts to cope with the general hyperbolic case and the difficulties that occur in doing so.

Appendix serves mostly auxiliary purposes: it briefly describes the theory of adic transformations.

The experienced reader will notice, of course, that some topics that might be in this survey paper, are missing. The reason for this is that either a nice exposition of the corresponding theory and results can be found elsewhere or the similarity with the actual framework of AD, however broad it is, is imaginary. Here is the list of some areas and subjects in question:

- *Odometers.* By this word is usually implied the adic transformation (see Appendix) on a general compact set (not necessarily a Markov compactum) with some extra "fullness" conditions – see, *e.g.*, [35]. I am still not convinced though that the odometers are really natural: as far as I am concerned, there are no interesting examples of symbolic codings of non-symbolic dynamical systems by means of odometers.
- *Algebraic codings of higher-rank actions on a torus.* The reader may refer to the paper by Einsiedler and Schmidt [22], in which a general construction (similar to the one from [62] discussed in Section 4) was suggested. There are also some partial results in this direction that can be found in the author's paper [65].

- *Measures of arithmetic nature.* By those we mean mainly Bernoulli-type convolutions parameterized by a Pisot number. The information about them can be found, for example, in [55] (see also references therein).

Since this is a survey paper, the proofs in the text are usually omitted; exception is made for a few new results, namely, Theorem 2.18 and all claims in Section 3.3 as well as some more minor claims.

The author is indebted to Anatoly Vershik for helpful suggestions, remarks and historical data.

2. BETA-EXPANSIONS

Let $\beta > 1$ and $x \geq 0$; we call any representation of the form

$$(2.1) \qquad x = \pi_\beta(\varepsilon) = \sum_{n=1}^{\infty} \varepsilon_n \beta^{-n}$$

a *β-expansion* (or *beta-expansion* if we do not want to specify the base). We need to make one historical remark. The matter is that in the literature the beta-expansion is often the one that uses the greedy algorithm for obtaining the "digits" ε_n (see Section 2.1); we acknowledge this tradition (which apparently dates back to B. Parry's pioneering work [56]), but think that within this text it would only cause confusion, because we intend to describe different types of algorithms for one and the same β.

2.1. **Greedy expansions and the beta-shift.**
This subsection is one of the few exceptions we have made from the rule that only the recent progress will be discussed. The reason for doing so is simple: without clear exposition of this model, it is impossible to explain the importance of ideas developed in recent papers. In this subsection we will confine ourselves mostly to the case of $x < 1$ (although beta-expansions can be extended to the positive half-axis as well – see Lemma 2.8).

So, let τ_β be the *β-transformation*, i.e., the map from $[0, 1)$ onto itself acting by the formula

$$\tau_\beta(x) = \beta x \bmod 1$$

(see Figure 1).

This map is very important in ergodic theory and as well as number theory, and the main tool for its study is its symbolic encoding which we are going to describe below. Note that if $\beta = d$ is an integer, then τ_β is isomorphic to the full shift on d symbols. The idea suggested in [60] was to generalize this construction to the non-integer β's.

As is well known, to "encode" it, one needs to apply the greedy algorithm in order to obtain the digits in (2.1), namely, $\varepsilon_n = [\beta \tau_\beta^{n-1} x]$, $n \geq 1$. Then the one-sided shift σ_β in the space X_β of all possible sequences ε that can be obtained this way, is clearly isomorphic to τ_β, with the conjugating map

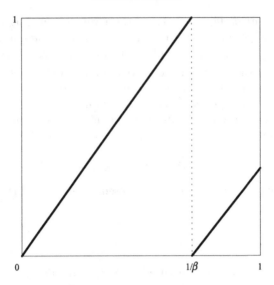

$$\text{FIGURE 1. The } \beta\text{-shift}$$

given by (2.1). We will call σ_β the *β-shift*. It is obviously a proper subshift of the full shift on $\prod_1^\infty \{0, 1, \ldots, [\beta]\}$. The question is, what kind of subshift is this – or equivalently – what is actually X_β?

This question was answered by B. Parry in the seminal paper [56]. The theorem he proved is the following. Let the sequence $(a_n)_1^\infty$ be defined as follows: let $1 = \sum_1^\infty a_k' \beta^{-k}$ be the greedy expansion of 1, i.e, $a_n' = [\beta \tau_\beta^{n-1} 1]$, $n \geq 1$; if the tail of the sequence (a_n') differs from 0^∞, then we put $a_n \equiv a_n'$. Otherwise let $k = \max\{j : a_j' > 0\}$; and $(a_1, a_2, \ldots) := (a_1', \ldots, a_{k-1}', a_k' - 1)^\infty$.

Theorem 2.1. [56]

1. *For any sequence $(a_n)_1^\infty$ described above, each power of its shift is lexicographically less than or equal to the sequence itself, and the equality occurs if and only if it is purely periodic. Conversely, each sequence having this property is $(a_n(\beta))_1^\infty$ for some $\beta > 1$.*
2. *For each greedy expansion $(\varepsilon_n)_1^\infty$ in base β, $(\varepsilon_n, \varepsilon_{n+1}, \ldots)$ is lexicographically less (notation: $\prec$) than $(a_1, a_2, \ldots)$ for every $n \geq 1$. Conversely, every sequence with this property is actually the greedy expansion in base β for some $x \in [0, 1)$.*

Let $\mathbf{a} = (a_n)_1^\infty$ and σ stands for the general one-sided shift on sequences. Thus, we have

$$(2.2) \qquad X_\beta = \left\{ \varepsilon \in \prod_1^\infty \{0, 1, \ldots, [\beta]\} \mid \sigma^n \varepsilon \prec \mathbf{a}, \ n \geq 0 \right\}$$

and the following diagram commutes:

$$
\begin{array}{ccc}
X_\beta & \xrightarrow{\;\sigma_\beta\;} & X_\beta \\
\pi_\beta \downarrow & & \downarrow \pi_\beta \\
[0,1) & \xrightarrow{\;\tau_\beta\;} & [0,1)
\end{array}
$$

Note that usually X_β is called the (one-sided) β-*compactum*.

Types of subshifts people know well how to deal with are mostly SFT (subshifts of finite type) and their factors, called *sofic subshifts* – see, *e.g.*, [51]. It is thus natural to ask whether X_β is such and if so, for which β? The following theorem gives a rather disappointing answer to this question. Roughly speaking, only certain algebraic β yield sofic beta-compacta. Recall that an algebraic integer is called a *Pisot number* if it is a real number greater than 1 and all its conjugates are less than 1 in modulus. A *Perron number* is an algebraic integer β greater than 1 whose conjugates are less than β in modulus.

Theorem 2.2. [56, 6, 50] *(see also the survey* [8]*)*

1. *The β-compactum X_β is an SFT if and only if the sequence a'_n (see above) is finite (i.e., its tail is 0^∞).*
2. *X_β is sofic if and only if (a_n) is ultimately periodic.*
3. *If β is a Pisot number, then X_β is sofic.*
4. *If X_β is sofic, then β is a Perron number.*

It has to be said that a more or less explicit description of X_β in case of transcendental β (as well as for $\beta = 3/2$, say) seems to be hopeless – see [8]. However, the ergodic-theoretic properties of the beta-shift are well studied and clearly understood by now for all $\beta > 1$. The following statement summarizes them.

Theorem 2.3. [60, 56, 71, 39]

1. *The β-transformation is topologically mixing, and its topological entropy is equal to $\log \beta$.*
2. *The β-shift is intrinsically ergodic[2] for any $\beta > 1$.*
3. *The unique measure of maximal entropy for τ_β is equivalent to the Lebesgue measure on $[0,1]$ and the corresponding density is bounded from both sides.*
4. *The natural extension of τ_β is Bernoulli. Moreover, the β-shift σ_β is weakly Bernoulli with respect to the natural (coordinate-wise) partition.*

Remark 2.4. The proof of item (2) given in [39] is rather complicated. In fact, I do not really understand why the proof of the intrinsic ergodicity of the transitive subshifts of finite type given by B. Parry (see, *e.g.*, [79, pp. 194–196]) cannot be applied to the β-shifts as well. The only property one needs

_____

[2]This means by definition that the map has a unique measure of maximal entropy.

apart from ergodicity, is the fact that the measure of any cylinder of length n divided by β^n is uniformly bounded. This is well known since [60].

Remark 2.5. In [7] a simplified proof of the above theorem was given. It is based on some auxiliary results on *coded systems* (which the β-shift is – see, *e.g.*, [8]). Unfortunately, this manuscript, rather helpful from the methodological point of view, is unpublished and not very easy to get hold of.[3]

The greedy expansion can be alternatively characterized as follows: assume that $n \geq 2$ and that the first $n-1$ digits of the expansion (2.1) are already chosen. Then if there is a choice for ε_n, we choose the largest possible number between 0 and $[\beta]$. Similarly, if we choose the smallest possible ε_n every time when we have a choice, this expansion is called the *lazy β-expansion*. Let us formally explain what we mean by the existence of a choice. Let $r_n(x,\beta) := x - \sum_1^{n-1} \varepsilon_k \beta^{-k}$; if $r_n(x,\beta) < \beta^{-n}$, then ε_n has be to equal to 0. If, on the contrary, $r_n(x,\beta) \geq [\beta]\beta^{-n}$, then inevitably $\varepsilon_n = [\beta]$. As is easy to see, in any other case there will be a choice for ε_n.

The following assertion is straightforward:

Lemma 2.6. *For a given $x \in [0,1)$, each of its β-expansions of the form (2.1) lies between its lazy and greedy β-expansions in the sense of lexicographic ordering of sequences.*

We will return to the case of "intermediate" expansions (i.e., those that lie strictly between the lazy and greedy ones) in Section 2.3.

Remark 2.7. As we have mentioned above, it is possible to expand any positive number (not necessarily from $(0,1)$) in base β by means of the greedy algorithm. Namely, let **a** be as above, and

$$(2.3) \qquad \widetilde{X}_\beta := \left\{ \varepsilon \in \prod_{-\infty}^{\infty} \{0, 1, \ldots, [\beta]\} \mid (\varepsilon_n, \varepsilon_{n+1}, \ldots) \prec \mathbf{a}, \ n \in \mathbb{Z} \right\},$$

i.e., the natural extension of the β-compactum. We will call it the *two-sided β-compactum*; it will be used extensively in Section 4.

Lemma 2.8. *Any $x \geq 0$ has the greedy β-expansion of the form*

$$(2.4) \qquad\qquad x = \sum_{n=-\infty}^{+\infty} \varepsilon_n \beta^{-n},$$

where (ε_n) is a sequence from $\widetilde{X}_\beta$ finite to the left, i.e., $\varepsilon_n \equiv 0$ for $n \leq N_0$ for some $N_0 \in \mathbb{Z}$.

The proof of this claim is similar to the one on the p-adic representations of positive reals, and we omit it (see also Section 4).

---

[3]In our days if a manuscript is unpublished, this is not necessarily that bad: a TEX file is even simpler to deal with than a hard copy. Unfortunately, this particular manuscript dates back the pre-TEX epoque ...

2.2. **Unique expansions and maps with gaps.** This subsection is aimed to describe a branch of a relatively new direction in the theory of arithmetic expansions, which deals with lifting all restrictions on "digits" leaving only the "Cartesian hull" (see Introduction).

2.2.1. $1 < \beta < 2$. Assume first that $\beta \in (1,2)$ and let $\Sigma = \prod_1^\infty \{0,1\}$. In the previous subsection we have described the specific (greedy) algorithm for choosing "digits" ε_n. As we have seen, the cost for this (very natural) approach is that the set of digits is quite complicated and unless β is an algebraic number, there is hardly any hope to describe it more or less explicitly (see Theorem 2.2).

In the 1990's a group of Hungarian mathematicians led by Paul Erdős began to investigate 0-1 sequences that provide **unique** representations of reals [24, 25, 26]. More precisely, let

$$\mathcal{A}'_\beta = \left\{ x \in \left(0, \frac{1}{\beta - 1}\right) \mid \exists! \ (\varepsilon_n)_1^\infty \in \Sigma : x = \sum_{n=1}^\infty \varepsilon_n \beta^{-n} \right\}$$

(it is obvious that the only representation for $x = 0$ is 0^∞ and the only representation for $x = 1/(\beta - 1)$ is 1^∞, so we will exclude both ends of the interval).

The first result about this set is given in [24]:

Proposition 2.9. *The set $\mathcal{A}'_\beta$ has Lebesgue measure zero for any $\beta \in (1,2)$. Moreover, if $\beta < G$, where $G = \frac{1+\sqrt{5}}{2}$, then in fact every x has $2^{\aleph_0}$ representations in the form (2.1).*

The question is, what can one say about the cardinality and – in case it is the continuum – about the Hausdorff dimension of this set. The answer to this question is given by P. Glendinning and the author in [32].

To present this result, we need some preliminaries. Let β_* denote the *Komornik-Loreti constant* introduced by V. Komornik and P. Loreti in [44], which is defined as the unique solution of the equation

$$\sum_1^\infty \mathfrak{m}_n x^{-n+1} = 1,$$

where $\mathfrak{m} = (\mathfrak{m}_n)_1^\infty$ is the Thue-Morse sequence

$$\mathfrak{m} = 0110 \ 1001 \ 1001 \ 0110 \ 1001 \ 0110 \ldots,$$

i.e., the fixed point of the substitution $0 \to 01$, $1 \to 10$. The Komornik-Loreti constant is known to be transcendental [4], and its numerical value is approximately as follows:

$$\beta_* = 1.787231650\ldots$$

The reason why this constant was introduced in [44] is that it proves to be the smallest number β such that $x = 1$ has a unique representation in the form (2.1). Now we are ready to formulate the result we mentioned above.

Theorem 2.10. [32] *The set $\mathcal{A}'_\beta$ is:*

- *empty if $\beta \in (1, G]$;*
- *countable for $\beta \in (G, \beta_*)$;*
- *an uncountable Cantor set of zero Hausdorff dimension if $\beta = \beta_*$; and*
- *a Cantor set of positive Hausdorff dimension for $\beta \in (\beta_*, 2)$.*

The proof of this result given in [32] is based on the observation that for ε to be a unique expansion for some $x \in \left(\frac{2-\beta}{\beta-1}, 1\right)$, it has to be its both greedy and lazy expansion – this is a direct consequence of Lemma 2.6. Let $\mathcal{A}_\beta = \mathcal{A}'_\beta \cap \left(\frac{2-\beta}{\beta-1}, 1\right)$ and $\mathcal{U}_\beta = \pi_\beta^{-1}(\mathcal{A}_\beta)$, where π_β is given by (2.1). It suffices to use (2.2) and a similar condition for the lazy expansion, which leads to the following lemma on the structure of the set $\mathcal{U}_\beta$:

Lemma 2.11. [32] *The set $\mathcal{U}_\beta$ can be described as follows:*

$$(2.5) \qquad \mathcal{U}_\beta = \{\varepsilon \in \Sigma : \overline{\mathbf{a}} \prec \sigma^n \varepsilon \prec \mathbf{a}, \ n \geq 0\}.$$

Remark 2.12. It is worth noting that although (2.2) looks similar to (2.5), the compacta X_β and $\mathcal{U}_\beta$ are completely different. In particular, the entropy of the β-shift is $\log \beta$ and in the case of the shift on $\mathcal{U}_\beta$ it is constant a.e. – see Theorem 2.14 below. Note also that restrictions like (2.5) are quite common in one-dimensional dynamics, and this is not a coincidence – see [33].

¿From Lemma 2.11 one can obtain the result on the cardinality of $\mathcal{U}_\beta$ and therefore of $\mathcal{A}'_\beta$ as well – see [32].

The problem with this proof is that it may be characterized as "a rabbit out of a hat" type of proof. Indeed, it does not say anything about the origin of the Komornik-Loreti constant as the main threshold between countable and uncountable set of uniquely representable points. This collision was overcome in the subsequent papers by the same authors [33, 34].

The key paper [33] is devoted to a "dynamical" version of the proof which will also involve various ergodic-theoretic and geometric applications for the shift on the space $\mathcal{U}_\beta$. Let us explain, where dynamics enters the game.

In a number of recent works *maps with gaps* or a *maps with holes* have been considered – see, *e.g.*, [12] and references therein. The model in question is as follows: let $X \subset \mathbb{R}^d$ and $f : X \to X$ be a map (invertible or not) with positive topological entropy. Let D be an open subset of X. The idea is to study the "dynamics of f on $X \setminus D$". More precisely, let

$$U = X \setminus \bigcup_{n \in \mathbb{Z}} f^n(D) \quad \text{if } f \text{ is invertible}$$

or

$$U = X \setminus \bigcup_{n \leq 0} f^n(D) \quad \text{if } f \text{ is non-invertible.}$$

One may ask two questions about this model.

Question 1. Is U empty? countable? uncountable? a Cantor set of positive Hausdorff dimension?

Question 2. If U has positive Hausdorff dimension, describe the dynamics of $f_U = f|_U$ (sometimes called the *exclusion map*). For instance, is f_U transitive? topologically mixing? intrinsically ergodic? etc.

Some (mostly, "generic" results) in this direction for Axiom A maps f on smooth manifolds can be found in [14, 15, 16, 12]. In particular, in [12] it is shown that if f is a hyperbolic algebraic automorphism of the torus $\mathbb{T}^m$ (see Section 4 for the relevant definitions) and $D = D(a_1, \ldots, a_m)$ is the parallelepiped built along the leaves of the stable and unstable foliation of f passing through $\mathbf{0}$ with the sides of length $a_1, \ldots, a_m$, then for a Lebesgue-generic m-tuple $(a_1, \ldots, a_m)$ the exclusion map $f|_U$ is a subshift of finite type.

Return to our situation. Let T_β be the following "map with a gap" acting from $[0, 1/(\beta - 1)]$ onto itself:

$$(2.6) \qquad T_\beta(x) = \begin{cases} \beta x, & x \in \left[0, \frac{1}{\beta}\right] \\ \text{not defined}, & x \in \left(\frac{1}{\beta}, \frac{1}{\beta(\beta-1)}\right) \\ \beta x - 1, & x \in \left[\frac{1}{\beta(\beta-1)}, \frac{1}{\beta-1}\right] \end{cases}$$

(see Figure 2).

The following simple lemma relates the dynamics of T_β to the original problem. Let

$$\Delta_\beta = [1/\beta, 1/\beta(\beta - 1)].$$

Lemma 2.13. [33]

$$\mathcal{A}_\beta' = \left\{ x \in \left(0, \frac{1}{\beta - 1}\right) : T_\beta^n(x) \notin \Delta_\beta \text{ for all } n \geq 0 \right\}.$$

Proof. Note first that if $x \in [0, 1/\beta)$, then necessarily ε_1 in (2.1) is 0 and if $x \in (1/\beta(\beta - 1), 1/(\beta - 1))$, then it is necessarily 1. If it belongs to Δ_β, then there is always at least two different representations.

Let S_β denote the shift on $\mathcal{U}_\beta$ and $R_\beta = T_\beta|_{\mathcal{A}_\beta}$. Then the following commutative diagram takes place:

$$\begin{array}{ccc} \mathcal{U}_\beta & \xrightarrow{S_\beta} & \mathcal{U}_\beta \\ \pi_\beta \downarrow & & \downarrow \pi_\beta \\ \mathcal{A}_\beta & \xrightarrow{R_\beta} & \mathcal{A}_\beta \end{array}$$

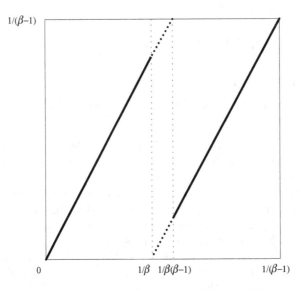

FIGURE 2. The map with a gap T_β

Thus, T_β acts as a shift on sequences providing unique representations. In-deed, it is either $x \mapsto \beta x$ or $\beta x - 1$ depending on whether we have $\varepsilon_1 = 0$ or 1, which is precisely the shift in (2.1). Hence for a T_β-orbit to stay out of Δ_β at any iteration is the same as keeping the representation (2.1) unique. $\square$

The important problem now is to describe the topological and ergodic properties of the shift S_β. The following theorem summarizes all we know at present about them.

Theorem 2.14. [33]

1. *For every* $\beta \in (\beta_*, 2)$ *the subshift* S_β *is* **essentially transitive**, *i.e., has a unique transitive component of maximal entropy.*
2. *The shift* S_β *is a subshift of finite type for a.e.* β.
3. *For a.e* $\beta \in (\beta_*, 2)$ *the subshift* S_β *is intrinsically ergodic and metrically isomorphic to a* **transitive** *SFT.*
4. *The function* $\beta \mapsto h_{top}(\sigma_\beta)$ *is continuous (but not Hölder continuous) and constant a.e. Every interval of constancy is naturally parameterized by an algebraic integer of a certain class.*

The open question is whether the shift S_β is "as good as" the beta-shift σ_β (see above). Namely, we *conjecture* that for **any** $\beta \in (\beta_*, 2)$ it is intrinsically ergodic and its natural extension is Bernoulli.

Remark 2.15. The above family of maps with gaps as a dynamical object might look artificial: we change not only the slope but the gap as well. How-ever, if one alters the size of gap only (which is more conventional), then the result on the symbolic level will be essentially the same. For example, let $Tx = 2x \bmod 1$ and $D_\delta = [\delta, 1 - \delta]$ for $\delta \in (0, 1/2)$. Let now

$$\mathcal{K}_\delta = \{x \in (0,1) : T^n(x) \notin D_\delta, \ n \geq 0\}.$$

It is known from the physical literature that $\dim_H(\mathcal{K}_\delta) > 0$ if and only if $\delta > \sum_1^\infty \mathfrak{m}_n 2^{-n} = 0.412\ldots$ (this has been apparently shown independently in [9, 81]). From the above results this claim follows almost immediately. Sketch of the proof of this fact is as follows: let $\mathbf{a}$ denote the binary expansion of 2δ; then in terms of the full 2-shift the set $\mathcal{K}_\delta$ is defined by (2.5); the only difference with $\mathcal{U}_\beta$ is that the sequence $\mathbf{a}$ does not necessarily satisfy the condition from Theorem 2.1 (1), but this is easy to deal with. So, the shifted Thue-Morse sequence is critical as well, which leads to the result in question. For a more general case see [33].

2.2.2. $\beta > 2$. Assume now that $\beta \in (N, N + 1)$ for some $N \geq 2$; we have similar results with some natural analog of the Thue-Morse sequence. Namely, let ρ denote the following substitution (morphism):

$$a \to ac, \ b \to ad, \ c \to da, \ d \to db.$$

Then

$$\rho^\infty(d) =: (\mathfrak{w}_n)_1^\infty = dbabacdb\ldots$$

The sequence $(\mathfrak{w}_n)_1^\infty$ is related to the Thue-Morse sequence in the following way:

$$\sigma(\mathfrak{m}) = \underbrace{1101}_{d} \ \underbrace{0011}_{b} \ \underbrace{0010}_{a} \ \underbrace{1101}_{d} \ \underbrace{0010}_{a} \ \underbrace{1100}_{c} \ \underbrace{1101}_{d} \ \underbrace{0011}_{b}\ldots$$

Let $P_n : \{0,1\} \to \{n-1, n\}$ be defined by

$$P_n(0) = n - 1, \ P_n(1) = n$$

and $Q_n : \{a, b, c, d\} \to \{n-1, n, n+1\}$:

$$Q_n(a) = n - 1, \ Q_n(b) = Q_n(c) = n, \ Q_n(d) = n + 1.$$

Then the critical value $x = \beta_*^{(N)}$ analogous to β_* is given by the equations

$$1 = \sum_{k=1}^\infty P_n(\mathfrak{m}_{k+1})x^{-k}, \quad N = 2n, \ n \geq 1$$

or

$$1 = \sum_{k=1}^\infty Q_n(\mathfrak{w}_k)x^{-k}, \quad N = 2n + 1, \ n \geq 1.$$

More precisely, let $\mathcal{A}_\beta^{(N)}$ denote the set of x which have a unique β-expansion with the digits $0, 1, \ldots, N - 1$.

Theorem 2.16. [34]

1. *The number $\beta_*^{(N)}$ is the smallest β for which $x = 1$ has a unique β-expansion with the digits $0, 1, \ldots, N-1$.*
2. *The set $\mathcal{A}_\beta^{(N)}$ has positive Hausdorff dimension if and only if $\beta > \beta_*^{(N)}$. If $\beta < \beta_*^{(N)}$, then it is at most countable.*

Remark 2.17. The method of proving this type of theorems comes from low-dimensional dynamics: it is called *renormalization*. For more detailed results in this direction and discussion see [33, 34].

2.3. **Intermediate beta-expansions.** Assume for simplicity that $\beta < 2$ and have another look at the map with a gap T_β defined above. More precisely, let

$$(2.7) \qquad T'_\beta(x) = \begin{cases} \beta x, & x \in \left[0, \frac{1}{\beta(\beta-1)}\right] \\ \beta x - 1, & x \in \left[\frac{1}{\beta}, \frac{1}{\beta-1}\right]. \end{cases}$$

Thus, we have a multivalued map on the middle interval Δ_β, and in order to get a "normal" (single-valued) map, one needs to make a choice for every $x \in \Delta_\beta$. From this point of view the β-shift corresponds to the choice of the lower branch and the "lazy" β-shift – the higher one for every x. On the other hand, as we have seen, if one removes the middle interval completely, then this corresponds to the unique expansions and eventually to the map that acts on at most a Cantor set, namely, $T_\beta|_{\mathcal{A}_\beta}$.

In [17] K. Dajani and C. Kraaikamp suggested the idea of considering "intermediate" beta-expansions, or how they called them, (β, α)-expansions. More precisely, let α be our parameter, $\alpha \in [0, (2-\beta)/(\beta-1)]$. We choose the upper branch if the ordinate is less than $1+\alpha$ and the lower branch otherwise, then restrict the resulting map to the interval $[\alpha, 1 + \alpha]$. Thus, we get

$$T_{\beta,\alpha}(x) = \begin{cases} \beta x, & x \in \left[\alpha, \frac{1+\alpha}{\beta}\right) \\ \beta x - 1, & x \in \left[\frac{1+\alpha}{\beta}, 1+\alpha\right]. \end{cases}$$

(see Figure 3).

As is easy to see, $T_{\beta,\alpha}$ is isomorphic to the map $S_{\beta,\gamma} : [0, 1) \to [0, 1)$ acting by the formula

$$S_{\beta,\gamma}(x) = \beta x + \gamma \bmod 1, \quad \gamma = (\beta - 1)\alpha \in [0, 2 - \beta].$$

The ergodic properties of the family $S_{\beta,\gamma}$ are well studied, see, *e.g.*, [57, 40, 27]. In particular, $S_{\beta,\gamma}$ is ergodic with respect to the Lebesgue measure, which is equivalent to the measure of maximal entropy.

2.4. **The realm of beta-expansions.** In the previous subsections we have studied the beta-expansions from the viewpoint of choosing a specific representation for a given x in (2.1) (and for $x \in \mathcal{A}'_\beta$ this choice was unique). We believe it is interesting to study the space of **all** possible representations of a given x, which this subsection will be devoted to.

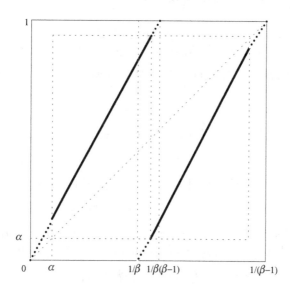

FIGURE 3. The pattern for the (β, α)-expansion

So, let $\beta > 1$ be fixed, and $\Sigma_q = \prod_1^\infty \{0, 1, \ldots, q-1\}$ for some fixed $q \geq 2$. We define for any $x \geq 0$,

$$\mathfrak{R}_{\beta, q}(x) := \left\{ \varepsilon \in \Sigma_q : x = \sum_{n=1}^\infty \varepsilon_n \beta^{-n} \right\}.$$

In the case when β is non-integer, and $q = [\beta] + 1$, we will simply write $\mathfrak{R}_\beta(x)$.

Not much is known about the general case yet. The following result was obtained by the author [67] by elementary means.[4] In this paper we would like to present an ergodic-theoretic proof that may possibly start the whole new line of research in this area (see Remark 2.20 below). The author is grateful to V. Komornik for his help with the history of the issue.

Theorem 2.18. *For any $\beta \in (1, 2)$ the cardinality of $\mathfrak{R}_\beta(x)$ is the continuum for a.e. $x \in (0, 1/(\beta - 1))$.*

Proof. Recall that if $\beta < G$, then **every** point x is known to have $2^{\aleph_0}$ β-expansions [25, Theorem 3]. So, let $\beta \geq G$.

---

[4]As with [32], this is a part of the author's *credo*: an elementary result deserves an elementary proof.

By the above, it suffices to show that

(2.8)
$$\mathrm{card}\left(\bigcup_{n=1}^{\infty}(T'_\beta)^n(x)\right) = 2^{\aleph_0}$$

for a.e. $x \in (0, 1/(\beta - 1))$ (where T'_β is the multivalued map given by (2.7)).
Let

$$T_0(x) = \begin{cases} \beta x, & x \in \left[0, \frac{1}{\beta(\beta-1)}\right) \\ \beta x - 1, & x \in \left[\frac{1}{\beta(\beta-1)}, \frac{1}{\beta-1}\right) \end{cases}$$

and

$$T_1(x) = \begin{cases} \beta x, & x \in \left[0, \frac{1}{\beta}\right) \\ \beta x - 1, & x \in \left[\frac{1}{\beta}, \frac{1}{\beta-1}\right) \end{cases}$$

(we omit the index β to simplify our notation). Then $T'_\beta x = T_0 x \cup T_1 x$ for
any $x \in \Delta_\beta$, whence any point y of the union in (2.8) is of the form

(2.9)
$$x_n = T_0^{k_n} T_1^{k_{n-1}} \ldots T_1^{k_2} T_0^{k_1}(x),$$

where $k_j \geq 0, 1 \leq j \leq n$. We are going to show that for a Lebesgue-generic x
and every x_n of the form (2.9) there exists $k_{n+1} \in \mathbb{N}$ such that $T_\beta^{k_{n+1}}(x_n) \in \Delta_\beta$
(where T_β is given by (2.6)), i.e., the branching for the multivalued map T'_β
has the form of the binary tree for a.e. x.

Fix the vector $(k_1, \ldots, k_n) \in \mathbb{Z}_+^n$ and let within this proof

$$T := T_0^{k_n} T_1^{k_{n-1}} \ldots T_1^{k_2} T_0^{k_1}.$$

Following the canonical proof of the corollary to the Poincaré Recurrence
Theorem (saying that a generic point returns to a given set of positive measure
infinitely many times), we will show that for a.e. x there exists $m \in \mathbb{N}$ such
that $T^m(x) \in \Delta_\beta$. This will prove the claim of our Theorem: let $E(k_1, \ldots, k_n)$
denote the generic set in question; then

$$E := \bigcap_{\substack{n \geq 1 \\ (k_1, \ldots, k_n) \in \mathbb{Z}_+^n}} E(k_1, \ldots, k_n)$$

will be a sought set of full measure.

So, it suffices to show that

$$\mathcal{L}\left(\bigcup_{j=1}^{\infty} T^{-j}\Delta_\beta\right) = 1$$

(where $\mathcal{L}$ is the **normalized** Lebesgue measure on $[0, 1/(\beta-1)]$). But since T_0
and T_1 are both extensions of T_β, we have $T^{-1}A \supset T_\beta^{-p}A$ for every measurable

A with $p = k_1 + \cdots + k_n$. It is thus left to show that

(2.10)
$$\mathcal{L}\left(\bigcup_{j=1}^{\infty} T_\beta^{-pj}\Delta_\beta\right) = 1$$

for every $p \geq 1$. Let

$$\widetilde{T}_\beta(x) = \begin{cases} \beta x, & x \in \left[0, \frac{1}{\beta}\right] \\ 1, & x \in \left(\frac{1}{\beta}, \frac{1}{\beta(\beta-1)}\right) \\ \beta x - 1, & x \in \left[\frac{1}{\beta(\beta-1)}, \frac{1}{\beta-1}\right]. \end{cases}$$

Thus, $\widetilde{T}_\beta$ is a "real map"; however, it is obvious that preimage-wise the maps T_β and $\widetilde{T}_\beta$ are the same. Moreover, the normalized Lebesgue measure on $[0, 1/(\beta - 1)]$ is quasi-invariant under $\widetilde{T}_\beta$; similarly to [60] and [57], one can easily show that there exists a unique $\widetilde{T}_\beta$-invariant measure $\mathcal{L}_\beta$ which is equivalent to the Lebesgue measure. Thus, the measure $\mathcal{L}_\beta$ is $(\widetilde{T}_\beta)^p$-invariant as well, which by the Poincaré Recurrence Theorem implies (2.10), and we are done. □

Remark 2.19. It is easy to generalize this theorem to the case of an arbitrary non-integer $\beta > 1$ and arbitrary $q > [\beta]$. We leave the details for the reader as a simple exercise (note: it is sufficient to consider the case $q = [\beta] + 1$).

Remark 2.20. The branching for T_β' described in the proof of the theorem, may be apparently studied in a more quantitative way. In particular, we *conjecture* that not only the cardinality of $\mathfrak{R}_\beta(x)$ is the continuum for a generic x but its Hausdorff dimension in the space Σ (provided with the natural (= binary) metric) is positive. This might be possibly shown by studying the average return times to the set Δ_β for the multivalued map T_β' and will be considered elsewhere.

Example 2.21. Let $\beta = G$; it is shown in [68, Appendix A] that $\mathfrak{R}_G(x)$ is always a continuum unless $x = nG \bmod 1$ for some $n \in \mathbb{Z}$. For example, $\mathfrak{R}_G(1/2) = \prod_1^\infty \{011, 100\}$, whence it is indeed a continuum, and its Hausdorff dimension equals $\frac{1}{3}$.

An important special case is when β is a Pisot number (see Section 2.1 for definition). Then one may study the combinatorics of the infinite space by means of the combinatorics of its finite approximations. More precisely, let us call two sequences ε and ε' from Σ_q *equivalent* if $\sum_1^\infty \varepsilon_k \beta^{-k} = \sum_1^\infty \varepsilon_k' \beta^{-k}$. There are a number of results about counting the cardinality of equivalence classes; most of them deal with random matrix products [46, 45, 55]. In particular, it is shown in [45] that for a generic 0-1 sequence $(x_n)_1^\infty$ the cardinality of the equivalence class of $(x_1 \ldots x_n)$ grows with n exponentially with an exponent greater than 1. The value of this exponent is given by the upper Lyapunov exponent of the random matrix product in question.

One special case is however worth mentioning on its own. Let $\beta = G$ (the golden ratio) and $q = 2$. We call a finite 0-1 word a *block* if it begins with 1 and ends by an even number of 0's. As is easy to show by induction, every block has the following form:

$$B = 1(01)^{a_1}(00)^{a_2}\ldots(00)^{a_r} \quad \text{or} \quad B = 1(00)^{a_1}(01)^{a_2}\ldots(00)^{a_r}.$$

Hence each block B is parameterized in a unique way by the natural numbers $a_1, \ldots, a_r$. We will write $B = B(a_1, \ldots, a_r)$. Let $\mathfrak{c}(w)$ denote the cardinality of the set of all 0-1 words equivalent to w.

In 1998 the author together with A. Vershik proved the following

Theorem 2.22. [68]

1. *Let $w = B_1 \ldots B_k$, where B_j is a block for all j. Then the space of all 0-1 words equivalent to w splits into the Cartesian product of the equivalence classes for B_j for $j = 1$ to k, and therefore, the function $\mathfrak{c}$ is blockwise multiplicative:*

$$\mathfrak{c}(B_1 \ldots B_k) = \mathfrak{c}(B_1) \ldots \mathfrak{c}(B_k).$$

2. *The cardinality of a block is given by the formula*

$$\mathfrak{c}(B) = p_r + q_r,$$

where $\frac{p_r}{q_r} = [a_1, \ldots, a_r]$ is the continued fraction.

The proof given in [68] is based on an induction argument; however, now it is clear that there exists a more direct and elegant way of proving this result. Let $X = X_G$ which in this case is the Markov compactum of all 0-1 sequences without two consecutive 1's and let $g : \prod_1^\infty \{a, b, c\} \to X$ act by the following rule:

$$g(a) = 00, \quad g(b) = 010, \quad g(c) = 10$$

and then by concatenation.

Lemma 2.23. *The map g is a bijection.*

Proof. It is a straightforward check that g^{-1} is well defined: any sequence from X can be split in a unique way into the concatenation of the blocks $00, 010$ and 10. $\qquad\square$

Let w be a finite word in X and $g^{-1}w = j_1 \ldots j_m$. In [55] it was shown that

$$\mathfrak{c}(w) = \begin{pmatrix} 1 & 0 \end{pmatrix} P_{j_1} P_{j_2} \ldots P_{j_m} \begin{pmatrix} 1 \\ 0 \end{pmatrix},$$

where

$$P_a = \begin{pmatrix} 1 & 1 \\ 0 & 1 \end{pmatrix}, \quad P_b = \frac{1}{2}\begin{pmatrix} 1 & 1 \\ 1 & 1 \end{pmatrix}, \quad P_c = \begin{pmatrix} 1 & 0 \\ 1 & 1 \end{pmatrix}.$$

This immediately yields both parts of Theorem 2.22, in view of the well-known relation between the matrix products of P_a and P_c and the finite continued fractions. The details are left to the reader.

The main consequence of Theorem 2.22 is the existence of the map called the *goldenshift* which acts on sequences from X starting with 1 as the shift by the length of the first block. This goldenshift is well defined a.e. and has a number of important properties that reveal a lot of information about the Bernoulli convolution parameterized by the golden ratio. For details see [68, §§2, 3].

Remark 2.24. In [68, Appendix A] A. Vershik and the author completely described all the possible patterns for $\mathfrak{R}_G(x)$ if $x \in (0, 1)$. Loosely speaking, these possibilities are as follows: either $\mathfrak{R}_G(x)$ is a Cartesian product (this is Lebesgue-generic – see also Example 2.21) or $\mathfrak{R}_G(x) = \mathcal{X}'_\alpha$ for a certain $\alpha = \alpha(x)$, where $\mathcal{X}'_\alpha$ is the "rotational" Markov compactum described in the next section. For more general Pisot numbers the analog of this theorem seems to be a delicate and interesting problem.

Remark 2.25. The combinatorics of the "integral" case $\beta \in \mathbb{N}$ is studied in detail by J.-M. Dumont, A. Thomas and the author in [19, §6]. In that case the cardinality function can be represented in terms of random matrix products as well. Specifically, if $\beta = 2$ and $q = 3$, then these matrices are precisely P_a and P_c.

Remark 2.26. There exists a class of singular measures (Bernoulli convolutions) that are based on the combinatorics we have just described. For more details see, *e.g.*, [3, 68] for the case of $\beta = G$ and [23, 31, 46, 45, 55] for more general cases of Pisot numbers (see also the survey article [58] for a general overview).

3. ROTATIONAL EXPANSIONS

In this section we are going to describe the model appeared first in [73] as an application of the general theorem by A. Vershik on adic realization – see Theorem 4.34 in Appendix, and in a more arithmetic form – in the joint paper [77] by A. Vershik and the author. Also, we will present another model which deals with more conventional base for arithmetic expansions.

3.1. General constructions.

3.1.1. *First model.* The problem we are going to consider in this section, is arithmetic codings of an irrational rotation of the circle $\mathbb{R}/\mathbb{Z}$ which we will identify with the interval $[0, 1)$. Let $\alpha \in (0, 1/2) \setminus \mathbb{Q}$ be the angle of rotation (if it is greater than $1/2$, simply take $1 - \alpha$), and let $\mathcal{R}_\alpha(x) = x + \alpha \bmod 1$.

Let the regular continued fraction expansion of α be

$$\alpha = \cfrac{1}{a_1 + \cfrac{1}{a_2 + \cfrac{1}{a_3 + \dots}}}$$

and $(p_n/q_n)_1^\infty$ be the sequence of convergents with $p_1 = 0, p_1 = 1, q_1 = 1, q_2 = a_1$ and $p_{n+1} = a_n p_n + p_{n-1}, q_{n+1} = a_n q_n + q_{n-1}$. Since $\alpha < 1/2$, we have $a_1 \geq 2$. Put $r_1 = a_1, r_n = a_n + 1, n \geq 2$ and,

$$M^{(n)} = M^{(n)}(\alpha) := \begin{pmatrix} 1 & \dots & 1 & 1 \\ 1 & \dots & 1 & 0 \\ \vdots & \vdots & \vdots & \vdots \\ 1 & \dots & 1 & 0 \end{pmatrix},$$

where the number of rows in $M^{(n)}$ is r_n, and the number of columns is r_{n+1}. Let $D_n = \{0, 1, \dots, r_n - 1\}$ be endowed with the natural ordering, i.e., $0 \prec 1 \prec \dots \prec r_n - 1$. Let now $\mathcal{X}_\alpha$ denote the Markov compactum determined by the sequence of matrices $(M^{(n)}(\alpha))_1^\infty$ and $\mathcal{T}_\alpha$ denote the adic transformation on $\mathcal{X}_\alpha$ with respect to the natural ordering (see Appendix for the definitions).

Note first that $\mathcal{T}_\alpha$ is well defined everywhere with the exception of two "maximal" sequences: $(a_1 - 1, 0, a_3, 0, a_5, \dots)$ and $(0, a_2, 0, a_4, 0, \dots)$. Thus, if we exclude the *cofinite* sequences (i.e., those whose tail is $(a_n, 0, a_{n+2}, 0, \dots)$ for some n), the positive part of the orbit of $\mathcal{T}_\alpha$ will be always well defined. To enable the whole trajectory of $\mathcal{T}_\alpha$ to be well defined, one has to remove the finite sequences as well.

We claim that the map $\mathcal{T}_\alpha$ is metrically isomorphic to $\mathcal{R}_\alpha$ and are going to present the conjugating map. Let $\psi_\alpha : \mathcal{X}_\alpha \to [0, 1)$ be defined as

(3.11) $$\psi_\alpha(x_1, x_2, \dots) := \alpha + \sum_{n=1}^\infty x_n (-1)^{n+1} \alpha_n,$$

where $\alpha_n = |q_n \alpha - p_n| = \|q_n \alpha\|$ (here $\| \cdot \|$ stands for the distance to the nearest integer).

Theorem 3.1. [77]

1. *The invertible dynamical system $(\mathcal{X}_\alpha, \mathcal{T}_\alpha)$ is uniquely ergodic. The unique invariant measure ν_α is Markov on $\mathcal{X}_\alpha$.*
2. *ψ_α is continuous and one-to-one except the cofinite sequences.*
3. *The map ψ_α metrically conjugates the automorphisms $(\mathcal{X}_\alpha, \nu_\alpha, \mathcal{T}_\alpha)$ and $([0, 1), \mathcal{L}, \mathcal{R}_\alpha)$, where $\mathcal{L}$ stands for the Lebesgue measure on the unit interval.*

Remark 3.2. As is well known, α_n decays at a very fast rate as n goes to the infinity, namely, $\alpha_n = O(q_{n+1}^{-1})$ (see, e.g., [43]). Thus, the n'th term of the sum in (3.11) is $O(q_n^{-1})$, i.e., decays at worst exponentially.

Remark 3.3. Expansion (3.11) was considered for the first time by Y. Dupain and V. Sos [20]; they also proved Theorem 3.1 (2), the fact A. Vershik and the author were unaware of when writing [77]. This however hardly undermines Theorem 3.1 as it appeared in [77], because the (most important) dynamical meaning of the rotational expansion was new.

Remark 3.4. A simple way to obtain these expansions is as follows: let $N \in \mathbb{N}$ and $(q_n)_1^\infty$ serve as a "base" for representations in the sense of [28]. Then there is a unique representation of N in the form

$$(3.12) \qquad N = 1 + \sum_k x_k q_k,$$

where $(x_1, x_2, \dots)$ is a finite sequence in $\mathcal{X}_\alpha$. All one has to do to get (3.11) is to make a *profinite completion* of (3.12) using the fact that the sequence $(N\alpha \bmod 1)_1^\infty$ is dense in $[0, 1)$ and the formula $q_k - p_k \alpha = (-1)^{k+1}\alpha_k$. For more details see [77, §2].

Remark 3.5. If one removes both the finite and cofinite sequences from $\mathcal{X}_\alpha$ and the $\mathcal{T}_\alpha$-trajectory of 0 from $[0, 1)$, then ψ_α becomes a homeomorphism and thus, acts as a conjugacy in the topological sense as well.

3.1.2. *Second model.* The price we pay for the natural ordering in the first model is that the base of the expansions is not always positive. The second model we are going to describe below, overcomes this problem but here there is the price to pay as well: the ordering is rather unusual. Apparently, it is impossible to take care of both issues simultaneously – this symbolizes the well-known fact that the convergents $(p_n/q_n) < \alpha$ if n is even and $> \alpha$ if n is odd.

Let $\alpha \in (0, 1) \setminus \mathbb{Q}$ and

$$\mathcal{M}^{(n)} = \mathcal{M}^{(n)}(\alpha) := \begin{pmatrix} 1 & 1 & \dots & 1 & 1 \\ \dots & \dots & \dots & \dots & \dots \\ 1 & 1 & \dots & 1 & 1 \\ 1 & 0 & \dots & 0 & 0 \end{pmatrix},$$

where the size of $\mathcal{M}^{(n)}$ is $(a_n + 1) \times (a_{n+1} + 1)$. Let $\mathcal{D}_n = \{0, 1, \dots, a_n\}$. We define the *alternating* ordering on $\mathcal{D}_n$ as follows: $0 \prec 1 \prec \cdots \prec a_n$ if n is odd and $0 \succ 1 \succ \cdots \succ a_n$ if n is even.

Let now $\mathcal{X}_\alpha'$ denote the corresponding Markov compactum, $\mathcal{T}_\alpha'$ – the adic transformation on it and $\psi_\alpha' : \mathcal{X}_\alpha' \to [0, 1)$ be defined as follows:

$$(3.13) \qquad \psi_\alpha'(x_1, x_2, \dots) := \sum_{n=1}^\infty x_n \alpha_n,$$

where α_n are as above. We have the following analog of Theorem 3.1:

Theorem 3.6. [77]

1. The map T'_α is well defined everywhere except the sequence $(a_1, 0, a_3, 0, \dots)$. Its inverse is not well defined only at $(0, a_2, 0, a_4, 0, \dots)$.
2. The dynamical system $(\mathcal{X}'_\alpha, T'_\alpha)$ is uniquely ergodic. The unique invariant measure μ_α is Markov on $\mathcal{X}'_\alpha$.
3. ψ'_α is continuous and one-to-one except the cofinite sequences.
4. The map ψ'_α metrically conjugates the automorphisms $(\mathcal{X}'_\alpha, \mu_\alpha, T'_\alpha)$ and $([0,1), \mathcal{L}, \mathcal{R}_\alpha)$.

Remark 3.7. The expansion (3.13) is a special case of the general class of *Cantor-Waterman expansions* [80]. More general systems of numeration are considered in [77] as well but without clear dynamical meaning.

Remark 3.8. An analog of (3.12) for the integers is as follows:

$$(3.14) \qquad N = \sum_n x_n(-1)^n q_n,$$

where $N \in \mathbb{Z}$ (not necessarily nonnegative!) and (x_n) is a finite sequence from $\mathcal{X}'_\alpha$ [77]. Thus, the alternating base is not for the reals but for the integers in the second model. Comparing (3.11) with (3.13) and (3.12) with (3.14), we see that the two models are in a way dual.

3.2. **Probabilistic properties of the "digits".** In this subsection we are going to mention briefly the results from [63] on Laws of Large Numbers (LLN and SLLN) and the Central Limit Theorem (CLT) for the sequences of digits for the expansions considered above. We will consider the metric space $(\mathcal{X}'_\alpha, \mu_\alpha)$; the results for $(\mathcal{X}_\alpha, \nu_\alpha)$ are very similar, so we will omit them.
    First, the initial distribution for μ_α is as follows:

$$\mu_\alpha(x_1 = i_1) = \begin{cases} \alpha, & x_1 < a_1 \\ \alpha_2, & x_1 = a_1, \end{cases}$$

the transition probabilities are given by

$$(3.15) \qquad \mu_\alpha\big(x_n = i_n \mid x_{n-1} = i_{n-1}\big) = \begin{cases} \frac{\alpha_n}{\alpha_{n-1}}, & i_{n-1} < a_{n-1}, \quad i_n < a_n \\ \frac{\alpha_{n+1}}{\alpha_{n-1}}, & i_{n-1} < a_{n-1}, \quad i_n = a_n \\ 1, & i_{n-1} = a_{n-1}, \quad i_n = 0 \\ 0, & \text{otherwise.} \end{cases}$$

Finally, the one-dimensional distributions are as follows:

$$(3.16) \qquad \mu_\alpha\big(x_n = i_n\big) = \begin{cases} (q_{n-1} + q_n)\alpha_n, & i_n = 0 \\ q_n\alpha_n, & 0 < i_n < a_n \\ q_n\alpha_{n+1}, & i_n = a_n. \end{cases}$$

    The following theorem shows that, roughly speaking, if the partial quotients a_n of α do no grow "too fast", then most of the probabilistic laws hold. More precisely, we have

Theorem 3.9. [63]

1. *If*

$$\sum_{k=1}^{n} a_k^2 = o(n^2), \quad n \to \infty,$$

then the LLN holds for $(\mathcal{X}'_\alpha, \mu_\alpha)$.

2. *The condition*

$$\sum_{n=1}^{\infty} \frac{a_n^2}{n^2} \ln^2 n < +\infty$$

is sufficient for the validity of SLLN for $(\mathcal{X}'_\alpha, \mu_\alpha)$.

3. *Finally, if the partial quotients for* α *are uniformly bounded, then the CLT holds for* $(\mathcal{X}'_\alpha, \mu_\alpha)$.

Remark 3.10. None of these conditions is Lebesgue-generic for α. We believe all of them can be improved but not significantly.

3.3. Unique rotational expansions.

Following the pattern of Section 2.2, it is interesting to study the combinatorics of expansions (3.13) with the lifted Markov restrictions (again, we will not consider the first model, where all the results is very similar and leave it to the interested reader). The results presented below are original (though some steps in this direction have been undertaken in [64]).

Let $Z_\alpha = \prod_{n=1}^{\infty} \{0, 1, \ldots, a_n\}$ and

$$\mathcal{V}_\alpha := \left\{ x \in (0, 1) \mid \exists! \ (x_1, x_2, \ldots) \in Z_\alpha : x = \sum_{n=1}^{\infty} x_n \alpha_n \right\}.$$

Our goal will be to study the properties of $\mathcal{V}_\alpha$.

Let us recall that $\alpha_{n-1} = a_n \alpha_n + \alpha_{n+1}$. Hence the triples $x_{n-1} = 1, x_n = 0, x_{n+1} = 0$ and $x_{n-1} = 0, x_n = a_n, x_{n+1} = 1$ give the same value in (3.13) provided all the other digits are the same. In a way, this claim is invertible, and this is what the proof will be based upon.

More precisely, in [64] it is shown that if $x \in (0, 1)$ has at least two different representations in the form (3.13), then in its *canonical* representation (the one with the digits in $\mathcal{X}'_\alpha$) there exists $n \in \mathbb{N}$ and a triple (i_{n-1}, i_n, i_{n+1}) with $i_{n-1} > 0, i_n = 0$ and $i_{n+1} < a_{n+1}$. We will call such triples *replaceable*. The question is, whether replaceable triples are generic with respect to the Lebesgue measure.[5]

We denote

$$\mathcal{V}'_\alpha = (\psi'_\alpha)^{-1}(\mathcal{V}_\alpha),$$

i.e., the set of admissible sequences providing unique rotational representations.

---

[5] The condition looks like something shift-invariant but there is no suitable ergodic theorem here, of course - the compactum $\mathcal{X}'_\alpha$ is non-stationary! (unless $a_n \equiv a$ for all n)

Lemma 3.11. *Let* $(x_1, x_2, \dots) \in \mathcal{V}'_\alpha$. *Then*

1. *if* $x_n = 0$ *for some* n, *then necessarily* $x_{n-1} = \cdots = x_1 = 0$ *as well;*
2. *it is impossible that* $x_n = a_n$ *for some* n.

Proof. It suffices to prove (1), because $x_n = a_n$ implies $x_{n+1} = 0$, which would contradict (1). Assume $x_n = 0$; if $x_{n-1} > 0$, then the triple (x_{n-1}, x_n, x_{n+1}) will be replaceable unless $x_{n+1} = a_{n+1}$. But then again, we have $x_{n+2} = 0$, which leads to the same problem! Since the tail $(x_n = a_n, 0, a_{n+2}, 0, \dots)$ is not admissible, we are done.   □

Proposition 3.12. $\mathcal{V}_\alpha = \emptyset$ *if and only if*

$$(3.17) \qquad\qquad \#\{n : a_n = 1\} = +\infty.$$

Proof. (1) Assume $a_{n_k} = 1$ for $k = 1, 2, \dots$ and $(x_1, x_2, \dots) \in \mathcal{V}'_\alpha$. Then for each k we have a choice between $x_{n_k} = 0$ and $x_{n_k} = 1$. The latter is impossible by Lemma 3.11, while the former leads to $x_j \equiv 0$ for all $j \le n_k$, which in turn leads to $x_j \equiv 0$ for all $j \in \mathbb{N}$. This is a contradiction, because $x > 0$.

(2) If the number of $a_n = 1$ is finite, we set

$$(3.18) \qquad\qquad n_0 = \sup\{n : a_n = 1\}.$$

Then the sequence with $x_j = 0$ for $1 \le j \le n_0$ and $x_j = 1$ otherwise, is a unique rotational representation.   □

Remark 3.13. The condition (3.17) is Lebesgue-generic for α. So, for a typical α our set is empty.

The following result may be regarded as an analog of Proposition 2.9 for the rotational expansions. The crucial difference is that it is not true that for every irrational α the set $\mathcal{V}_\alpha$ has zero Lebesgue measure – this depends on how fast the partial quotients grow.

Theorem 3.14. *The set* $\mathcal{V}_\alpha$ *has Lebesgue measure zero if and only if*

$$\sum_{n=1}^\infty \frac{1}{a_n} = +\infty. (*)$$

Proof. Assume first that $(*)$ is **not** satisfied. Then there exists only a finite number of n such that $a_n = 1$. Let n_0 be given by (3.18) and

$$K_\alpha := \{(x_1, x_2, \dots) \in \mathcal{X}'_\alpha : 0 < x_n < a_n,\ n \ge n_0 + 1\}.$$

By the above, each sequence from K_α is a unique representation, whence it would suffice to show that $\mu_\alpha(K_\alpha) > 0$. This measure can be computed explicitly: by (3.16) and (3.15) and in view of $x_1 = \cdots = x_{n_0} = 0$ (see Lemma 3.11),

$$\mu_\alpha(K_\alpha) = \alpha_{n_0} \cdot \prod_{n=n_0+1}^\infty (a_n - 1)\frac{\alpha_n}{\alpha_{n-1}} > \alpha_{n_0} \prod_{n=n_0+1}^\infty \frac{a_n - 1}{a_n + 1} > 0$$

(the first inequality follows from the fact that $(a_n + 1)\alpha_n > \alpha_{n-1}$ and the second one is a consequence of failing of $(*)$).

Assume now that $(*)$ holds. Our goal is to show that $\mu_\alpha(\mathcal{V}'_\alpha) = 0$, and our first remark consists in the observation that by Proposition 3.12, it suffices to consider α such that $n_0 < \infty$. Furthermore, each sequence from $\mathcal{V}'_\alpha$ that contains at least one zero, is of the form $(0, 0, \ldots, 0, x_n, x_{n+1}, \ldots)$, where $0 < x_j < a_j$ for $j \geq n$ (see Lemma 3.11). Thus, we come again to the set similar to K_α – see above. We have

$$\mu_\alpha(\mathcal{V}'_\alpha) = \sum_{k=0}^\infty \alpha_k \prod_{n=k+1}^\infty (a_n - 1)\frac{\alpha_n}{\alpha_{n-1}} \leq \sum_{k=0}^\infty \alpha_k \prod_{n=k+1}^\infty \frac{a_n - 1}{a_n} = 0,$$

because by $(*)$, each infinite product in the last-mentioned sum equals 0. $\square$

Remark 3.15. In [64] it is shown that if $(*)$ is not satisfied, then the image of the uniform measure on Z_α (i.e., $\prod_{n=1}^\infty \{1/a_n, \ldots, 1/a_n\}$) under the map ψ'_α given by (3.13), is an absolutely continuous measure. In the opposite direction the result is incomplete: apart from $(*)$ for this measure to be singular, there is one (apparently, parasite) condition, which at the time we have not been able to get rid of. Of course, if, for instance, $a_n \equiv a$ for all $n \geq 1$, then the measure in question is singular, which is the famous Erdös Theorem [23].

What is left if we wish to follow the pattern of Section 2.2, is the Hausdorff dimension of $\mathcal{V}_\alpha$ when $(*)$ is satisfied.

Proposition 3.16. *Assume that the number of n such that $a_n = 1$, is finite. Then the cardinality of $\mathcal{V}_\alpha$ is the continuum if and only if the tail of $(a_1, a_2, \ldots)$ is different from $(2, 2, 2, \ldots)$. Otherwise $\mathcal{V}_\alpha$ is a finite set.*

Proof. Let again n_0 be given by (3.18). If $a_n \equiv 2$ for $n \geq n_1$, then we must have $x_n \equiv 1$ for $n \geq n_1$. If, on the contrary, there exists a subsequence (m_k) such that $a_{m_k} \geq 3$, then we will have a choice of $x_{m_k} = 1$ or 2, which yields a continuum. $\square$

Remark 3.17. Thus, here we also have some kind of monotonicity, namely, the cardinality and Hausdorff dimension of $\mathcal{V}_\alpha$ are nondecreasing functions with respect to the partial quotients. The difference with Section 2.2 is that $\mathcal{V}_\alpha$ is never infinite countable.

Theorem 3.18. *Under the assumption of Proposition 3.16, the Hausdorff dimension of $\mathcal{V}_\alpha$ is positive if and only if*

$$(3.19) \qquad \liminf_{n \to +\infty} \frac{\sum_{k=n_0+1}^n \log(a_k - 1)}{\log q_{n+1}} > 0,$$

where n_0 is given by (3.18).

Proof. Assume for the simplicity of notation that $a_n > 1$ for all $n \geq 1$. Let $\mathcal{V}_\alpha^{(n)}$ denote the set of all cylinders of length n in $\mathcal{V}'_\alpha$. By the above, we will have the following choice for $\mathcal{V}_\alpha^{(n)}$: if $a_k = 2$, then necessarily $x_k = 1$; otherwise $x_k \in \{1, 2, \ldots, a_k - 1\}$. Hence by (3.16) and (3.15),

$$\mu_\alpha(\mathcal{V}_\alpha^{(n)}) = \alpha_n \prod_{k=1}^{n}(a_k - 1)$$

(all the transitional measures at the k'th step are the same), and the condition for the positivity of the Hausdorff dimension of $\mathcal{V}_\alpha$ is a follows:

$$\liminf_n \frac{\log \prod_{k=1}^{n}(a_k - 1)}{-\log \alpha_n} > 0,$$

which, in view of the inequality $1/2 < q_{n+1}\alpha_n < 1$ is equivalent to (3.19). □

Corollary 3.19. *If n_0 given by (3.18) is less than infinity, and*

$$\liminf_{n \to +\infty} \frac{\sum_{k=n_0+1}^{n} \log(a_k - 1)}{\sum_{k=n_0+1}^{n} \log(a_k + 1)} > 0,$$

then $\dim_H(\mathcal{V}_\alpha) > 0$.

Proof. It suffices to use the relation $q_{n+1} = a_n q_n + q_{n-1}$, from which it follows that $q_{n+1} \leq \prod_{k=1}^{n}(a_k + 1)$. □

Corollary 3.20. *If $a_n \leq C$ for all $n \geq 1$ and $n_0 < \infty$, then $\dim_H(\mathcal{V}_\alpha) > 0$ if and only if*

$$\Delta = \liminf_{n \to \infty} \frac{1}{n} \#\{1 \leq k \leq n : a_k = 2\} < 1.$$

Proof. Again, for simplicity we assume that $a_n \neq 1$ for all $n \geq 1$. We have: $\sum_{k=1}^{n} \log(a_k - 1) \geq (1 - \Delta)n \log 2$, and $\sum_{k=1}^{n} \log(a_k + 1) \leq n \log(1 + C)$, whence by the previous corollary, the "if" part follows. The proof of the "only if" part is left to the reader. □

4. ARITHMETIC CODINGS OF TORAL AUTOMORPHISMS

This section is devoted to the arithmetic codings of hyperbolic automorphisms of a torus. The idea of a coding is to expand the points of a torus in power series in base its homoclinic point. It was suggested by A. Vershik in special cases [74, 75] and developed by the author and A. Vershik in [68, 69, 78] in dimension 2 and in higher dimensions (chronologically) – by R. Kenyon and A. Vershik [42], S. Le Borgne in his Ph. D. Thesis [47] and subsequent works [48, 49], K. Schmidt [62] and finally by the author [66].

4.1. An important example: the Fibonacci automorphism. We begin with the example that was studied in detail in 1991–92 and has eventually led to the theory described in the rest of the section.

We are going to expose it just the way it appeared. The initial motivation has come from the theory of p-adic numbers: let p be a prime, and Z_p denote the group of p-adic integers, i.e., one-sided formal series in powers of p:

$$Z_p = \left\{ \sum_{n=-\infty}^{-1} x_n p^{-n} : 0 \le x_n \le p-1,\ n \le -1 \right\}.$$

Let Q_p denote the field of p-adic numbers, i.e.,

$$Q_p = \left\{ \sum_{n=-\infty}^{\infty} x_n p^{-n} \mid 0 \le x_n \le p-1,\ n \in \mathbb{Z},\ \exists N \in \mathbb{Z} : x_n \equiv 0, n \ge N \right\}.$$

Thus, Q_p is the space of two-sided p-adic expansions finite to the right.[6] Finally, if one considers the "full-scale" two-sided p-adic expansions

$$S_p = \left\{ \sum_{n=-\infty}^{\infty} x_n p^{-n} \mid 0 \le x_n \le p-1,\ n \in \mathbb{Z} \right\},$$

then we obtain the p-adic solenoid.

Question 4.1. What will all the above objects become if one replaces p by an algebraic unit $\beta > 1$ and the full p-adic compactum – by the two-sided β-compactum $\widetilde{X}_\beta$?

The obvious candidate to start investigation seemed $\beta = \frac{1}{2}(1 + \sqrt{5})$, in which case, we recall, $\widetilde{X} := \widetilde{X}_\beta$ is the set of two-sided 0-1 sequences without two consecutive 1's (see Section 2). There is another good reason for considering the golden ratio. Let $F_1 = 1, F_2 = 2, \ldots$ be the Fibonacci sequence; as was explained in Section 3, every natural number N has a unique representation in base $(F_n)_1^\infty$ with the digits from $X = X_\beta$ – see (3.12). Furthermore, as we know, the profinite completion of (3.12) turns $\mathbb{N}$ into S^1, whence the analog of Z_p is S^1. This suggests that unlike the p-adic case, the *fibadic* case, as we will call it, produces the topology of the real line instead of the p-adic topology.

Recall also that the set of p-adic expansions as well as β-expansions finite to the **left**, is simply $\mathbb{R}_+$ (Lemma 2.8). The situation with the spaces that involve *formal* power series in base β (infinite to the left) is completely different and strongly depends on β, as we will see below.

To deal with the problems regarding the formal power series, we notice that in the p-adic case the key to the structure of Z_p, Q_p and S_p is just the following relation: $p v_n = v_{n-1}$, where $v_n = p^{-n}$. In the fibadic case the analog of this relation is

(4.20) $$u_{n-1} = u_n + u_{n+1}.$$

---

[6]I have heard some people call them "1.5-sided expansions". Informally, of course.

Thus, for instance, the analog of Q_p is as follows:

$$Q_\beta := \left\{ \sum_{n=-\infty}^{\infty} \varepsilon_n u_n : (\varepsilon_n)_{-\infty}^{\infty} \in \widetilde{X},\ \varepsilon_n \equiv 0,\ n \geq N \text{ for some } N \in \mathbb{Z} \right\},$$

where the sequence (u_n) satisfies (4.20).

Proposition 4.2. [74] *After identification of a countable number of certain pairs of sequences Q_β becomes a field isomorphic to $\mathbb{R}$.*

The pairs in question arise because, loosely speaking, unlike the p-adic case, where $-v_0 = (p-1)v_{-1} + (p-1)v_{-2} + \ldots$, in the fibadic pattern we have two different representations: $-u_0 = u_{-1} + u_{-3} + u_{-5} + \cdots = u_1 + u_{-2} + u_{-4} + u_{-6} + \ldots$. The pairwise identification in question thus concerns certain sequences that are finite to the left and cofinite to the right. A formal way to establish this fact given in [74] is as follows: while the standard representation of the generators u_n is $\pi(u_n) = \beta^{-n}$, there is another one, namely $\pi'(u_n) = (-\beta)^n$. Then $\pi'(Q_\beta) \subset \mathbb{R}$, and it suffices to show that every real number does have a representation in base $((-\beta)^{-n})_{n \in \mathbb{Z}}$, and this representation is unique everywhere except a certain countable set. This claim follows from the results of Section 3 (see (3.14)).

Remark 4.3. It is interesting to find out what will correspond to different subsets of $\mathbb{R}$ in $\widetilde{X}$. Since $2 = 1 + 1 = 1 + \beta^{-1} + \beta^{-2} = \beta + \beta^{-2}$, and similarly, $3 = \beta^2 + \beta^{-2}$, etc., it is easy to see that $\mathbb{N} \subset \widetilde{X}$ consists of finite sequences only.[7] However, there are a lot of finite sequences that do not yield a natural number, for example, $1 + \beta^{-2}$. Moreover, it is shown in [74] that if one takes the union of the finite sequences and the sequences that are finite to the right and cofinite to the left, then after the identification mentioned above, this set becomes naturally isomorphic to the ring $\mathbb{Z}[\beta] \simeq \mathbb{Z} + \mathbb{Z}$ (and the finite sequences are of course isomorphic to $\mathbb{Z}[\beta] \cap \mathbb{R}_+$). This is again the crucial difference with the p-adic case, where the analogs are respectively $\mathbb{Z}$ and $\mathbb{Z} \cap \mathbb{R}_+$.

Remark 4.4. More detailed results about the embedding of different subsets of $\mathbb{R}_+$ into $\widetilde{X}$ as well as about relations with finite automata can be found in [30]. Note also that by the theorem proven independently by A. Bertrand [5] and K. Schmidt [61], $\pi^{-1}(\mathbb{Q}(\beta) \cap \mathbb{R}_+)$ is precisely the set of all sequences finite to the left and eventually periodic to the right (this is very similar to the p-adic case and is true for all Pisot numbers).

The most important discovery made in [74] was the fact that the fibadic analog of the solenoid $\mathcal{S}_p$ is actually the 2-torus $\mathbb{T}^2 = \mathbb{R}^2/\mathbb{Z}^2$. Let us explain this in detail as it appeared in subsequent works [68, 69]. Let $\mathcal{L}_m$ stand for the

---

[7] "Finite" henceforward will mean "finite in both directions".

Haar (= Lebesgue) measure on $\mathbb{T}^m$, and Φ denote the *Fibonacci automorphism* of $\mathbb{T}^2$, namely, the algebraic automorphism given by the matrix

$$M_\Phi = \begin{pmatrix} 1 & 1 \\ 1 & 0 \end{pmatrix}.$$

As is well known since the pioneering work by R. L. Adler and B. Weiss [1], Φ is metrically isomorphic to the two-sided β-shift σ_β. So, this is nothing new that $\widetilde{X}$ as a set is essentially the torus; what **is** new, however, is that the natural arithmetic of $\widetilde{X}$ is the same as the natural arithmetic of $\mathbb{T}^2$. Our goal is thus dual: to show that $\widetilde{X}$ is indeed *arithmetically* isomorphic to the 2-torus (i.e., not only in the ergodic-theoretic sense but in the arithmetic sense as well) and also to give a proof of the Adler-Weiss Theorem cited above that reveals the arithmetic structure of $\widetilde{X}_\beta$. Both problems will be discussed simultaneously.

We denote by X_f the set of all sequences from $\widetilde{X}$ finite to the left (recall that $X_f \simeq \mathbb{R}_+$). Consider $x \geq 0$ and its greedy expansion $x = \sum_{k=-\infty}^{\infty} \varepsilon_k \beta^{-k}$ given by (2.4) with $\varepsilon_k \equiv 0$ for $k \leq N(x)$. Consider now the map $f_\beta : X_f \to \mathbb{T}^2$ acting by the formula

$$f_\beta(\varepsilon) = \{(\{x\}, \{\beta^{-1}x\}) \mid x \geq 0\},$$

where $\{\cdot\}$ denotes the fractional part of a number. Let $\mathcal{R}_\beta \subset \mathbb{T}^2$ denote the image of X_f under f_β. Since $(1, \beta^{-1})$ is an eigenvector of M_Φ corresponding to the eigenvalue β, the set W_β is the half-leaf of the unstable foliation for the Fibonacci automorphism passing through **0**. Hence

(4.21) $$(f_\beta \sigma_\beta)(\varepsilon) = \Phi f_\beta(\varepsilon)$$

for any $\varepsilon = (\varepsilon_n)$ finite to the left.

Since the set W_β is dense in the 2-torus, as well as the set of sequences finite to the right is dense in $\widetilde{X}$, we can extend the relation (4.21) to the whole compactum $\widetilde{X}$, i.e. $(f_\beta \sigma_\beta)(\varepsilon) = \Phi f_\beta(\varepsilon)$ everywhere on $\widetilde{X}$. Besides, f_β is surjective and can be written in a very "arithmetic" sort of way, namely

(4.22) $$f_\beta(\varepsilon) = \left(\sum_{k=-\infty}^{\infty} \varepsilon_k \beta^{-k} \bmod 1, \ \sum_{k=-\infty}^{\infty} \varepsilon_k \beta^{-k-1} \bmod 1 \right),$$

where the expression $\sum_{n=-\infty}^{\infty} x_n = x \bmod 1$ means that $\lim_N \| \sum_{n=-N}^{N} x_n - x \| = 0$. The number-theoretic reason why these series do converge modulo 1 is that β is a Pisot number, whence $\|\beta^n\| \to 0$ as an exponential rate.

Lemma 4.5. [68] *The map f_β semiconjugates the automorphisms $(\widetilde{X}, m, \sigma_\beta)$ and $(\mathbb{T}^2, \mathcal{L}_2, \Phi)$, where m denotes the (Markov) measure of maximal entropy for σ_β. Moreover, after an identification on $\widetilde{X}$ that concerns a set of sequences of zero measure, $\widetilde{X}$ becomes an additive group $\widetilde{X}'$, and f_β becomes a group homomorphism of $\widetilde{X}'$ and $\mathbb{T}^2$.*

Thus, we seemed to have succeeded in our attempt to insert the arithmetic compactum $\widetilde{X}$ into $\mathbb{T}^2$. However, this is not that simple; the issue with f_β is that it is **not** bijective a.e. and thus cannot be regarded as an actual isomorphism. In fact, in [68] it was shown that it is 5-to-1 a.e.[8] This is not a coincidence – in Section 4.2 we will see that the discriminant of an irrational in question plays an important role in this theory (see Proposition 4.17). The deep reason why f_β has failed is because of the wrong choice of a homoclinic point – see below.

The way to construct an actual isomorphism is a slight modification of f_β. Namely, let $F_\beta : \widetilde{X} \to \mathbb{T}^2$ be defined by the formula

$$(4.23) \qquad F_\beta(\varepsilon) = \left(\sum_{k=-\infty}^{\infty} \varepsilon_k \frac{\beta^{-k}}{\sqrt{5}} \bmod 1, \ \sum_{k=-\infty}^{\infty} \varepsilon_k \frac{\beta^{-k-1}}{\sqrt{5}} \bmod 1 \right).$$

Similarly to the above, the convergence of both series is a consequence of the fact that $\|\beta^n/\sqrt{5}\| = \beta^{-n}/\sqrt{5}$, $n \geq 0$.

Theorem 4.6. [68] *The map F_β is 1-to-1 a.e. It is both a metric isomorphism of the automorphisms $(\widetilde{X}, m, \sigma_\beta)$ and $(\mathbb{T}^2, \mathcal{L}_2, \Phi)$ and of the groups $\widetilde{X}'$ and $\mathbb{T}^2$.*

The question is, why F_β succeeded where f_β failed? The reason becomes more transparent if we rewrite both maps. To do so, we need to recall some basic notions and facts from hyperbolic dynamics. Let T be a hyperbolic automorphism of the torus $\mathbb{T}^m = \mathbb{R}^m/\mathbb{Z}^m$, L_s and L_u denote respectively the leaves of the stable and unstable foliations passing through $\mathbf{0}$. Recall that a point homoclinic to $\mathbf{0}$ (or simply a *homoclinic point*) is a point which belongs to $L_s \cap L_u$. In other words, $\mathbf{t}$ is homoclinic iff $T^n \mathbf{t} \to \mathbf{0}$ as $n \to \pm\infty$. The homoclinic points are a group under addition isomorphic to $\mathbb{Z}^m$, and we will denote it by $H(T)$. Each homoclinic point $\mathbf{t}$ can be obtained as follows: take some $\mathbf{n} \in \mathbb{Z}^m$ and project it onto L_u along L_s and then onto $\mathbb{T}$ by taking the fractional parts of all coordinates of the vector (see [75]).

We claim that both (4.22) and (4.23) can be written in the form

$$(4.24) \qquad h_\mathbf{t}(\varepsilon) = \sum_{n \in \mathbb{Z}} \varepsilon_n T^{-n} \mathbf{t},$$

where $T = \Phi$ and $\mathbf{t} = \mathbf{t}_1 = (1, \beta^{-1})$ in the case of f_β and $\mathbf{t} = \mathbf{t}_0 = (1/\sqrt{5}, \beta^{-1}/\sqrt{5})$ in the case of F_β.

The reason why $\mathbf{t}_0$ is "better" than $\mathbf{t}_1$ is because it is a *fundamental* homoclinic point, i.e., the one for which the linear span of its orbit is the whole group $H(\Phi)$.

Remark 4.7. Any fundamental homoclinic point for Φ is of the form $\Phi^n \mathbf{t}_0$ for some $n \in \mathbb{Z}$. In other words, $\mathbf{t}$ is fundamental iff $\mathbf{t} = (\beta^n/\sqrt{5}, \beta^{n-1}/\sqrt{5})$ mod

---

[8]This means that $\mathcal{L}_2$-a.e. $x \in \mathbb{T}^2$ has exactly 5 f_β-preimages.

$\mathbb{Z}^2$ for some $n \in \mathbb{Z}$. In the next subsection we will have a generalization of this fact.

4.2. Pisot automorphisms.

The next step was made by the author and A. Vershik in [69, 78] – it concerned the general case of dimension 2. In this paper however we will jump to the next stage, which will completely cover the two-dimensional case, namely to the hyperbolic automorphisms of the m-torus ($m \geq 2$) whose stable (unstable) foliation is one-dimensional.

Let T be an algebraic automorphism of the torus $\mathbb{T}^m$ given by a matrix $M \in GL(m, \mathbb{Z})$ with the following property: the characteristic polynomial for M is irreducible over $\mathbb{Q}$, and a Pisot number $\beta > 1$ is one of its roots (we recall that an algebraic integer is called *a Pisot number*, if it is greater than 1 and all its Galois conjugates are less than 1 in modulus). Since $\det M = \pm 1, \beta$ is a *unit*, i.e., an invertible element of the ring $\mathbb{Z}[\beta] = \mathbb{Z}[\beta^{-1}]$. We will call such an automorphism a *Pisot automorphism*. Note that since none of the eigenvalues of M lies on the unit circle, T is hyperbolic. It is obvious that any hyperbolic automorphism T of $\mathbb{T}^2$ or $\mathbb{T}^3$ is either Pisot or one of the automorphisms of the form $\pm T, \pm T^{-1}$ is such.

Our goal is, as above, to present a symbolic coding of T which, roughly speaking, reveals not just the structure of T itself but the natural arithmetic of the torus as well. Let us give a precise definition.

Definition 4.8. An *arithmetic coding* h of T is a map from $\widetilde{X}_\beta$ onto $\mathbb{T}^m$ that satisfies the following set of properties:

1. h is continuous and bounded-to-one;
2. $h\sigma_\beta = Th$;
3. $h(\varepsilon + \varepsilon') = h(\varepsilon) + h(\varepsilon')$ for any pair of sequences finite to the left.

Thus, unlike the classical symbolic dynamics, where one has to "encode" the action of T itself, our goal is to give a **simultaneous** encoding of T and the action of $\mathbb{T}^m$ on itself by addition. This makes the choice of h much more restricted; in fact, there are only a countable number of arithmetic codings, as the following lemma shows:

Lemma 4.9. [69, 65] *Any arithmetic coding of a Pisot automorphism of $\mathbb{T}^m$ is $h_{\mathbf{t}}$ given by (4.24), where $\mathbf{t} \in H(T)$.*

The issue is to find (if possible) an arithmetic coding of a Pisot automorphism which is one-to-one a.e. We will call it a *bijective arithmetic coding* or BAC. Before we formulate a necessary and sufficient condition for T to admit a BAC, we need some auxiliary definitions. Let first the characteristic equation for β be

$$\beta^m = k_1\beta^{m-1} + k_2\beta^{m-2} + \cdots + k_m, \quad k_m = \pm 1,$$

and T_β denote the toral automorphism given by the *companion matrix* M_β
for β, i.e.,

$$M_\beta = \begin{pmatrix} k_1 & k_2 & \ldots & k_{m-1} & k_m \\ 1 & 0 & \ldots & 0 & 0 \\ 0 & 1 & \ldots & 0 & 0 \\ \ldots & \ldots & \ldots & \ldots & \ldots \\ 0 & 0 & \ldots & 1 & 0 \end{pmatrix}.$$

We need one more (arithmetic) condition on β to discuss. Let $Fin(\beta)$ denote
the set of all $x \geq 0$ having finite greedy β-expansion. It is obvious that
$Fin(\beta) \subset \mathbb{Z}[\beta]_+ = \mathbb{Z}[\beta] \cap \mathbb{R}_+$. However, the inverse inclusion does not holds
for some Pisot units; those for which it does hold, are called *finitary*. For
examples see, *e.g.*, [66, §2]. The property of β to be finitary helps in many
Pisot-related issues, but our goal here is to present a more general result,
which is based on a more general property.

Definition 4.10. A Pisot unit β is called *weakly finitary* if for any $\delta > 0$
and any $x \in \mathbb{Z}[\beta]_+$ there exists $f \in Fin(\beta) \cap (0, \delta)$ such that $x + f \in Fin(\beta)$
as well.

This notion has appeared in different contexts and is related to different
problems – see [2, 41, 65]. The following conjecture (apparently, very difficult
to prove) is shared by most experts.

Conjecture 4.11. Any Pisot unit is weakly finitary.

To find out more about this property and about the algorithm how to
verify that a **given** Pisot unit is weakly finitary, see [2].

Return to our setting. We assume the following conditions to be satisfied:

1. T is algebraically conjugate to T_β, i.e., there exists a matrix $C \in GL(m, \mathbb{Z})$
   such that $CM = M_\beta C$ (notation: $T \sim T_\beta$).
2. A homoclinic point **t** is fundamental.
3. β is weakly finitary.

Theorem 4.12. *(1) If a Pisot automorphism T admits a BAC, then T is
algebraically conjugate to T_β.*
*(2) Assume that the three conditions above are satisfied. Then T admits an
arithmetic coding bijective a.e.*

Remark 4.13. Theorem 4.12 (2) for the case of finitary Pisot eigenvalue has
been proven by Le Borgne in his Ph. D. Thesis [47] (see also [62] for some
cases).

Remark 4.14. If $T \sim T_\beta$, then a fundamental homoclinic point always exists.
Thus, modulo Conjecture 4.11, the algebraic conjugacy to the companion
matrix is the necessary and sufficient condition for a Pisot automorphism to
admit a BAC.

For the rest of the subsection we assume β to be weakly finitary. Similarly to the Fibonacci case, the set $\widetilde{X}_\beta$ is an *almost group* in the following sense.

Proposition 4.15. [65] *Let $\sim$ denotes the identification on $\widetilde{X}_\beta$ defined as follows: $\varepsilon \sim \varepsilon'$ iff $h_{\mathbf{t}}(\varepsilon) = h_{\mathbf{t}}(\varepsilon')$, where $\mathbf{t}$ is fundamental. Then it touches only a set of measure zero, and $\widetilde{X}'_\beta = \widetilde{X}_\beta / \sim$ is a group isomorphic to $\mathbb{T}^m$.*

A natural question to ask is as follows: what is the number of preimages of a generic point if $\mathbf{t}$ is not fundamental? (for instance, if $T \not\sim T_\beta$) In [65] this question is answered completely.

We start with the case $T = T_\beta$ and show how this problem is related to Algebraic Number Theory. Let

$$\mathcal{P}_\beta = \{\xi \in \mathbb{R} : \|\xi\beta^n\| \to 0, \ n \to +\infty\}.$$

It is well-known that $\mathcal{P}_\beta \subset \mathbb{Q}(\beta)$ (see, *e.g.*, [13]). Let $\mathrm{Tr}(\xi)$ denote the trace of ξ, i.e. the sum of ξ and all its conjugates. It is shown in [66] that the set $\mathcal{P}_\beta$ is a commutative group under addition containing $\mathbb{Z}[\beta]$ and also that it can be characterized as follows:

$$\mathcal{P}_\beta = \{\xi \in \mathbb{Q}(\beta) : \mathrm{Tr}(a\xi) \in \mathbb{Z} \text{ for any } a \in \mathbb{Z}[\beta]\}.$$

Lemma 4.16. [66] *There exists a one-to-one correspondence between the homoclinic points and the elements of $\mathcal{P}_\beta$. Namely, $\mathbf{t} \in H(T)$ if and only if*

$$\mathbf{t} = (\xi, \xi\beta^{-1}, \dots, \xi\beta^{-m+1}) \bmod \mathbb{Z}^m$$

for some $\xi \in \mathcal{P}_\beta$.

Thus, any arithmetic coding of T_β is of the form

$$(4.25) \qquad h_\xi(\varepsilon) = \sum_{k \in \mathbb{Z}} \varepsilon_k T^{-k} \mathbf{t} = \lim_{N \to +\infty} \left(\sum_{k=-N}^{\infty} \varepsilon_k \beta^{-k} \right) \begin{pmatrix} \xi \\ \xi\beta^{-1} \\ \vdots \\ \xi\beta^{-m+1} \end{pmatrix},$$

where $\xi = \xi(\mathbf{t}) \in \mathcal{P}_\beta$. Let $N(\cdot)$ denote the norm in $\mathbb{Q}(\beta)$ and $D = D(\beta)$ stand for the discriminant of β.

Proposition 4.17. [65] *The map h_ξ is K-to-1 a.e., where $K = |DN(\xi)|$.*

Remark 4.18. Thus, h_ξ is a BAC if and only if $N(\xi) = \pm 1/D$, which is equivalent to the fact that ξ/ξ_0 is a unit in $\mathbb{Q}(\beta)$. If $\xi = 1$, then we come to the historically the first attempt to encode a Pisot automorphism undertaken by A. Bertrand-Mathis in [6]. Now we see that h_1 is in fact $|D|$-to-1 (provided β is weakly finitary).

Consider now the general case. We will be interested in the **minimal** number of preimages of $h_{\mathbf{t}}$ that one can attain for a given T. Let $M \in GL(m, \mathbb{Z})$ denote the matrix which determines T. To answer the above question, we are going to describe **all** integral square matrices that semiconjugate M and

M_β. Let for $\mathbf{n} \in \mathbb{Z}^m$ the matrix $B_M(\mathbf{n})$ be defined as follows (we write it column-wise):

$$B_M(\mathbf{n}) = (M\mathbf{n}, (M^2 - k_1 M)\mathbf{n}, (M^3 - k_1 M^2 - k_2 M)\mathbf{n}, \dots,$$
$$M^{m-1} - k_1 M^{m-2} - \dots - k_{m-2}M)\mathbf{n}, k_m\mathbf{n}).$$

Lemma 4.19. [65] *Any integral square matrix satisfying the relation*

$$BM_\beta = MB$$

is $B = B_M(\mathbf{n})$ for some $\mathbf{n} \in \mathbb{Z}^m$.

Let

$$f_M(\mathbf{n}) := \det B_M(\mathbf{n})$$

(an m-form of m variables).

Proposition 4.20. *Let $\mathbf{t} \in H(T)$. Then there exists $\mathbf{n} \in \mathbb{Z}^m$ such that*

$$\#\varphi_{\mathbf{t}}^{-1}(x) \equiv |f_M(\mathbf{n})|$$

for $\mathcal{L}_m$-a.e. point $x \in \mathbb{T}^m$.

Corollary 4.21. *The minimal number of preimages for an arithmetic coding of T equals the arithmetic minimum of the form f_M.*

Thus, T admits a BAC iff the Diophantine equation

(4.26) $f_M(\mathbf{n}) = \pm 1$

is solvable. In the case $m = 2$, (4.26) is especially natural: if $M = \begin{pmatrix} a & b \\ c & d \end{pmatrix}$, then it is

$$cx^2 - (a - d)xy - by^2 = \pm 1$$

and therefore, belongs to the class of well-known quadratic Diophantine equations. For more details about the two-dimensional case see [69, 78].

4.3. General case. The previous subsection has covered the case when one of the eigenvalues of the matrix of an automorphism is outside (inside) the unit disc and all the others are inside (resp. outside). The model explained above looks rather natural, explicit and canonical. What can be done in case when at least two eigenvalues are outside the unit disc and at least two – inside it? The main difficulty here lies in the fact that unlike the Pisot case, where the entropy is $\log \beta$ and the β-compactum is the obvious candidate for a coding space, in the general case this choice is not at all obvious.

There are several constructions that cover the general hyperbolic (or even ergodic) case, and each of them has its own advantages and disadvantages. Before we describe all of them in detail, let us try to understand what is that we actually want from an arithmetic coding. Obviously, there are no new properties of algebraic toral automorphisms that can be revealed this way – simply because they all are so well known.[9] What then? The unclear situation with this has, in my opinion, led to a certain impasse in this theory. No model seems to be canonical, and until we find an appropriate application, any theory will be a 𝔇ing-an-sich.

Let us also note that there are two main challenges any general arithmetic encoding has to meet:

1. it has to be bounded-to-one and, if possible, one-to-one a.e.;
2. the alphabet - it should be as simple as possible (preferably integers).

Which one is more important (if one cannot achieve both aims)? Here is one possible application that might measure the value of different constructions.

We have already mentioned the theorem on maps with holes proven by S. Bundfuss, T. Krueger and S. Troubetzkoy in [12] (see Section 2.2). Recall that this theorem claims that a if one cuts out a "typical" parallelepiped from $\mathbb{T}^m$ along the directions of the stable and unstable foliations with a vertex at 0 and the sides of length $a_1, \ldots, a_m$, then the corresponding exclusion map will be a subshift of finite type. This nice result however does not give any conditions on a_i for this subshift to be **nondegenerate**. At the same time, it is possible to show that if a_i are very small, then its entropy will be positive, and it is obvious that for "large" a_i the images of the hole will cover the whole torus, so it will be degenerate. Thus, if we make a natural assumption that similarly to the one-dimensional case, the entropy of the exclusion subshift is a continuous function of $(a_1, \ldots, a_m)$, then there exists a threshold similar to the Komornik-Loreti constant for the map T_β (see Section 2.2). In other words, we will have the surface Π in the space $(a_1, \ldots, a_m)$, underneath which the entropy of the exclusion subshift parameterized by (a_i) is positive, and it is zero above Π.

We do not know how the surface $\Pi = \Pi(T)$ looks like even in the case of the Fibonacci automorphism (where it in fact must be a curve). Nonetheless, we believe the exact simple formula for the symbolic encoding like (4.24) with an explicitly described symbolic compactum will probably help to reformulate the problem in terms symbolic sequences and to treat it in a way similar to the one described in [32]. In particular, let us ask the following question: is there any multidimensional analog of the Thue-Morse sequence (cf. Section 2)?

For this problem it is obvious that a bounded-to-one encoding map will be sufficient, as long as the set of digits and the map itself are explicit (because the entropy is preserved). We plan to return to this problem in our subsequent

---

[9]For instance, the construction of Markov partitions for the hyperbolic automorphisms of a torus (even for more general Axiom A diffeomorphisms [70, 11]) was revolutionary in the sense that although it was practically implicit, it nonetheless allowed to show "for free" (with the help of the famous Ornstein Theorem, of course) that they are all Bernoulli, which completely justified all the hard efforts and technicalities.

papers. Now it is time to present all the models known to date and to compare them.

4.3.1. *The construction of Kenyon and Vershik.* Historically the first general arithmetic symbolic model for the hyperbolic automorphisms was suggested by R. Kenyon and A. Vershik [42] (published in 1998 but written in 1995). This model is based on certain constructions that intensively use Algebraic Number Theory. We refer the reader to the textbooks, *e.g.*, [10, 29] for the relevant notions and results. We will keep the original notation of [42] and hope this will not make any confusion with the notation of the rest of the present paper.

Alphabet. Let $\lambda_1, \ldots, \lambda_m$ be the eigenvalues of M, where $|\lambda_i| > 1$ if and only if $i = 1, \ldots, k$. Let $K = \mathbb{Q}[x]/p(x)$, where $p(x)$ is the characteristic polynomial for M, and $\mathcal{O}$ denote the ring of integers in K. The ring K (and therefore, $\mathcal{O}$ as well) is naturally embedded into $\mathbb{R}^m$ via the standard coordinate-wise embeddings. The set $\mathcal{O}$ becomes a full-rank lattice in $\mathbb{R}^m$.

We denote by B the closed ball centered at $\mathbf{0}$ with the radius r defined as the smallest t such that its any translation has a nonempty intersection with $\mathcal{O}$. Finally, $D := \mathcal{O} \cap (B + xB)$, where multiplication by x symbolizes the multiplication by the companion matrix for M. The set D is shown to be finite, and this is precisely the set of digits for the model of [42].

Coding. Let σ denote the shift on $D^{\mathbb{N}}$ and (Σ_u, σ) denote the subshift defined as follows: assume D is endowed with some full order $\prec$; this creates the lexicographic ordering on $D^{\mathbb{N}}$. If $(\varepsilon_1, \ldots, \varepsilon_j)$ is a finite sequence, we say it is non-minimal if there exists a word $(\varepsilon'_1, \ldots, \varepsilon'_j) \prec (\varepsilon_1, \ldots, \varepsilon_j)$ such that $\sum_{i=1}^{j} \varepsilon_i x^{j-i} = \sum_{i=1}^{j} \varepsilon'_i x^{j-i}$. If a sequence is not non-minimal, we call it *minimal*.

The space Σ_u is thus the closed shift-invariant subset of $D^{\mathbb{N}}$ consisting of those sequences whose finite subsequences are all minimal. The coding space will be (Σ, σ), where Σ is the natural extension of Σ_u.

Proposition 4.22. [42] *The subshift (Σ, σ) is sofic.*

Now let us follow the authors of [42] in their construction of the encoding map. Define for $d = (d_0, d_1, \ldots) \in D^{\mathbb{N}}$,

$$S_i(d) = \sum_{j=0}^{\infty} \rho_i(d_j) \lambda_i^{-j}$$

if $i = 1, \ldots, k$ and

$$S_i(d) = \sum_{j=0}^{\infty} \rho_i(d_j) \lambda_i^{j+1}$$

otherwise. Furthermore, let $R_u : D^{\mathbb{N}} \to W_u$ (the unstable eigenspace of the companion matrix) act as follows: $R_u(d) = (S_1(d), \ldots, S_k(d))$ and similarly

$R_s(d) = (S_{k+1}(d), \ldots, S_m(d))$. Finally, let

$$R(\ldots, d_{-1}, d_0, d_1, \ldots) := R_u(d_0, d_1, \ldots) - R_s(d_{-1}, d_{-2}, \ldots)$$

be the map from Σ to $\mathbb{R}^m$, and π denote the natural projection from $\mathbb{R}^m$ to $\mathbb{T}^m$.

Theorem 4.23. [42] *The map πR is a factor map from (Σ, σ) to $(\mathbb{T}^m, T)$. It is bounded-to-one everywhere and constant-to-one a.e.*

Examples. The authors consider in detail the Fibonacci and similar quadratic cases as well as some cubic cases. Unfortunately, none of them uses the original set of digits D described above (in the Fibonacci case, for example, they take the conventional $D = \{0, 1\}$). Thus, it is difficult to assess the effectiveness of this model; nonetheless, the authors show how to deal with the "reasonable" choice of digits in specific cases. Note also that E. Hirsch proved in [38] that it is impossible for a general case to use this model with D containing just nonnegative integers.

4.3.2. *The construction of Le Borgne.* The model suggested by S. Le Borgne in his Ph. D. Thesis [47] (see also [48, 49]) is in fact a generalization (map-wise) of the Pisot model described above. As usual, we preserve the author's notation.

Alphabet. Let F_u, F_s denote the unstable and stable foliations for T and π_u stand for the projection from $\mathbb{R}^m$ onto F_u along F_s, and we define π_s in a similar way. Let M_u denote the restriction of M to F_u.

Assume $E \subset \pi_u(\mathbb{Z}^m)$ to be a finite set, and

(4.27) $$W_E = \left\{ \sum_{j=0}^{\infty} M_u^{-j} e_j \mid e_j \in E \right\}.$$

Lemma 4.24. [47, 48] *It is always possible to choose E in such a way that the interior of W_E is nonempty.*

Henceforward we assume E to be such, and $W = W_E$. Let now Y denote the set of all sequences that appear in the expansion (4.27) and let Z be its natural extension. Finally, denote by X the maximal transitive subshift of Z.

Lemma 4.25. [47, 48] *The shift (X, σ) is sofic and has a unique measure of maximal entropy (ν, say).*

The set X is the sought symbolic compactum. The "digits" thus are in fact vectors, and the actual choice is hidden in Lemma 4.24; see below how to convert vectors into (more conventional) integers in the case of M algebraically conjugate to its companion matrix.

Coding. Let X be as above, and $\varphi : X \to \mathbb{T}^m$ be defined by the formula

$$(4.28) \qquad \varphi(\varepsilon) = \sum_{j=1}^{\infty} \pi_u(M^{-j}\varepsilon_j) \bmod \mathbb{Z}^m - \sum_{j=-\infty}^{0} \pi_s(M^{-j}\varepsilon_j) \bmod \mathbb{Z}^m.$$

Theorem 4.26. [47, 48] *The map φ given by (4.28) is surjective, Hölder continuous and p-to-one a.e for a certain $p \in \mathbb{N}$. It semiconjugates the transitive sofic shift (X, ν, σ) and $(\mathbb{T}^m, \mathcal{L}_m, T)$.*

The main issue is to make it one-to-one a.e. (by an appropriate choice of E) as well as to make the alphabet more canonical. In the case when M is algebraically conjugate to its companion matrix, this has been partially done in the thesis [47]. Let $\Xi = \pi_u^{-1}(E) \subset \mathbb{Z}^m$.

Proposition 4.27. *Let $u_0 \in \mathbb{Z}^m$ be such that $\langle M^j u_0 \mid j = 0, 1 \ldots, m - 1 \rangle = \mathbb{Z}^m$. There exists $N \geq 1$ such that Ξ may be chosen in the form $\{-Nu_0, \ldots, Nu_0\}$.*

Thus, in a way, one might say that the digits are integers. The author also shows how (theoretically) the alphabet can be constructed but gives no non-Pisot examples.

4.3.3. *The construction of Schmidt.* The paper [62] by K. Schmidt appeared right after [69] and used the map defined by (4.24). More precisely, the case considered in [62] was more general than the hyperbolic toral automorphisms: the author deals with expansive group automorphisms of compact abelian groups. We will not be concerned with the general case though and will confine ourselves to the setting in question.

Theorem 4.28. [62] *For a given hyperbolic automorphism T of $\mathbb{T}^m$ whose matrix is algebraically conjugate to its companion matrix there exists a topologically mixing sofic subshift V of $l^\infty(\mathbb{Z}, \mathbb{Z})$ such that*

1. *$h_t(V) = \mathbb{T}^m$, where h_t is given by (4.24);*
2. *The restriction of h_t to V is one-to-one everywhere except the set of doubly transitive points of T.*

Remark 4.29. The proof given in [62] is non-constructive. As the author himself states, the above theorem only asserts the *existence* of a sofic shift V with the properties described above.

4.3.4. *Conclusions.* Let us compare all models by gathering all we know about them in the following table:

	Sidorov-Vershik	Kenyon-Vershik	Le Borgne	Schmidt
Automorphisms covered	2D and generalized Pisot (modulo arithmetic conjecture)	Hyperbolic	Hyperbolic	Hyperbolic cyclic
Is the subshift explicit?	Yes	No	No	No
Is the coding canonical?	Yes	Yes	No	No
"Digits"	Nonnegative integers	Algebraic numbers	Vectors	Integers
The encoding map is	K-to-one a.e. and one-to-one a.e. for the cyclic	K-to-one a.e.	K-to-one a.e.	Bounded-to-one and one-to-one a.e.

Remark 4.30. Here a *generalized Pisot automorphism* means that its stable (unstable) foliation is one-dimensional. They all can be arithmetically encoded using the construction for the Pisot automorphisms – see [65] for details. The expression "K-to-one a.e." implies that there exists $K \in \mathbb{N}$ such that almost every point of the torus has K preimages and "cyclic" means "the matrix is algebraically conjugate to its companion matrix".

It is also worth noting that an attempt to deal with the general case has been undertaken by the author in [65]. The idea is as follows: assume S is a hyperbolic automorphism of $\mathbb{T}^m$ and T is a generalized Pisot automorphism of $\mathbb{T}^m$ that commutes with S. Then $S = \sum_{j=0}^{m-1} c_j T^j$ with $c_j \in \mathbb{Q}$. Actually, the denominators of c_j are known to be bounded, and we assume that $c_j \in \mathbb{Z}$ for all j. Recall that the map h_t given by (4.24) semiconjugates (or conjugates if T is cyclic) the shift $(\widetilde{X}_\beta, \sigma_\beta)$ and $(\mathbb{T}^m, T)$. Hence the same map semiconjugates the linear combination of the powers of σ_β, namely, $\sum_{j=0}^{m-1} c_j \sigma_\beta^j$, and S (recall that by Proposition 4.15 the set $\widetilde{X}_\beta$ is an "almost group", whence any fixed finite integral combination of the powers of the shift is well defined a.e.). Thus, if we do not require that it must be necessarily a shift that encodes S, we are practically done.

The main issue is number-theoretic: the question is whether in a given algebraic field $K \simeq \mathbb{Q}(\lambda)$ there exists a Pisot unit β, and if it exists, whether it can be found in such a way that λ is an **integral** linear combination of powers of β. Of course, if, for instance, K is totally real (which leads to the *Cartan action*, i.e., the $\mathbb{Z}^{m-1}$-action by algebraic automorphisms), then it always contains a Pisot unit but the second property seems to be more difficult to prove – it requires some knowledge about the structure of the

Pisot units in an algebraic field, which is apparently missing in the classical Algebraic Number Theory.

Example 4.31. [65] Let $M = \begin{pmatrix} 4 & 0 & -3 & 1 \\ 1 & 0 & 0 & 0 \\ 0 & 1 & 0 & 0 \\ 0 & 0 & 1 & 0 \end{pmatrix}$. Note M is a companion matrix, and its spectrum is purely real. Now take the action generated by $M_1 = M, M_2 = M + E$ and $M_3 = M - E$. It is easy to check that they all belong to $GL(4, \mathbb{Z})$ and that this will yield a Cartan action on $\mathbb{T}^4$ as well as the fact that the dominant eigenvalue β of M is indeed weakly finitary. We leave the details to the reader. Therefore, the usual mapping h_t conjugates the action generated by $(\sigma_\beta, \sigma_\beta + id, \sigma_\beta - id)$ on the compactum X_β and the Cartan action generated by (T_1, T_2, T_3). Furthermore, T_3 has two eigenvalues strictly inside the unit disc and two strictly outside it. Perhaps, this is the first ever explicit bijective a.e. encoding of a non-generalized Pisot automorphism (though not by means of a shift).

Is this model any good application-wise? I am not sure; in particular, for the maps with holes the fact that instead of a shift we have this modified map, does not help a lot. However, it might be worth trying to apply it, when the Pisot case becomes clear. The author is grateful to A. Manning, M. Einsiedler and K. Schmidt for helpful discussions and number-theoretic insights regarding this question.

APPENDIX: ADIC TRANSFORMATIONS

In this appendix we are going to describe the class of maps on symbolic spaces which is in a way transversal to the shifts.[10] Let us give the precise definition.

Let $(D_k)_{k=1}^\infty$ be a sequence of finite sets, $r_k = \#D_k$, and let $\mathfrak{X}' := \prod_1^\infty D_k$ endowed with the weak topology. A closed subset $\mathfrak{X}$ of $\mathfrak{X}'$ is called a *Markov compactum* if there exists a sequence of 0-1 matrices $(M^{(k)})_{k=1}^\infty$, where $M^{(k)}$ is an $r_k \times r_{k+1}$ matrix, such that

$$\mathfrak{X} = \mathfrak{X}(\{M^{(k)}\}) = \{(x_1, x_2, \dots) \in \mathfrak{X}' : M^{(k)}_{x_k x_{k+1}} = 1\}.$$

In other words, $\mathfrak{X}$ is a (generally speaking, non-stationary) analog of topological Markov chain, and the $M^{(k)}$ are its incidence matrices. Assume that there is a full ordering $\prec_k$ on each set D_k. Then this sequence of orderings induces the partial lexicographic order on $\mathfrak{X}$ in a standard way: two distinct sequences x and x' are comparable iff there exists $n \geq 1$ such that $x_n \neq x'_n$ and $x_k = x'_k$ for all $k \geq n + 1$. Then $x \prec x'$ iff $x_n \prec_n x'_n$.

---

[10]Actually, this statement can be made precise whenever the symbolic space is stationary (= shift-invariant) – see, *e.g.*, [76] for some cases. As we will see, the adic transformations cover a much wider class of spaces.

Definition 4.32. The *adic transformation* S on $\mathfrak{X}$ is defined as a map that assigns to a sequence x its immediate successor in the sense of the lexicographic ordering defined above (if exists).

Remark 4.33. If $M^{(k)}_{x_k x_{k+1}} = 1$ for **all** pairs (x_k, x_{k+1}), then we have the *full odometer* or the **r**-*adic transformation* $S_\mathbf{r}$. If we identify D_k with $\{0, 1, \dots, r_k - 1\}$, then $S_\mathbf{r}$ acts on the the set of **r**-adic integers in the following way: every finite sequence x can be associated with a nonnegative integer as usual, i.e., $N = \sum_k x_k r_k$; then $S_\mathbf{r}(N) = N + 1$. The map $S_\mathbf{r}$ on the whole space is thus the profinite completion of the operation $N \mapsto N + 1$. It is well defined everywhere except the sequence $(r_1 - 1, r_2 - 1, \dots)$.

As is well known, this map has purely discrete spectrum for any $\mathbf{r} = (r_k)_1^\infty$. Thus, the adic transformation on an arbitrary Markov compactum may be regarded as a Poincaré map for the full odometer and some Markov subcompactum.

The adic transformation is known to be well defined a.e. for a large class of systems (see, *e.g.*, [53]). The importance of this model is confirmed by the following theorem proved by A. Vershik in the seminal paper [73], where the notion in question was first introduced (see also [72]).

Theorem 4.34. [73] *Each ergodic automorphism of the Lebesgue space is metrically isomorphic to some adic transformation.*

Remark 4.35. A "topological" version of this theorem have been obtained by M. Herman, I. Putnam and C. Skau [37]. Recently A. Dooley and T. Hamachi obtained a version of this theorem for the quasi-invariant measures of type III [18]. Note also that the adic realization in a special case were earlier considered by M. Pimsner and D. Voiculescu [59] in connection with approximations of certain operator algebras.

Note that although the proof of Theorem 4.34 is based on Rokhlin's Lemma and is thus to some extent constructive, there are very few explicit examples of "adic realization". The irrational rotations of the circle are among those rare exceptions (see Section 3); unfortunately, even for a general ergodic shift on the 2-torus the model seems to be hardly constructible.

One more fact worth noting is that if a Markov compactum is stationary (i.e., if $M^{(k)} \equiv M$ for any $k \geq 1$), then, as was shown by Livshits [52], the adic transformation on it is isomorphic to a *substitution* or, as it is more appropriate to call it, a *substitutional dynamical system*. The converse is also true, i.e., any primitive substitution has a stationary adic realization. For instance, the Fibonacci substitution $0 \rightarrow 01$, $1 \rightarrow 0$ is isomorphic to the adic transformation on X_G, while the Morse substitution $0 \rightarrow 01$, $1 \rightarrow 10$ leads to the the adic transformation on $\prod_1^\infty \{0, 1\}$ with the alternating ordering similar to the one described in Section 3.1 (see the second model). A good exposition of this theory can be found in [76].

REFERENCES

[1] R. L. Adler and B. Weiss, *Entropy, a complete metric invariant for automorphisms of the torus*, Proc. Nat. Acad. Sci. USA **57** (1967), 1573–1576.

[2] Sh. Akiyama, *On the boundary of self-affine tiling generated by Pisot numbers*, to appear in J. Math. Soc. Japan.

[3] J. C. Alexander and D. Zagier, *The entropy of a certain infinitely convolved Bernoulli measure* J. London Math. Soc. **44** (1991), 121–134.

[4] J.-P. Allouche and M. Cosnard, *The Komornik-Loreti constant is transcendental*, Amer. Math. Monthly **107** (2000), 448–449.

[5] A. Bertrand, *Développement en base de Pisot et répartition modulo 1*, C. R. Acad. Sci. Paris **385** (1977), 419–421.

[6] A. Bertrand-Mathis, *Développement en base θ, répartition modulo un de la suite $(x\theta^n)_{n\geq 0}$; langages codés et θ-shift*, Bull. Soc. Math. Fr. **114** (1986), 271–323.

[7] A. Bertrand-Mathis, *Le θ-shift sans peine*, unpublished manuscript.

[8] F. Blanchard, *β-expansions and symbolic dynamics*, Theoret. Comp. Sci. **65** (1989), 131–141.

[9] P. H. Borcherds and G. P. McCauley, *The digital tent map and the trapezoidal map*, Chaos Solitons Fractals **3** (1993), 451–466.

[10] Z. Borevich and I. Shafarevich, Number Theory, Acad. Press, NY, 1986.

[11] R. Bowen, *Markov partitions for Axiom A diffeomorphisms*, Amer. J. Math. **92** (1970), 725–747.

[12] S. Bundfuss, T. Krueger and S. Troubetzkoy, *Symbolic dynamics for Axiom A diffeomorphisms with holes*, preprint (2001) – see http://xxx.lanl.gov

[13] J. Cassels, An Introduction in Diophantine Approximation, Cambridge Univ. Press, 1957.

[14] N. Chernov and R. Markarian, *Ergodic properties of Anosov maps with rectangular holes*, Bol. Soc. Bras. Mat. **28** (1997), 271–314.

[15] N. Chernov and R. Markarian, *Anosov maps with rectangular holes*, Bol. Soc. Bras. Mat. **28** (1997), 315–342.

[16] N. Chernov, R. Markarian and S. Troubetzkoy, *Conditionally invariant measures for Anosov maps with small holes*, Erg. Th. Dyn. Sys. **18** (1998), 1049–1073.

[17] K. Dajani and C. Kraaikamp, *From greedy to lazy expansions and their driving dynamics*, preprint.

[18] A. H. Dooley and T. Hamachi, *Markov odometer actions not of product type*, to appear in Trans. Amer. Math. Soc.

[19] J.-M. Dumont, N. Sidorov and A. Thomas, *Number of representations related to a linear recurrent basis*, Acta Arith. **88** (1999), 371–394.

[20] Y. Dupain and V. Sos, *On the one-sided boundedness of discrepancy-function of the sequence $\{n\alpha\}$*, Acta Arith. **37** (1980), 363–374.

[21] M. Einsiedler, G. Everest and T. Ward, *Canonical heights and entropy in arithmetic dynamics*, preprint.

[22] M. Einsiedler and K. Schmidt, *Markov partitions and homoclinic points of algebraic $\mathbb{Z}^d$-actions* in: Dynamical Systems and Related Topics, Proc. Steklov Inst. Math., vol. 216, Interperiodica Publishing, Moscow, 1997, 259–279.

[23] P. Erdős, *On a family of symmetric Bernoulli convolutions*, Amer. J. Math. **61** (1939), 974–976.

[24] P. Erdős and I. Joó, *On the number of expansions $1 = \sum q^{-n_i}$*, Ann. Univ. Sci. Budapest Eötvös Sect. Math. **35** (1992), 129–132.

[25] P. Erdős, I. Joó and V. Komornik, *Characterization of the unique expansions $1 = \sum_{i=1}^{\infty} q^{-n_i}$ and related problems*, Bull. Soc. Math. Fr. **118** (1990), 377–390.

[26] P. Erdös, I. Joó and V. Komornik, *On the number of q-expansions*, Ann. Univ. Sci. Budapest Eötvös Sect. Math. **37** (1994), 109–118.

[27] L. Flatto and J. Lagarias, *The lap-counting function for linear mod 1 transformations I. Explicit formulas and renormalizability*, Erg. Theory Dynam. Systems **16** (1996), 451–491.

[28] A. Fraenkel, *Systems of numeration*, Amer. Math. Monthly **92** (1985), 105–114.

[29] A. Frölich and M. Taylor, Algebraic Number Theory, Cambridge Univ. Press, 1991.

[30] Ch. Frougny and J. Sakarovitch, *Automatic conversion from Fibonacci representation to representation in base φ, and a generalization*, Internat. J. Algebra Comput. **9** (1999), 351–384.

[31] A. Garsia, *Entropy and singularity of infinite convolutions*, Pac. J. Math. **13** (1963), 1159–1169.

[32] P. Glendinning and N. Sidorov, *Unique representations of real numbers in non-integer bases*, Math. Res. Letters **8** (2001), 535–543.

[33] P. Glendinning and N. Sidorov, *Uniqueness of β-expansions, a dynamical systems approach*, in preparation.

[34] P. Glendinning and N. Sidorov, *Unique representations of numbers as power series with coefficients in $\{0, 1, \ldots, N - 1\}$*, in preparation.

[35] P. Grabner, P. Liardet and R. Tichy, *Odometers and systems of numeration*, Acta Arith. **80** (1995), 103–123.

[36] B. M. Gurevich and Ya. G. Sinai, *Algebraic toral automorphisms and Markov chains*, supplement to the Russian translation of P. Billingsley, *Ergodic Theory and Information*, Izdat. "Mir", Moscow, 1969, 205–233 (in Russian).

[37] M. Herman, I. Putnam and C. Skau, *Ordered Bratteli diagrams, dimension groups and topological dynamics*, Int. J. Math. **3** (1992), 827–864.

[38] E. Hirsch, *On the construction of a symbolic realization of a hyperbolic automorphism of a torus*, Zap. Nauchn. Semin. POMI **223** (1995), 137–139 (in Russian); English transl. J. Math. Sci. **87** (1997), 4065–4066.

[39] F. Hofbauer, *β-shifts have unique maximal measure*, Monatsh. Math. **85** (1978), 189–198.

[40] F. Hofbauer, *Maximal measures for simple piecewise monotonic transformations*, Z. Wahrsch. Verw. Gebiete **52** (1980), 289–300.

[41] M. Hollander, Linear Numeration Systems, Finite Beta Expansions, and Discrete Spectrum of Substitution Dynamical Systems, Ph.D. Thesis, University of Washington, 1996.

[42] R. Kenyon and A. Vershik, *Arithmetic construction of sofic partitions and hyperbolic toral automorphisms*, Ergodic Theory Dynam. Systems **18** (1998), 357–372.

[43] A. I. Khinchin, Continued Fractions, New York Chelsea Pub. Co., 1963.

[44] V. Komornik and P. Loreti, *Unique developments in non-integer bases*, Amer. Math. Monthly **105** (1998), 636–639.

[45] S. Lalley, *Beta expansions with deleted digits for Pisot numbers beta*, Trans. Amer. Math. Soc. **349** (1997), 4355–4365.

[46] S. Lalley, *Random series in powers of algebraic integers: Hausdorff dimension of the limit distribution*, J. London Math. Soc. (2) **57** (1998), 629–654.

[47] S. Le Borgne, Dynamique Symbolique et Propriét'es Stochastiques des Automorphismes du Tore : Cas Hyperbolique et Quasi-hyperbolique, Thèse de Doctorat, 1997.

[48] S. Le Borgne, *Un codage sofique des automorphismes hyperboliques du tore*, C. R. Acad. Sci. Paris Sér. I Math. **323** (1996), 1123–1128.

[49] S. Le Borgne, *Un codage sofique des automorphismes hyperboliques du tore*, Bol. Soc. Bras. Mat. **30** (1999), 61–93.

[50] D. Lind, *The entropies of topological Markov shifts and a related class of algebraic integers*, Erg. Theory Dynam. Systems **4** (1984), 283–300.

[51] D. Lind and B. Marcus, An Introduction to Symbolic Dynamics and Coding, Cambridge University Press, 1995.

[52] A. Livshits, *Sufficient conditions for weak mixing of substitutions and of stationary adic transformations* (Russian), Mat. Zametki **44** (1988), 785–793, 862; English transl. in Math. Notes **44** (1988), 920–925.

[53] A. N. Livshits and A. M. Vershik, *Adic models of ergodic transformations, spectral theory and related topics*, Adv. in Soviet Math. **9** (1992), 185–204.

[54] R. Miles, Arithmetic Dynamical Systems, Ph. D. Thesis, University of East Anglia, 2000.

[55] E. Olivier, N. Sidorov and A. Thomas, *On the Gibbs properties of Bernoulli convolutions, and related problems in fractal geometry*, preprint.

[56] W. Parry, *On the β-expansions of real numbers*, Acta Math. Acad. Sci. Hung. **11** (1960), 401–416.

[57] W. Parry, *Representations for real numbers*, Acta Math. Acad. Sci. Hung. **15** (1964), 95–105.

[58] Y. Peres, W. Schlag and B. Solomyak, *Sixty years of Bernoulli convolutions*, Fractal geometry and stochastics, II (Greifswald/Koserow, 1998), 39–65, Progr. Probab., 46, Birkhauser, Basel, 2000.

[59] M. Pimsner and D. Voiculescu, *Embedding the irrational rotation C^*-algebra into an AF-algebra*, J. Operator Theory **4** (1980), 201–210.

[60] A. Rényi, *Representations for real numbers and their ergodic properties*, Acta Math. Acad. Sci. Hung. **8** (1957) 477–493.

[61] K. Schmidt, *On periodic expansions of Pisot numbers and Salem numbers*, Bull. London Math. Soc. **12** (1980), 269–278.

[62] K. Schmidt, *Algebraic codings of expansive group automorphisms and two-sided beta-shifts*, Monatsh. Math. **129** (2000), 37–61.

[63] N. Sidorov, *Laws of large numbers and the central limit theorem for sequences of coefficients of rotational expansions*, Zapiski Nauchn. Seminarov POMI **223** (1995), 313–322 (in Russian); English transl. J. Math. Sci. **87** (1997), 4180–4186.

[64] N. Sidorov, *Singularity and absolute continuity of measures associated with a rotation of the circle*, Zapiski Nauchn. Seminarov POMI **223** (1995), 323–336 (in Russian); English transl. J. Math. Sci. **87** (1997), 4187–4198.

[65] N. Sidorov, *Bijective and general arithmetic codings for Pisot toral automorphisms*, J. Dynam. Control Systems **7** (2001), 447–472.

[66] N. Sidorov, *An arithmetic group associated with a Pisot unit, and its symbolic-dynamical representation*, Acta Arith. **101** (2002), 199–213.

[67] N. Sidorov, *Almost every number has a continuum of β-expansions*, preprint, http://www.ma.umist.ac.uk/nikita

[68] N. Sidorov and A. Vershik, *Ergodic properties of Erdös measure, the entropy of the goldenshift, and related problems*, Monatsh. Math. **126** (1998), 215–261.

[69] N. Sidorov and A. Vershik, *Bijective arithmetic codings of hyperbolic automorphisms of the 2-torus, and binary quadratic forms*, J. Dynam. Control Systems **4** (1998), 365–399.

[70] Ya. Sinai, *Markov partitions and U-diffeomorphisms*, Funct. Anal. Appl. **2** (1968), 64–89.

[71] M. Smorodinsky, *β-automorphisms are Bernoulli shifts*, Acta Math. Acad. Sci. Hung. **24** (1973), 273–278.

[72] A. Vershik, *Uniform algebraic approximation of shift and multiplication operators*, Sov. Math. Doklady **24** (1981), 97–100.

[73] A. Vershik, *A theorem on Markov periodic approximation in ergodic theory*, Zapiski Nauchn. Sem. LOMI **115** (1982), 72–82 (in Russian); English transl. in J. Soviet Math. **28** (1985), 667–673.

[74] A. Vershik, *The fibadic expansions of real numbers and adic transformation*, Prep. Report Inst. Mittag-Leffler, no 4, 1991/1992, pp. 1–9, unpublished.

[75] A. Vershik, *Arithmetic isomorphism of the toral hyperbolic automorphisms and sofic systems*, Functional. Anal. Appl. **26** (1992), 170–173.

[76] A. Vershik, *Locally transversal symbolic dynamics*, St. Petersburg J. Math. **6** (1995), 526–540.

[77] A. Vershik and N. Sidorov, *Arithmetic expansions associated with a rotation of the circle and with continued fractions*, Algebra i Analiz **5** (1993), 97–115 (in Russian); English transl. St. Petersburg Math. J. **5** (1994), 1121–1136.

[78] A. Vershik and N. Sidorov, *Bijective codings of automorphisms of the torus, and binary quadratic forms*, Uspekhi Mat. Nauk **53** (1998), 231–233 (in Russian); English transl. Russian Math. Surveys **53** (1998), 1106–1107.

[79] P. Walters, An Introduction to Ergodic Theory, Springer, 1982.

[80] M. Waterman, *Cantor series for vectors*, Amer. Math. Monthly **82** (1975), 622–625.

[81] K. Zyczkowski and E. Bollt, *On the entropy devil's staircase in a family of gap-tent maps*, Physica D **132** (1999), 392–410.

DEPARTMENT OF MATHEMATICS, UMIST, P.O. BOX 88, MANCHESTER M60 1QD, UNITED KINGDOM

E-mail address: Nikita.A.Sidorov@umist.ac.uk

THE DEFECT OF FACTOR MAPS AND FINITE EQUIVALENCE OF DYNAMICAL SYSTEMS

KLAUS THOMSEN

Contents

1. Introduction

The defect, $D(\pi)$, of a factor map $\pi : (Y, \psi) \to (X, \varphi)$ between dynamical systems was defined in [Th1] under the assumption that X is a totally disconnected compact metric space, and it was calculated in a series of specific cases. The defect gives a numerical indication of how far π is from being injective; an indication which is particularly sensitive to the ambiguity of π over periodic orbits of φ. In [Th2] a variational principle for the defect was established:

$$D(\pi) = \sup_{\mu} \int_X \log \#\pi^{-1}(x) \, d\mu(x),$$

where we take the supremum over all φ-invariant Borel probability measures on X. In this paper the definition of the defect is extended to the general case, i.e. we drop the assumption that Y is totally disconnected and define the defect in a way which is analogous to - and generalizes - the case when Y is totally disconnected. We prove the variational principle in the general case and show how almost all the general properties of the defect follow from this principle. In particular, we obtain the subadditivity

$$D(\pi_2 \circ \pi_1) \leq D(\pi_2) + D(\pi_1)$$

for the composition of factor maps between invertible dynamical systems. This property fails dramatically for general non-invertible dynamical systems and this has effects for the notion of finite equivalence between dynamical systems which the defect suggests in a natural way. For this reason we consider a slight variation in the definition of the defect which makes no difference for factor maps between invertible dynamical systems, but results in a smaller number in general. We call this *the reduced defect* of the factor map, and denote it by $D_r(\pi)$. The reduced defect is sub-additive in general and relates directly to the defect via the notion of natural invertible extensions of dynamical systems. Recall that the (inverse of the) natural invertible extension is the invertible dynamical system which arises as the shift acting on the inverse limit of the given space with the given map as bonding maps. A factor map, π, between (non-invertible) dynamical systems induces in a natural way a factor map, $\widehat{\pi}$, between the natural invertible extensions, and it turns out that

$$D_r(\pi) = D(\widehat{\pi}).$$

This makes it possible to transfer general properties of the defect to properties of the reduced defect. For example, we use it to obtain a variational principle for the reduced defect of a factor map $\pi : (Y, \psi) \to (X, \varphi)$;

$$D_r(\pi) = \sup_\mu \int_X \log A_\pi(x) \, d\mu(x),$$

where we take the supremum over all φ-invariant Borel probability measures on X and A_π is a Borel function $A_\pi : X \to \mathbb{N} \cup \{\infty\}$ canonically associated to π. It turns out that, unlike the defect itself, the reduced defect is always the logarithm of a natural number. It follows therefore that to any factor map between arbitrary dynamical systems there is associated a natural number (or $+\infty$) which carries substantial information about how well the factor map relates the dynamical systems.

As indicated above the subadditivity of the defect (or the reduced defect), combined with the variational principle, leads to a natural generalization of the notion of finite equivalence first introduced by Parry, cf. [P], and used by him to give a classification of irreducible sofic shifts in terms of topological entropy. Namely we say that two invertible dynamical systems, $(X, \varphi), (Y, \psi)$, are *finitely equivalent* when there is an invertible dynamical system (Z, κ) and factor maps $\pi_1 : (Z, \kappa) \to (X, \varphi)$, $\pi_2 : (Z, \kappa) \to (Y, \psi)$ such that $D(\pi_1) + D(\pi_2) < \infty$. This equivalence relation generalizes the notion of finite equivalence of irreducible sofic shifts, and by using the reduced defect instead we obtain a further generalization to arbitrary dynamical systems. In the remaining part of the paper we make a first investigation of this equivalence relation. Specifically, we determine the finite equivalence classes of a series of dynamical systems which are all quite well-understood: Irreducible sofic shifts (two-sided as well as one-sided), hyperbolic toral automorphisms and expansive endomorphisms, periodic maps, homeomorphisms of the circle with an irrational rotation number, minimal rotations of tori and certain classes of unimodal maps of the interval. The most important invariant for finite equivalence is the topological entropy, and for some sufficiently restricted classes of dynamical systems (such as hyperbolic automorphisms of tori or expansive endomorphisms of manifolds) it is also the only invariant. But in general it is not. For example we show that two orientation preserving homeomorphisms of the circle $\mathbb{T}$ with irrational rotation numbers, $\alpha, \beta \in \mathbb{R}$, are finitely equivalent if and only if $1, \alpha$ and β are rationally dependent.

2. THE DEFECT OF FACTOR MAPS: DEFINITION AND THE VARIATIONAL PRINCIPLE

Let (X, φ) be a dynamical system[1] acting on a compact space X. Let Y be another compact space and $\pi : Y \to X$ be a continuous surjection. Let $\mathcal{U} = \{U_i : i \in I\}$ be a finite cover of Y. For any subset $F \subseteq Y$ we let $C(F, \mathcal{U})$ denote the minimal number of elements in $\mathcal{U}$ needed to cover F, i.e.

$$C(F, \mathcal{U}) = \min\{\#J : J \subseteq I, \bigcup_{j \in J} U_j \supseteq F\} .$$

For $x \in X$, set

$$a_k(x, \varphi, \mathcal{U}) = C(\pi^{-1}(x), \mathcal{U}) C(\pi^{-1}(\varphi(x)), \mathcal{U}) \cdots C(\pi^{-1}(\varphi^{k-1}(x)), \mathcal{U}) . \quad (2.1)$$

When $\mathcal{U}$ is a partition of Y, $a_k(x, \varphi, \mathcal{U}) = q_k(x, \pi(\mathcal{U}))$, where the last quantity was one of the fundamental entities used to define the defect in [Th1]. Set

$$a_k(\varphi, \mathcal{U}) = \sup_{x \in X} a_k(x, \varphi, \mathcal{U}),$$

so that $a_k(\varphi, \mathcal{U}) = q_k(\varphi, \pi(\mathcal{U}))$ when $\mathcal{U}$ is a partition, cf. [Th1]. Then

1) $a_{k+n}(\varphi, \mathcal{U}) \leq a_k(\varphi, \mathcal{U}) a_n(\varphi, \mathcal{U})$,
2) $a_k(\varphi, \mathcal{U}) \leq a_k(\varphi, \mathcal{V})$ when $\mathcal{V}$ is a refinement of $\mathcal{U}$.

It follows from 1) that we can consider the limit

$$A(\varphi, \mathcal{U}) = \lim_{n \to \infty} \frac{1}{n} \log a_n(\varphi, \mathcal{U}).$$

We define the defect of π to be

$$D(\pi) = \sup_{\mathcal{U}} A(\varphi, \mathcal{U})$$

where we take the supremum over all finite open covers $\mathcal{U}$ of Y. It follows from 2) that

$$D(\pi) = \lim_{k \to \infty} A(\varphi, \mathcal{U}_k),$$

for any sequence $\mathcal{U}_k, k \in \mathbb{N}$, of open covers of Y for which the maximal diameter of any set in the cover $\mathcal{U}_k$ goes to zero as k tends to infinity. In particular, $D(\pi)$ agrees with the defect defined in [Th1] when Y is totally disconnected.

---

[1] Here and in the following all dynamical systems are implicitly assumed to act on a compact *metric* space.

Theorem 2.1. *(The variational principle.) The function $x \mapsto \#\pi^{-1}(x)$ is Borel, and*

$$D(\pi) = \sup_{\mu} \int_X \log \#\pi^{-1}(x) \, d\mu(x),$$

where we take the supremum over all φ-invariant Borel probability measures on X. In fact, it suffices to take the supremum over all φ-ergodic Borel probability measures on X.

Proof. Let $\mathcal{U}_n = \{U_i^n : i = 1, 2, \cdots, I_n\}$ be a sequence of finite open covers of Y such that

$$\lim_{n \to \infty} \max\{\mathrm{diam}\, U_i^n : i = 1, 2, \cdots, I_n\} \;=\; 0.$$

For each n, set $\tilde{U}_1^n = U_1^n$ and

$$\tilde{U}_i^n = U_i^n \backslash (U_1^n \cup U_2^n \cup \cdots \cup U_{i-1}^n),$$

for $i \geq 2$. Note that $\tilde{U}_i^n$ is an F_σ-set so that $\pi(\tilde{U}_i^n)$ is an F_σ-set and hence also a Borel set for all n, i. For each n, define a Borel function $f_n : X \to \mathbb{N}$ by

$$f_n(x) = \#\{i : x \in \pi(\tilde{U}_i^n)\} = \sum_{i=1}^{I_n} 1_{\pi(\tilde{U}_i^n)}(x).$$

We claim that

$$\#\pi^{-1}(x) = \lim_{n \to \infty} f_n(x) \tag{2.2}$$

for all $x \in X$. To see this, let $k \in \mathbb{N}$ satisfy that $k \leq \#\pi^{-1}(x)$. There are then k distinct elements $y_1, y_2, \cdots, y_k \in \pi^{-1}(x)$. Let $N \in \mathbb{N}$ satisfy that $\max_i \mathrm{diam}\, \tilde{U}_i^n$ is smaller than any distance between y_k and y_l when $k \neq l$, for all $n \geq N$. Then $k \leq f_n(x) \leq \#\pi^{-1}(x)$ for all $n \geq N$, proving (2.2). In particular, we see that $x \mapsto \#\pi^{-1}(x)$ is a Borel function. By Fatou's lemma,

$$\int_X \log \#\pi^{-1}(x) \, d\mu(x) \leq \liminf_n \int_X \log f_n \, d\mu \tag{2.3}$$

for any φ-invariant Borel probability measure μ on X. Let $t < \int_X \log \#\pi^{-1}(x) \, d\mu$. It follows from (2.3) that we can choose n so large that

$$t < \int_X \log f_n \, d\mu.$$

Since each $\tilde{U}_i^n$ is an F_σ-set we can find sequences $F_i^1 \subseteq F_i^2 \subseteq F_i^3 \subseteq \cdots$ of closed sets such that $\tilde{U}_i^n = \bigcup_k F_i^k$. Then

$$\lim_{k \to \infty} \max\{\#\{i : x \in \pi(F_i^k)\}, 1\} = f_n(x),$$

non-decreasingly, for all $x \in X$, so Lebesgue's monotone convergence theorem gives us a k such that

$$t < \int_X \log(\max\{\#\{i : x \in \pi(F_i^k)\}, 1\}) \, d\mu(x). \tag{2.4}$$

When $\mathcal{F} = \{F_i : i \in I\}$ is a collection of subsets of X (not necessarily a cover), we set

$$q'_k(x, \mathcal{F}) = \prod_{j=0}^{k-1} \max\{1, \#\{i : \varphi^j(x) \in F_i\}\}$$

for $x \in X, k \in \mathbb{N}$. Set $q'_k(\varphi, \mathcal{F}) = \sup_{x \in X} q'_k(x, \mathcal{F})$. Then the limit $Q'(\varphi, \mathcal{F}) = \lim_{n \to \infty} \frac{1}{n} \log q'_n(\varphi, \mathcal{F})$ exists and is equal to $\inf_n \frac{1}{n} \log q'_n(\varphi, \mathcal{F})$. In particular, there is, for $\epsilon > 0$, an m such that

$$\frac{1}{m} \log q'_m(\varphi, \mathcal{G}) < Q'(\varphi, \mathcal{G}) + \epsilon$$

when $\mathcal{G} = \{\pi(F_i^k) : i = 1, 2, \cdots, I_n\}$. For each i we choose a decreasing sequence $U_i^1 \supseteq U_i^2 \supseteq \cdots$ of open sets in X such that $\overline{U_i^{l+1}} \subseteq U_i^l$ for all l and $\bigcap_l U_i^l = \pi(F_i^k)$. Since

$$\bigcap_l U_{i_1}^l \cap \varphi^{-1}(U_{i_2}^l) \cap \varphi^{-2}(U_{i_3}^l) \cap \cdots \cap \varphi^{-n+1}(U_{i_m}^l)$$

$$= \pi(F_{i_1}^k) \cap \varphi^{-1}(\pi(F_{i_2}^k)) \cap \varphi^{-2}(\pi(F_{i_3}^k)) \cap \cdots \cap \varphi^{-n+1}(\pi(F_{i_m}^k))$$

for each tuple $(i_1, i_2, \cdots, i_m) \in I_n^m$, there is an l so large that

$$q'_m(\varphi, \mathcal{U}^l) = q'_m(\varphi, \mathcal{G}),$$

when we set $\mathcal{U}^l = \{U_i^l : i = 1, 2, \cdots, I_n\}$. It follows that

$$Q'(\varphi, \mathcal{U}^l) \leq Q'(\varphi, \mathcal{G}) + \epsilon. \tag{2.5}$$

For each $d \in \mathbb{N}$, let

$$L_d = \bigcup_J \bigcap_{j \in J} U_j^l,$$

where we take the union over all subsets J of $\{1, 2, \cdots, I_n\}$ of cardinality $\leq d$. Take continuous functions $g_d : X \to [0, 1]$ with support in L_d. We claim that

$$Q'(\varphi, \mathcal{G}) \geq \int_X \log(\max\{\sum_{d=1}^{I_n} g_d(x), 1\}) \, d\mu(x) - 2\epsilon. \tag{2.6}$$

To prove (2.6) it suffices, since μ is the weak*-limit of a convex combination of φ-ergodic Borel probability measures, to consider the case when μ is φ-ergodic. In that case

$$\int_X \log(\max\{\sum_{d=1}^{I_n} g_d(x), 1\}) \, d\mu(x) = \lim_{m \to \infty} \frac{1}{m} \sum_{i=0}^{m-1} \log(\max\{\sum_{d=1}^{I_n} g_d(\varphi^i(z)), 1\})$$

for μ-almost all $z \in X$. There is therefore a point $z \in X$ such that

$$\frac{1}{m}\sum_{i=0}^{m-1} \log(\max\{\sum_{d=1}^{I_n} g_d(\varphi^i(z)), 1\}) \geq \int_X \log(\max\{\sum_{d=1}^{I_n} g_d(x), 1\}) \, d\mu(x) - \epsilon$$

for all sufficiently large m. Since $\max\{\sum_{d=1}^{I_n} g_d(\varphi^i(z)), 1\} \leq \max\{1, \#\{d \in I_n : \varphi^i(z) \in U_d^l\}\}$, we deduce that

$$\frac{1}{m} \log q'_m(\varphi, \mathcal{U}^l) \geq \int_X \log(\max\{\sum_{d=1}^{I_n} g_d(x), 1\}) \, d\mu(x) - \epsilon$$

for all large enough m. (2.6) follows from this and (2.5). Let $g_d^1 \leq g_d^2 \leq g_d^3 \leq \cdots$ be an increasing sequence of continuous functions such that $\lim_{n\to\infty} g_d^n = 1_{L_d}$. Then

$$\lim_{m\to\infty} \log(\max\{\sum_{d=1}^{I_n} g_d^m(x), 1\}) = \log(\max\{\#\{i : x \in U_i^l\}, 1\})$$

for all $x \in X$. It follows therefore from (2.4) and (2.6) that

$$t - 2\epsilon < \int_X \log(\max\{\#\{i : x \in \pi(F_i^k)\}, 1\}) \, d\mu(x) - 2\epsilon$$

$$\leq \int_X \log(\max\{\#\{i : x \in U_i^l\}, 1\}) \, d\mu(x) - 2\epsilon \leq Q'(\varphi, \mathcal{G}).$$

Since $F_i^k \cap F_j^k = \emptyset$ when $i \neq j$, we can easily construct an open cover $\mathcal{V} = \{V_i : i = 1, 2, \cdots, I_n + 1\}$ of Y such that $F_i^k \subseteq V_i \backslash \bigcup_{j\neq i} V_j$ for all $i = 1, 2, \cdots, I_n$. Then $C(\pi^{-1}(x), \mathcal{V}) \geq \#\{i : x \in \pi(F_i^k)\}$ for all $x \in X$ and hence

$$A(\varphi, \mathcal{V}) \geq Q'(\varphi, \mathcal{G}) > t - 2\epsilon.$$

It follows that $D(\pi) > t - 2\epsilon$, proving that

$$D(\pi) \geq \sup_\mu \int_X \log \#\pi^{-1}(x) \, d\mu(x).$$

To prove the reversed inequality, let $t \in \mathbb{R}$ be a number such that $t < D(\pi)$. There is an open cover $\mathcal{U}$ of Y such that $A(\varphi, \mathcal{U}) > t$. For each n choose a point $x_n \in X$ such that $\frac{1}{n} \log a_n(x_n, \varphi, \mathcal{U}) > \frac{1}{n} \log a_n(\varphi, \mathcal{U}) - \epsilon$. Then

$$\frac{1}{n} \log a_n(x_n, \varphi, \mathcal{U}) > A(\varphi, \mathcal{U}) - \epsilon > t - \epsilon \tag{2.7}$$

for all n. Let μ_n be the measure

$$\mu_n = \frac{1}{n}\sum_{i=0}^{n-1} \delta_{\varphi^i(x_n)}.$$

There is then a sequence $\{n_j\}$ in $\mathbb{N}$ and a φ-invariant Borel probability measure μ_∞ on X such that $\lim_{j\to\infty}\mu_{n_j} = \mu_\infty$. Let $\delta > 0$ be a Lebesgue number for $\mathcal{U}$. We need the following

Observation 2.2. There is an open cover $\mathcal{V} = \{U_i : i = 1, 2, \cdots, m\}$ of Y such that $\mathrm{diam}(U_i) < \delta$ for all i and such that

$$\mu_n(\pi(\overline{U_1})\backslash\pi(U_1)) = 0,$$

and

$$\mu_n(\pi(\overline{U_j\backslash(U_1 \cup U_2 \cup \cdots \cup U_{j-1})})\backslash\pi(U_j\backslash(U_1 \cup U_2 \cup \cdots \cup U_{j-1}))) = 0$$

for $j = 2, 3, \cdots, m$, and for all $n \in \mathbb{N} \cup \{\infty\}$.

To prove this observation we need some notation. For every set $B \subseteq Y$ and every $\epsilon > 0$, let $B^\epsilon = \{y \in Y : \mathrm{dist}(y, B) < \epsilon\}$. Let $\{S_i : i = 1, 2, \cdots, m\}$ be an open cover of Y such that $\mathrm{diam}\, S_i < \frac{\delta}{2}$ for all i. Since $\overline{S_1^t} \subseteq S_1^s$ when $t < s$, we see that the sets

$$\pi(\overline{S_1^t})\backslash\pi(S_1^t), \; t \in]0, \frac{\delta}{2}[,$$

are mutually disjoint. There must therefore be an $\epsilon_1 \in]0, \frac{\delta}{2}[$ such that $\mu_n(\pi(\overline{S_1^{\epsilon_1}})\backslash\pi(S_1^{\epsilon_1})) = 0$ for all $n \in \mathbb{N} \cup \{\infty\}$. Note that

$$\pi(\overline{S_2^t\backslash S_1^{\epsilon_1}}) \subseteq \pi(S_2^s\backslash S_1^{\epsilon_1})$$

when $t < s$. We can therefore repeat the above argument to find a $\epsilon_2 \in]0, \frac{\delta}{2}[$ such that

$$\mu_n(\pi(\overline{S_2^{\epsilon_2}\backslash S_1^{\epsilon_1}})\backslash\pi(S_2^{\epsilon_2}\backslash S_1^{\epsilon_1})) = 0$$

for all $n \in \mathbb{N}\cup\{\infty\}$. Continuing in this way we find $\epsilon_j \in]0, \frac{\delta}{2}[, j = 1, 2, \cdots, m$, such that

$$\mu_n(\pi(\overline{S_j^{\epsilon_j}\backslash(S_1^{\epsilon_1} \cup S_2^{\epsilon_2} \cup \cdots \cup S_{j-1}^{\epsilon_{j-1}})})\backslash\pi(S_j^{\epsilon_j}\backslash(S_1^{\epsilon_1} \cup S_2^{\epsilon_2} \cup \cdots \cup S_{j-1}^{\epsilon_{j-1}}))) = 0$$

for all $n \in \mathbb{N} \cup \{\infty\}$. Set $U_j = S_j^{\epsilon_j}, j = 1, 2, \cdots, m$. Then $\mathcal{V} = \{U_i : i = 1, 2, \cdots, m\}$ is an open cover with the desired property.

Set $V_1 = U_1, V_j = U_j\backslash(U_{j-1} \cup U_{j-2} \cup \cdots \cup U_1)$. Then $\mathcal{W} = \{V_i : i = 1, 2, \cdots, m\}$ is a partition of Y which refines $\mathcal{U}$. In particular, $a(x, \varphi, \mathcal{U}) \leq a(x, \varphi, \mathcal{W})$ for all x and hence

$$\frac{1}{n} \log \, a_n(x_n, \varphi, \mathcal{U}) \leq \frac{1}{n} \log \, a_n(x_n, \varphi, \mathcal{W}) \qquad (2.8)$$

for all $n \in \mathbb{N}$. Note that

$$\frac{1}{n} \log a_n(x_n, \varphi, \mathcal{W}) = \frac{1}{n} \sum_{j=0}^{n-1} \log \, C(\pi^{-1}(\varphi^j(x_n)), \mathcal{W})$$

$$= \frac{1}{n} \sum_{j=0}^{n-1} \log \#\{l : \varphi^j(x_n) \in \pi(V_l)\} = \int_X \log \#\{l : x \in \pi(V_l)\} \, d\mu_n(x)$$

for all n. The special properties of $\mathcal{V}$ ensure that

$$\int_X \log \#\{l : x \in \pi(V_l)\} \, d\mu_n(x) = \int_X \log \#\{l : x \in \pi(\overline{V_l})\} \, d\mu_n(x)$$

for all n. So when we combine with (2.7) and (2.8) we find that

$$\int_X \log \#\{l : x \in \pi(\overline{V_l})\} \, d\mu_n(x) > t - \epsilon \qquad (2.9)$$

for all n. Since $\pi(\overline{V_l})$ is closed for all l there is a decreasing sequence $h_1 \geq h_2 \geq h_3 \geq \cdots$ of continuous functions such that $\lim_{n\to\infty} h_n(x) = \log \#\{l : x \in \pi(\overline{V_l})\}$ for all $x \in X$. For each n we have that

$$t - \epsilon \leq \int_X h_k \, d\mu_n$$

for all k. By restricting to $\{n_j\}$ and taking the limit over j it follows that

$$t - \epsilon \leq \int_X h_k \, d\mu_\infty$$

for all k, and by taking the limit over k, that $t - \epsilon \leq \int_X \log \#\{l : x \in \pi(\overline{V_l})\} \, d\mu_\infty(x) = \int_X \log \#\{l : x \in \pi(V_l)\} \, d\mu_\infty(x)$. Since each V_l is an F_σ-set, an application of Lebesgue's theorem on monotone convergence gives us closed subsets $F_l \subseteq V_l$, $l = 1, 2, \cdots, m$, such that

$$t - 2\epsilon \leq \int_X \log \#\{l : x \in \pi(F_l)\} \, d\mu_\infty(x). \qquad (2.10)$$

Being the infimum of continuous functions, affine the map W given by

$$W(\nu) = \int_X \log \#\{l : x \in \pi(F_l)\} \, d\nu(x)$$

is upper semi-continuous on the compact convex set of φ-invariant Borel probability measures on X and hence it attains it maximum at an extreme point. It follows therefore from (2.10) that there is φ-ergodic measure ν on X such that $t - 2\epsilon \leq \int_X \log \#\{l : x \in \pi(F_l)\} \, d\nu(x)$. Since $\#\{l : x \in \pi(F_l)\} \leq \#\pi^{-1}(x)$ we conclude that

$$D(\pi) \leq \sup\{\int_X \log \#\pi^{-1}(x) \, d\mu(x) : \mu \text{ is a } \varphi\text{-ergodic Borel probability measure}\}.$$

$\square$

3. GENERAL PROPERTIES OF THE DEFECT

With the variational principle established we can now quickly generalize the general properties of the defect from [Th1].

Proposition 3.1.   1) $D(\pi) \leq \log d$ when $\#\pi^{-1}(x) \leq d$ for all $x \in X$, and $D(\pi) = \log d$ when $\#\pi^{-1}(x) = d$ for all $x \in X$.

2) $D(\pi) \geq \frac{1}{p} \sum_{i=0}^{p-1} \log \#\pi^{-1}(\varphi^i(x))$ when $x \in X$ is p-periodic.

3) When $A_i \subseteq X, i \in I$, is a family of closed φ-invariant subsets such that $\bigcup_{i \in I} A_i = X$, $D(\pi) = \sup_i D(\pi|_{\pi^{-1}(A_i)})$.
4) $D(\pi) = D(\pi|_{\pi^{-1}(\Omega)})$, where Ω is the set of non-wandering points for φ.
5) $D(\pi) = D(\pi|_{\pi^{-1}(\cap_{k=0}^{\infty} \varphi^k(X))})$.
6) $D(\pi_2 \circ \pi_1) \geq D(\pi_1)$ when $\pi_1 : (Y, \psi) \to (X, \varphi)$ and $\pi_2 : (X, \varphi) \to (Z, \lambda)$ are factor maps, and $D(\pi_2 \circ \pi_1) = D(\pi_1)$ when π_2 is a conjugacy.

Proof. All items follow straightforwardly from Theorem 2.1. $\square$

Theorem 3.2. *Let* $\pi_n : Y_n \to X_n$, $\varphi_n : X_n \to X_n, n \in \mathbb{N}$, *be continuous maps between compact metric spaces. Assume that each* π_n *is surjective. Consider the dynamical system* $(\prod_{n=1}^{\infty} X_n, \prod_{n=1}^{\infty} \varphi_n)$ *and the continuous surjection* $\prod_{n=1}^{\infty} \pi_n : \prod_{n=1}^{\infty} Y_n \to \prod_{n=1}^{\infty} X_n$. *Then*

$$D(\prod_{n=1}^{\infty} \pi_n) = \sum_{n=1}^{\infty} D(\pi_n).$$

Proof. Let μ be a $\prod_{n=1}^{\infty} \varphi_n$-invariant Borel probability measure on $\prod_{n=1}^{\infty} X_n$ and set $\pi_\infty = \prod_{n=1}^{\infty} \pi_n$. Let $\rho_k : \prod_{n=1}^{\infty} X_n \to X_k$ be the projection. Then

$$\int_{\prod_{n=1}^{\infty} X_n} \log \#\pi_\infty^{-1}(x) \, d\mu(x) = \int_{\prod_{n=1}^{\infty} X_n} \sum_{k=1}^{\infty} \log \#\pi_k^{-1}(\rho_k(x)) \, d\mu(x)$$

$$= \sum_{k=1}^{\infty} \int_{X_k} \log \#\pi_k^{-1}(z) \, d\mu \circ \rho_k^{-1}(z) \leq \sum_{k=1}^{\infty} D(\pi_k),$$

proving that $D(\pi_\infty) \leq \sum_{k=1}^{\infty} D(\pi_k)$. To obtain the reversed inequality, choose for each k a $t_k \in \mathbb{R}$ such that $t_k < D(\pi_k)$ and a φ_k-invariant Borel probability measure μ_k on X_k such that

$$\int_{X_k} \log \#\pi_k^{-1}(x) \, d\mu_k(x) \geq t_k.$$

Let μ be the product measure $\prod_{k=1}^{\infty} \mu_k$ on $\prod_{k=1}^{\infty} X_k$ and note that

$$\int_{\prod_{n=1}^{\infty} X_n} \log \#\pi_\infty^{-1}(x) \, d\mu(x) = \sum_{k=1}^{\infty} \int_{X_k} \log \#\pi_k^{-1}(z) \, d\mu_k(z) \geq \sum_{k=1}^{\infty} t_k.$$

It follows that $D(\pi_\infty) \geq \sum_{k=1}^{\infty} D(\pi_k)$. $\square$

Consider a commuting diagram

$$
\begin{array}{ccccccccc}
Y_1 & \xleftarrow{\lambda_1^Y} & Y_2 & \xleftarrow{\lambda_2^Y} & Y_3 & \xleftarrow{\lambda_3^Y} & Y_4 & \xleftarrow{\lambda_4^Y} & \cdots \\
\downarrow{\scriptstyle\pi_1} & & \downarrow{\scriptstyle\pi_2} & & \downarrow{\scriptstyle\pi_3} & & \downarrow{\scriptstyle\pi_4} & & \\
X_1 & \xleftarrow{\lambda_1^X} & X_2 & \xleftarrow{\lambda_2^X} & X_3 & \xleftarrow{\lambda_3^X} & X_4 & \xleftarrow{\lambda_4^X} & \cdots \\
\downarrow{\scriptstyle\varphi_1} & & \downarrow{\scriptstyle\varphi_2} & & \downarrow{\scriptstyle\varphi_3} & & \downarrow{\scriptstyle\varphi_4} & & \\
X_1 & \xleftarrow{\lambda_1^X} & X_2 & \xleftarrow{\lambda_2^X} & X_3 & \xleftarrow{\lambda_3^X} & X_4 & \xleftarrow{\lambda_4^X} & \cdots
\end{array}
$$

of compact metric spaces and continuous maps such that each π_n is surjective. The φ_n's give rise to a dynamical system $\varphi_\infty : \varprojlim(X_n, \lambda_n^X) \to \varprojlim(X_n, \lambda_n^X)$ and the π_n's to a continuous surjection $\pi_\infty : Y_\infty = \varprojlim(Y_n, \lambda_n^Y) \to \varprojlim(X_n, \lambda_n^X) = X_\infty$. In fact, when we set $X_{\infty,k} = \bigcap_{j>k} \lambda_k^X \circ \lambda_{k-1}^X \circ \cdots \circ \lambda_j^X(X_{j+1})$ and $Y_{\infty,k} = \bigcap_{j>k} \lambda_k^Y \circ \lambda_{k-1}^Y \circ \cdots \circ \lambda_j^Y(Y_{j+1})$, we have that

$$\pi_k(Y_{\infty,k}) = X_{\infty,k}$$

for all $k \in \mathbb{N}$.

Proposition 3.3.

$$D(\pi_\infty) \le \liminf_{k \to \infty} D(\pi_k|_{Y_{\infty,k}}).$$

Proof. Let $\rho_i : Y_\infty \to Y_i$ and $\rho_i' : X_\infty \to X_i$ be the projections to the i'th coordinate. By definition of the topology, for every finite open cover $\mathcal{U}$ of Y_∞ there is an $N \in \mathbb{N}$ such that for all $k \ge N$ there is a refinement of $\mathcal{U}$ of the form $\rho_k^{-1}(\mathcal{V})$ where $\mathcal{V}$ is a finite open cover of $\rho_k(Y_\infty) = Y_{\infty,k}$. Let $x \in X_{\infty,k}$. Since $\rho_k(\pi_\infty^{-1}(x)) \subseteq \pi_k^{-1}(\rho_k'(x)) \cap \rho_k(Y_\infty)$, we find that $C(\pi_\infty^{-1}(x), \rho_k^{-1}(\mathcal{V})) \le C(\pi_k^{-1}(\rho_k'(x)), \mathcal{V})$. It follows that $A(\varphi_\infty, \mathcal{U}) \le A(\varphi_\infty, \rho_k^{-1}(\mathcal{V})) \le A(\varphi_k|_{X_{\infty,k}}, \mathcal{V}) \le D(\pi_k|_{Y_{\infty,k}})$. Since this is true for all $k \ge N$ we find that $A(\varphi_\infty, \mathcal{U}) \le \inf_{k \ge N} D(\pi_k|_{Y_{\infty,k}}) \le \liminf_k D(\pi_k|_{Y_{\infty,k}})$. $\square$

In general equality fails in Proposition 3.3, except under appropriate additional assumptions, like condition (A) of Theorem 1.9 in [Th1].

Lemma 3.4. *Let $(Y, \psi), (X, \varphi)$ be dynamical systems on compact metric spaces X and Y. Let $\pi : (Y, \psi) \to (X, \varphi)$ be a factor map. Then*

$$\sup_{x \in X} h(\psi, \pi^{-1}(x)) \le D(\pi).$$

Proof. Choose $t \in \mathbb{R}$ such that $t < \sup_{x \in X} h(\psi, \pi^{-1}(x))$ and let $x \in X$ be a point such that $t < h(\psi, \pi^{-1}(x))$. There is then an $\epsilon > 0$ such that, in the notation of [B2],

$$t < \overline{s}_{\psi,d}(\epsilon, \pi^{-1}(x)).$$

Let $\mathcal{V} = \{V_i : i \in I\}$ be an open cover of Y by balls of radius $< \frac{\epsilon}{2}$. Consider an $n \in \mathbb{N}$ and let E_n be an (n, ϵ)-separated subset of $\pi^{-1}(x)$ of maximal cardinality. If

$$\prod_{j=0}^{n-1} C(\pi^{-1}(\varphi^j(x)), \mathcal{V})) < \#E_n,$$

there would have to be two different elements, s_1 and s_2, of E_n such that $\psi^j(s_1)$ and $\psi^j(s_2)$ were contained in the same element of $\mathcal{V}$ for all $j = 0, 1, 2, \cdots, n - 1$. These two elements would not be (n, ϵ)-separated, contradicting the choice of E_n. So we see that $\prod_{j=0}^{n-1} C(\pi^{-1}(\varphi^j(x)), \mathcal{V}) \geq \#E_n$. Thus $\log a_n(\varphi, \mathcal{V}) \geq \log a_n(x, \varphi, \mathcal{V}) \geq \log \#E_n = \log s_n(\epsilon, \pi^{-1}(x))$. Since n was arbitrary we conclude that

$$D(\pi) \geq A(\varphi, \mathcal{V}) \geq \overline{s}_{\psi, d}(\epsilon, \pi^{-1}(x)) > t.$$

$\square$

Lemma 3.5. *Let* $(Y, \psi), (X, \varphi)$ *be dynamical systems on compact metric spaces* X *and* Y. *Let* $\pi : (Y, \psi) \to (X, \varphi)$ *be a factor map. Then*

$$h(\psi) \leq h(\varphi) + D(\pi).$$

Proof. Combine Lemma 3.4 with Theorem 17 of [B2]. $\square$

Theorem 3.6. *Let* $(Y, \psi), (X, \varphi)$ *be dynamical systems on compact metric spaces* X *and* Y. *Let* $\pi : (Y, \psi) \to (X, \varphi)$ *be a factor map. Then*

$$D(\pi) < \infty \implies h(\psi) = h(\varphi).$$

Proof. With Lemma 3.5 substituting for Lemma 3.6 of [Th2] and Proposition 3.3 for Remark 1.10 of [Th1] the proof of Theorem 3.5 in [Th2] can be used ad verbatim. $\square$

Lemma 3.7. *Let* $\pi : (Y, \psi) \to (X, \varphi)$ *be a factor map. Assume that* ψ *is surjective and* φ *injective. Let* μ *be a* φ-*ergodic Borel probability measure. There is then a natural number* $k_\mu^\pi \in \mathbb{N}$ *or* $k_\mu^\pi = \infty$ *and a Borel set* $B \subseteq X$ *of full measure such that* $\#\pi^{-1}(x) = k_\mu^\pi$ *for all* $x \in B$.

Proof. Under the present assumptions on φ and ψ the identity $\pi^{-1}(\varphi(x)) = \psi(\pi^{-1}(x))$ is valid and shows that $\varphi^{-1}(\{x \in X : \#\pi^{-1}(x) \geq k\}) \subseteq \{x \in X : \#\pi^{-1}(x) \geq k\}$ for all $k \in \mathbb{N}$. By ergodicity, this implies that

$$\mu(\{x \in X : \#\pi^{-1}(x) \geq k\}) \in \{0, 1\}.$$

Let k_μ^π be the supremum of all $k \in \mathbb{N}$ for which $\mu(\{x \in X : \#\pi^{-1}(x) \geq k\}) = 1$ and set $B = \bigcap_{k < k_\mu^\pi} \{x \in X : \#\pi^{-1}(x) > k\}$. $\square$

Theorem 3.8. *Let* $\pi : (Y, \psi) \to (X, \varphi)$ *be a factor map. Assume that* ψ *is surjective and* φ *injective and that* $D(\pi) < \infty$. *There is then a* $k \in \mathbb{N}$ *and a* φ-*ergodic probability measure* μ *on* X *such that* $\#\pi^{-1}(x) = k$ *for* μ-*almost all* x, *and*

$$D(\pi) = \log k.$$

Proof. Combine Lemma 3.7 with Theorem 2.1.

$\square$

As in [Th2], the variational principle implies a certain subadditivity of the defect between invertible dynamical systems. This fact will be exploited below.

Theorem 3.9. *(Subadditivity of the defect.) Let* $(X, \varphi), (Y, \psi)$ *and* (Z, λ) *be invertible dynamical systems. Let* $\pi_1 : (X, \varphi) \rightarrow (Y, \psi)$ *and* $\pi_2 : (Y, \psi) \rightarrow (Z, \lambda)$ *be factor maps. It follows that*

$$D(\pi_2 \circ \pi_1) \leq D(\pi_1) + D(\pi_2).$$

Proof. The proof of Theorem 3.3 in [Th2] can be adopted ad verbatim. $\square$

It follows from Theorem 3.7 that the defect of a factor map $\pi : (Y, \psi) \rightarrow (X, \varphi)$ between invertible dynamical systems is infinite or the logarithm of a natural number. This number has other interpretations: When ν is a ψ-ergodic Borel probability measure and $\mathcal{B}$ and $\mathcal{B}_0$ denote the Borel σ-algebras of Y and X, respectively, the relative entropy $H_\nu(\mathcal{B}|\pi^{-1}(\mathcal{B}_0))$ equals $\int_X \log \#\pi^{-1}(x) \, d\nu \circ \pi^{-1}(x)$ by Lemma 1 of [NP] and hence

$$D(\pi) = \sup_\nu H_\nu(\mathcal{B}|\pi^{-1}(\mathcal{B}_0)), \tag{3.1}$$

where we take the supremum over all ψ-ergodic Borel probability measures.

Remark 3.10. Mike Boyle, Doris and Ulf Fiebig have introduced a notion of conditional entropy for factor maps between invertible dynamical systems, [BFF], and shown, among others, that this quantity is related to the defect. As pointed out in Proposition B.4 of [BFF], it follows from the variational principle for the defect that whenever the identity in their variational principle holds, see Theorem 6.6 of [BFF], finite defect implies that the conditional entropy is zero.

The defect is also related, via the crossed product construction (or group-measure space construction), to the Jones index for sub-factors, [J]. See [DT1] and [DT2].

4. THE REDUCED DEFECT

For general (non-invertible) dynamical systems the subadditivity of the defect, Theorem 3.9, fails as shown by example in Remark 3.4 of [Th2]. In fact, as the next example shows, there are factor maps

$$(X, \varphi) \xrightarrow{\pi_1} (Y, \psi) \xrightarrow{\pi_2} (Z, \sigma)$$

such that $D(\pi_1) = 0$, $D(\pi_2) < \infty$ and $D(\pi_2 \circ \pi_1) = \infty$.

Example 4.1. We elaborate first on the examples from Example 2.5 of [Th1] and Remark 3.4 of [Th2] as follows. Let $m \in \mathbb{N}$. Consider finite sets A and B such that $\#A \geq \#B \geq 1$. Define $\varphi : A \cup \{1, 2, \cdots, m\} \to A \cup \{1, 2, \cdots, m\}$ such that $\varphi(A \cup \{1\}) = \{2\}$, $\varphi(i) = i + 1$, modulo m, $2 \leq i \leq m$, and define $\psi : B \cup \{1, 2, \cdots, m\} \to B \cup \{1, 2, \cdots, m\}$, such that $\psi(B \cup \{1\}) = \{2\}$ and $\psi(i) = i + 1$, modulo m, $2 \leq i \leq m$. Then $(A \cup \{1, 2, \cdots, m\}, \varphi)$ and $(B \cup \{1, 2, \cdots, m\}, \psi)$ are both non-invertible dynamical systems. Define $\pi_1 : A \cup \{1, 2, \cdots, m\} \to B \cup \{1, 2, \cdots, m\}$ such that $\pi_1(A) = B$ and $\pi_1(i) = i$ for all i. Then π_1 is a factor map with defect $D(\pi_1) = 0$, cf. Remark 3.4 of [Th2]. Let $\sigma : \{1, 2, \cdots, m\} \to \{1, 2, \cdots, m\}$ be cyclic permutation and define $\pi_2 : B \cup \{1, 2, \cdots, m\} \to \{1, 2, \cdots, m\}$ such that $\pi_2(B \cup \{1\}) = \{1\}$ and $\pi_2(i) = i$ when $i \geq 2$. Then π_2 is a factor map and $D(\pi_2) = \frac{\log(\#B+1)}{m}$ while $D(\pi_2 \circ \pi_1) = \frac{\log(\#A+1)}{m}$, cf. Remark 3.4 of [Th2].

By using the freedom in this construction we can find sequences of factor maps,

$$(X_n, \varphi_n) \xrightarrow{\pi_1^n} (Y_n, \psi_n) \xrightarrow{\pi_2^n} (Z_n, \sigma_n),$$

such that $D(\pi_1^n) = 0$ for all n, $\sum_{n=1}^{\infty} D(\pi_2^n) < \infty$ and $\sum_{n=1}^{\infty} D(\pi_2^n \circ \pi_1^n) = \infty$. Then, by Theorem 3.2 above or Theorem 1.11 of [Th2],

$$\left(\prod_{n=1}^{\infty} X_n, \prod_{n=1}^{\infty} \varphi_n\right) \xrightarrow{\prod_{n=1}^{\infty} \pi_1^n} \left(\prod_{n=1}^{\infty} Y_n, \prod_{n=1}^{\infty} \psi_n\right) \xrightarrow{\prod_{n=1}^{\infty} \pi_2^n} \left(\prod_{n=1}^{\infty} Z_n, \prod_{n=1}^{\infty} \sigma_n\right)$$

are factor maps such that $D(\prod_{n=1}^{\infty} \pi_1^n) = 0$, $D(\prod_{n=1}^{\infty} \pi_2^n) < \infty$, while $D((\prod_{n=1}^{\infty} \pi_2^n) \circ (\prod_{n=1}^{\infty} \pi_1^n)) = D(\prod_{n=1}^{\infty} \pi_2^n \circ \pi_1^n) = \infty$.

To obtain a notion of defect which is also subadditive for factor maps between non-invertible dynamical systems, we take the definition of the defect up for a slight revision. Let (X, φ) and (Y, ψ) be dynamical systems acting on compact metric spaces, and let $\pi : (Y, \psi) \to (X, \varphi)$ be a factor map. Let $\mathcal{U} = \{U_i : i \in I\}$ be a finite open cover of Y. For each $k \in \mathbb{N}$, $H \subseteq X$, set

$b_k(H, \pi, \mathcal{U}) =$

$C(\pi^{-1}(H), \mathcal{U})C(\psi(\pi^{-1}(H)), \mathcal{U})C(\psi^2(\pi^{-1}(H)), \mathcal{U}) \cdots C(\psi^{k-1}(\pi^{-1}(H)), \mathcal{U}).$

For $x \in X$, set $b_k(x, \pi, \mathcal{U}) = b_k(\pi^{-1}(x), \pi, \mathcal{U})$, i.e.

$$b_k(x, \pi, \mathcal{U}) = C(\pi^{-1}(x), \mathcal{U})C(\psi(\pi^{-1}(x)), \mathcal{U}) \cdots C(\psi^{k-1}(\pi^{-1}(x)), \mathcal{U}).$$

(Compare with (2.1).) Note that $b_{k+n}(x, \pi, \mathcal{U}) \leq b_k(x, \pi, \mathcal{U})b_n(\varphi^k(x), \pi, \mathcal{U})$ and that $b_k(x, \pi, \mathcal{U}) \leq b_k(x, \pi, \mathcal{V})$ when $\mathcal{V}$ refines $\mathcal{U}$. We set

$$b_k(\psi, \mathcal{U}) = \sup_{x \in X} b_k(x, \pi, \mathcal{U}).$$

Then

1) $b_{k+n}(\psi, \mathcal{U}) \leq b_k(\psi, \mathcal{U})b_n(\psi, \mathcal{U}),$

2) $b_k(\psi, \mathcal{U}) \leq b_k(\psi, \mathcal{V})$ when $\mathcal{V}$ is a refinement of $\mathcal{U}$.

It follows from 1) that we can consider the limit

$$B(\psi, \mathcal{U}) = \lim_{n \to \infty} \frac{1}{n} \log b_n(\psi, \mathcal{U}).$$

We define *the reduced defect of π* to be

$$D_r(\pi) = \sup_{\mathcal{U}} B(\psi, \mathcal{U}),$$

where we take the supremum over all finite open covers $\mathcal{U}$ of Y. It follows from 2) that

$$D_r(\pi) = \lim_{k \to \infty} B(\psi, \mathcal{U}_k),$$

for any sequence $\mathcal{U}_k, k \in \mathbb{N}$, of open covers of Y for which the maximal diameter of any set in the cover $\mathcal{U}_k$ goes to zero as k tends to infinity.

Lemma 4.2. 1) $D_r(\pi) \leq D(\pi)$,
2) $D_r(\pi) = D(\pi)$ *when ψ is surjective and φ injective.*

Proof. In general $\psi^j(\pi^{-1}(x)) \subseteq \pi^{-1}(\varphi^j(x))$ for all j, x, and this gives 1). Under the assumptions of 2) we have that $\psi^j(\pi^{-1}(x)) = \pi^{-1}(\varphi^j(x))$. $\square$

Lemma 4.3. *Let $\pi : (Y, \psi) \to (X, \varphi)$ be a factor map. Then $\pi(\bigcap_{j \in \mathbb{N}} \psi^j(Y)) = \bigcap_{j \in \mathbb{N}} \varphi^j(X)$ and*

$$B(\psi, \mathcal{U}) = B(\psi|_{\bigcap_{j \in \mathbb{N}} \psi^j(Y)}, \mathcal{U})$$

for every finite open cover $\mathcal{U}$ of Y. In particular,

$$D_r(\pi) = D_r(\pi|_{\bigcap_{j \in \mathbb{N}} \psi^j(Y)}).$$

Proof. The first statement is straightforward to check. Let $\mathcal{U}$ be a finite open cover of Y. We claim that

$$B(\psi|_{\bigcap_j \psi^j(Y)}, \mathcal{U}) = \lim_{k \to \infty} B(\psi|_{\psi^k(Y)}, \mathcal{U}). \tag{4.1}$$

To prove (4.1), let $\epsilon > 0$ and choose $n \in \mathbb{N}$ such that

$$\frac{1}{n} \log b_n(\psi|_{\bigcap_j \psi^j(Y)}, \mathcal{U}) \leq B(\psi|_{\bigcap_j \psi^j(Y)}, \mathcal{U}) + \epsilon.$$

Recall that

$$b_n(\psi|_{\bigcap_j \psi^j(Y)}, \mathcal{U})$$

$$= \sup_{x \in \bigcap_j \varphi^j(X)} \prod_{l=0}^{n-1} C(\psi^l(\pi^{-1}(x) \cap \bigcap_j \psi^j(Y)), \mathcal{U}).$$

An easy compactness argument gives us for each $x \in \bigcap_j \varphi^j(X)$ a $\delta_x > 0$ and a $j_x \in \mathbb{N}$ such that

$$\prod_{l=0}^{n-1} C(\psi^l(\pi^{-1}(x) \cap \bigcap_j \psi^j(Y)), \mathcal{U})$$

$$= \prod_{l=0}^{n-1} C(\psi^l(\pi^{-1}(B_{\delta_x}(x)) \cap \psi^i(Y)), \mathcal{U})$$

for all $i \geq j_x$. The cover $B_{\delta_x}(x), x \in \bigcap_j \varphi^j(X)$, of $\bigcap_j \varphi^j(X)$ has a finite subcover $\{B_{\delta_i}(x_i) : i \in I\}$. By (the proof of) Lebesgue's covering lemma, (cf. Theorem 0.20 of [W]), there is $\delta > 0$ so small that every ball $B_\delta(x), x \in \bigcap_j \varphi^j(X)$, is contained in $B_{\delta_r}(x_r)$ for some $r \in I$. Hence

$$\sup_{x \in \bigcap_j \varphi^j(X)} \prod_{l=0}^{n-1} C(\psi^l(\pi^{-1}(B_\delta(x)) \cap \psi^i(Y)), \mathcal{U})$$

$$\leq \max_r \prod_{l=0}^{n-1} C(\psi^l(\pi^{-1}(B_{\delta_r}(x_r)) \cap \psi^i(Y)), \mathcal{U}) \qquad (4.2)$$

$$\leq b_n(\psi|_{\bigcap_j \psi^j(Y)}, \mathcal{U})$$

for all $i \geq \max_r j_{x_r}$. Choose $k \geq \max_r j_{x_r}$ so large that every element of $\varphi^k(X)$ has distance less than δ to an element of $\bigcap_j \varphi^j(X)$. Then (4.2) shows that

$$\sup_{x \in \varphi^k(X)} \prod_{l=0}^{n-1} C(\psi^l(\pi^{-1}(x) \cap \psi^k(Y)), \mathcal{U})$$

$$\leq b_n(\psi|_{\bigcap_j \psi^j(Y)}, \mathcal{U}).$$

Hence

$$B(\psi|_{\psi^k(Y)}, \mathcal{U}) \leq \frac{1}{n} \log b_n(\psi|_{\psi^k(Y)}, \mathcal{U})$$

$$\leq \frac{1}{n} \log b_n(\psi|_{\bigcap_j \psi^j(Y)}, \mathcal{U}) \leq B(\psi|_{\bigcap_j \psi^j(Y)}, \mathcal{U}) + \epsilon.$$

Since $B(\psi|_{\psi^k(Y)}, \mathcal{U})$ decreases with k and $B(\psi|_{\psi^k(Y)}, \mathcal{U}) \geq B(\psi|_{\bigcap_j \psi^j(Y)}, \mathcal{U})$ for all k, this proves (4.1). To complete the proof, we need only show that

$$B(\psi, \mathcal{U}) = B(\psi|_{\psi^m(Y)}, \mathcal{U}) \qquad (4.3)$$

for all $m \in \mathbb{N}$. To establish (4.3) observe that

$$\psi^{m+i}(\pi^{-1}(x)) \subseteq \psi^i(\pi^{-1}(\varphi^m(x)) \cap \psi^m(Y))$$

for all $x \in X$ and all $i \in \mathbb{N}$. It follows that there is an $L \in \mathbb{N}$ which only depends on $\mathcal{U}$ and m such that

$$b_k(x, \pi, \mathcal{U}) \leq L b_{k-m}(\varphi^m(x), \pi|_{\psi^m(Y)}, \mathcal{U})$$

for all $k > m$ and all $x \in X$. Hence

$$b_k(\psi|_{\psi^m(Y)}, \mathcal{U}) \leq b_k(\psi, \mathcal{U}) \leq Lb_{k-m}(\psi|_{\psi^m(Y)}, \mathcal{U})$$

all $k > m$, proving (4.3). $\qquad\square$

Proposition 4.4. *In the setting of Proposition 3.3,*

$$D_r(\pi_\infty) \leq \liminf_{k \to \infty} D_r(\pi_k|_{Y_{\infty,k}}).$$

Proof. Let $\rho_i : Y_\infty \to Y_i$ and $\rho'_i : X_\infty \to X_i$ be the projections to the i'th coordinate. By definition of the topology, for every finite open cover $\mathcal{U}$ of Y_∞ there is an $N \in \mathbb{N}$ such that for all $k \geq N$ there is a refinement of $\mathcal{U}$ of the form $\rho_k^{-1}(\mathcal{V})$ where $\mathcal{V} = \{V_i : i \in I\}$ is a finite open cover of Y_k. Since

$$\psi_\infty^j(\pi_\infty^{-1}(x)) \subseteq \bigcup_{j \in J} \rho_k^{-1}(V_j)$$

$$\Leftrightarrow \qquad\qquad\qquad\qquad\qquad (4.4)$$

$$\psi_k^j(\pi_k|_{Y_{\infty,k}}^{-1}(\rho'_k(x))) \subseteq \bigcup_{j \in J} V_j$$

for all $x \in X_\infty$, all $J \subseteq I$ and all $j \in \mathbb{N}$, we see that $b_k(\psi_\infty, \rho_k^{-1}(\mathcal{V})) = b_k(\psi_k|_{Y_{\infty,k}}, \mathcal{V})$ for all k. It follows that $B(\psi_\infty, \mathcal{U}) \leq D_r(\pi_k|_{Y_{\infty,k}})$ for all $k \geq N$. Hence $B(\psi_\infty, \mathcal{U}) \leq \liminf_k D_r(\pi_k|_{Y_{\infty,k}})$ and by taking the supremum over $\mathcal{U}$ we get that $D_r(\pi_\infty) \leq \liminf_k D_r(\pi_k|_{Y_{\infty,k}})$. $\qquad\square$

For $\epsilon > 0$, let $B_\epsilon(x)$ denote the closed ball of radius ϵ centered at x. Set

$$b_k^\epsilon(\psi, \mathcal{U}) = \sup_{x \in X} b_k(B_\epsilon(x), \mathcal{U}).$$

Clearly $b_k^\epsilon(\psi, \mathcal{U}) \geq b_k(\psi, \mathcal{U})$ for all $\epsilon > 0$.

Lemma 4.5. *For each k there is an $\epsilon > 0$ for which $b_k^\epsilon(\psi, \mathcal{U}) = b_k(\psi, \mathcal{U})$.*

Proof. For any given $x \in X$ there is clearly an $\epsilon_x > 0$ such that

$$b_k(B_{\epsilon_x}(x), \pi, \mathcal{U}) = b_k(x, \pi, \mathcal{U}).$$

Let $\{B_{\epsilon_{x_i}}(x_i) : i \in I\}$ be a finite subcover of $\{B_{\epsilon_x}(x) : x \in X\}$ and let $\epsilon > 0$ be a Lebesgue number for $\{B_{x_i}(x_i) : i \in I\}$. Let $z \in X$. Then $B_\epsilon(z) \subseteq B_{\epsilon_{x_i}}(x_i)$ for some i and hence

$$b_k(B_\epsilon(z), \pi, \mathcal{U}) \leq b_k(B_{\epsilon_{x_i}}(x_i), \pi, \mathcal{U}) = b_k(x_i, \pi, \mathcal{U}) \leq b_k(\psi, \mathcal{U}).$$

It follows that $b_k^\epsilon(\psi, \mathcal{U}) = b_k(\psi, \mathcal{U})$. $\qquad\square$

Consider a factor map $\pi : (Y, \psi) \to (X, \varphi)$. Then

$$
\begin{array}{ccccccccc}
Y & \xleftarrow{\psi} & Y & \xleftarrow{\psi} & Y & \xleftarrow{\psi} & Y & \xleftarrow{\psi} & \cdots \\
\downarrow{\scriptstyle \pi} & & \downarrow{\scriptstyle \pi} & & \downarrow{\scriptstyle \pi} & & \downarrow{\scriptstyle \pi} & & \\
X & \xleftarrow{\varphi} & X & \xleftarrow{\varphi} & X & \xleftarrow{\varphi} & X & \xleftarrow{\varphi} & \cdots
\end{array}
$$

commutes and we get therefore a factor map $\widehat{\pi} : (\widehat{Y}, \widehat{\psi}) \to (\widehat{X}, \widehat{\varphi})$, where $\widehat{Y} = \varprojlim(Y, \psi)$ and $\widehat{X} = \varprojlim(X, \varphi)$. $\widehat{\psi}$ is the homeomorphism of $\widehat{Y}$ given by $\widehat{\psi}((y_i)) = (\psi(y_i))$. $\widehat{\varphi}$ is defined similarly. The invertible dynamical system $(\widehat{Y}, \widehat{\psi})$ is *the natural invertible extension* of (Y, ψ), and we call the factor map $\widehat{\pi}$ *the natural extension of* π.

Theorem 4.6. *Let* $\pi : (Y, \psi) \to (X, \varphi)$ *be a factor map. Then*

$$D_r(\pi) = D(\widehat{\pi}),$$

where $\widehat{\pi} : (\widehat{Y}, \widehat{\psi}) \to (\widehat{X}, \widehat{\varphi})$ *is the natural extension of* π.

Proof. It follows from Proposition 4.4 that $D_r(\widehat{\pi}) \leq D_r(\pi)$, so by combining with 2) of Lemma 4.2, we have that $D(\widehat{\pi}) \leq D_r(\pi)$. To prove the reversed inequality observe first that $\widehat{\pi} : (\widehat{Y}, \widehat{\psi}) \to (\widehat{X}, \widehat{\varphi})$ is also induced by the commuting diagram

$$
\begin{array}{ccccccccc}
\bigcap_j \psi^j(Y) & \xleftarrow{\psi} & \bigcap_j \psi^j(Y) & \xleftarrow{\psi} & \bigcap_j \psi^j(Y) & \xleftarrow{\psi} & \bigcap_j \psi^j(Y) & \xleftarrow{\psi} & \cdots \\
\downarrow{\scriptstyle \pi} & & \downarrow{\scriptstyle \pi} & & \downarrow{\scriptstyle \pi} & & \downarrow{\scriptstyle \pi} & & \\
\bigcap_j \varphi^j(X) & \xleftarrow{\varphi} & \bigcap_j \varphi^j(X) & \xleftarrow{\varphi} & \bigcap_j \varphi^j(X) & \xleftarrow{\varphi} & \bigcap_j \varphi^j(X) & \xleftarrow{\varphi} & \cdots
\end{array}
$$

We can therefore, by Lemma 4.3, substitute $\pi : (Y, \psi) \to (X, \varphi)$ by $\pi : (\bigcap_j \psi^j(Y), \psi) \to (\bigcap_j \varphi^j(X), \varphi)$, and hence assume that φ and ψ are both surjective.

Let $t < D_r(\pi)$ and choose a finite open cover $\mathcal{V}$ of Y such that $B(\psi, \mathcal{V}) > t$. Let $\delta > 0$ and choose $m \in \mathbb{N}$ so large that $\frac{1}{m} \log b_m(\widehat{\psi}, \rho_1^{-1}(\mathcal{V})) \leq B(\widehat{\psi}, \rho_1^{-1}(\mathcal{V})) + \delta$. (Remember that $\rho_k : \widehat{Y} \to Y$ and $\rho_k' : \widehat{X} \to X$ are the projections to the k'th coordinate.) By Lemma 4.5 there is an $\epsilon > 0$ so small that $b_m(\widehat{\psi}, \rho_1^{-1}(\mathcal{V})) = b_m^\epsilon(\widehat{\psi}, \rho_1^{-1}(\mathcal{V}))$. There is a $k \in \mathbb{N}$ so large that the diameter of every subset of $\widehat{X}$ of the form $\rho_k^{-1}(z)$ for some $z \in X$ has diameter less than ϵ. Since $\rho_k(\pi_\infty^{-1}(\rho_k'^{-1}(z))) = \pi^{-1}(z)$ (because of the surjectivity of ψ), we have that $\rho_k(\psi_\infty^j(\pi_\infty^{-1}(\rho_k'^{-1}(z)))) = \psi^j(\pi^{-1}(z))$ for all $j \in \mathbb{N}$ and all $z \in X$. Since $\rho_k'^{-1}(z) \subseteq B_\epsilon(z_\infty)$ for some $z_\infty \in \widehat{X}$ we can combine this with (4.4) to see that

$$b_m(\psi, \mathcal{U}) \leq b_m^\epsilon(\widehat{\psi}, \rho_k^{-1}(\mathcal{U}))$$

for every finite open cover $\mathcal{U}$ of Y. If we set $\mathcal{U} = \psi^{-k+1}(\mathcal{V})$, we have that $\rho_k^{-1}(\mathcal{U}) = \rho_1^{-1}(\mathcal{V})$ and hence that

$$
\begin{aligned}
\frac{1}{m} \log b_m(\psi, \psi^{-k+1}(\mathcal{V})) &\leq \frac{1}{m} \log b_m^\epsilon(\widehat{\psi}, \rho_1^{-1}(\mathcal{V})) \\
&= \frac{1}{m} \log b_m(\widehat{\psi}, \rho_1^{-1}(\mathcal{V})) \leq B(\widehat{\psi}, \rho_1^{-1}(\mathcal{V})) + \delta.
\end{aligned}
\tag{4.5}
$$

But it is easy to see, directly from the definition, that $B(\psi, \psi^{-k+1}(\mathcal{V})) = B(\psi, \mathcal{V})$, so we see from (4.5) that $B(\psi, \mathcal{V}) \leq B(\widehat{\psi}, \rho_1^{-1}(\mathcal{V})) + \delta$. It follows that $D_r(\pi) \leq D_r(\widehat{\pi})$. Since $D_r(\widehat{\pi}) = D(\widehat{\pi})$ by 2) of Lemma 4.2, the proof is complete. $\qquad\square$

Corollary 4.7. *(Subadditivity of the reduced defect.)* Let $\pi_1 : (Y, \psi) \to (X, \varphi)$ and $\pi_2 : (X, \varphi) \to (Z, \lambda)$ be factor maps. Then

$$
D_r(\pi_2 \circ \pi_1) \leq D_r(\pi_1) + D_r(\pi_2).
$$

Proof. Combine Theorem 4.6 with Theorem 3.9. $\qquad\square$

Corollary 4.8. Let $\pi : (Y, \psi) \to (X, \varphi)$ be a factor map. Then

$$
D_r(\pi) < \infty \;\Rightarrow\; h(\psi) = h(\varphi).
$$

Proof. Combine Proposition 5.2 of [B1] with Theorem 4.6 and Theorem 3.6. $\qquad\square$

There is also a variational principle for the reduced defect. It is, however, somewhat more complicated. Let $\pi : (Y, \psi) \to (X, \varphi)$ be a factor map. For $\epsilon > 0$ and any closed subset $F \subseteq Y$ we let $\#^\epsilon F$ denote the largest number of elements in an ϵ-separated subset of F, i.e. if d denotes the metric of Y,

$$
\#^\epsilon F = \max\{n \in \mathbb{N} : \exists \{x_1, x_2, \cdots, x_n\} \subseteq F \text{ such that } d(x_i, x_j) \geq \epsilon,\ i \neq j\}.
$$

For $F = \emptyset$ we set $\#^\epsilon F = 0$.

Lemma 4.9. For every $\epsilon > 0$ and every $k \in \mathbb{N}$, the function

$$
x \mapsto \#^\epsilon \psi^k \Big(\pi^{-1}(x) \cap \bigcap_j \psi^j(Y) \Big)
$$

is Borel.

Proof. For $F \subseteq Y$ closed, set

$$
\#_\epsilon F = \max\{n \in \mathbb{N} : \exists \{x_1, x_2, \cdots, x_n\} \subseteq F \text{ such that } d(x_i, x_j) > \epsilon,\ i \neq j\}.
$$

Then

$$
\#^\epsilon \psi^k \Big(\pi^{-1}(x) \cap \bigcap_j \psi^j(Y) \Big) = \inf_q \#_q \psi^k \Big(\pi^{-1}(x) \cap \bigcap_j \psi^j(Y) \Big)
$$

for all $x \in X$, when we take the infimum over all rational $q < \epsilon$. It suffices therefore to show that

$$x \mapsto \#_\epsilon \psi^k \big(\pi^{-1}(x) \cap \bigcap_j \psi^j(Y) \big)$$

is Borel. Let $\mathcal{F} = \{F_1, F_2, \cdots, F_M\}$ be closed subsets of Y. Set

$$B(\mathcal{F})(x) = \#\{j : \psi^k \big(\pi^{-1}(x) \cap \bigcap_l \psi^l(Y) \big) \bigcap F_j \neq \emptyset\},$$

for $x \in X$. Since

$$\#\{j : \psi^k \big(\pi^{-1}(x) \cap \bigcap_l \psi^l(Y) \big) \bigcap F_j \neq \emptyset\} = \sum_{i=1}^{M} 1_{\pi(\psi^{-k}(F_i) \cap \bigcap_l \psi^l(Y))}(x),$$

we see that $B(\mathcal{F})$ is Borel. For each $t > 0$ and $y \in Y$, let $B_t(y)$ denote the open ball of radius t centered at y. Let $\{y_i\}$ be a dense sequence in Y. Then

$$\#_\epsilon \psi^k \big(\pi^{-1}(x) \cap \bigcap_l \psi^l(Y) \big) = \sup_{\mathcal{F}} B(\mathcal{F})(x),$$

where we take the supremum over all collections $\mathcal{F}$ of the form

$$\mathcal{F} = \{\overline{B_q(y_{i_j})} : j = 1, 2, \cdots, K\},$$

where $q > 0$ is rational and $d(y_{i_k}, y_{i_j}) > \epsilon + 2q$ for $k \neq j$. Since this is a countable collection of functions, we are done. $\qquad \square$

Since $\#^\epsilon \psi^k \big(\pi^{-1}(x) \cap \bigcap_j \psi^j(Y) \big)$ is decreasing in ϵ we can define

$$A_\pi(x) = \lim_{\epsilon \to 0} \big[\limsup_k \#^\epsilon \psi^k \big(\pi^{-1}(x) \cap \bigcap_j \psi^j(Y) \big) \big],$$

which is a Borel function $A_\pi : X \to \mathbb{N} \cup \{\infty\}$ by Lemma 4.9. Observe that $A_\pi(x) = 0$ for $x \notin \bigcap_j \varphi^j(X)$, and that A_π does not depend on the metric.

Theorem 4.10. *(The variational principle for the reduced defect.) Let $\pi : (Y, \psi) \to (X, \varphi)$ be a factor map. Then*

$$D_r(\pi) = \sup_\mu \int_X \log A_\pi(x) \, d\mu(x), \qquad (4.6)$$

where we take the supremum over all φ-invariant Borel probability measures on X and use the convention $\log 0 = 0$. In fact, it suffices to take the supremum over all φ-ergodic Borel probability measures on X.

Proof. We will prove (4.6) by combining Theorem 4.6 with the variational principle for the defect, Theorem 2.1. Let $\rho_i : \widehat{Y} \to Y$ and $\rho_i' : \widehat{X} \to X$ be the projections to the i'th coordinate. Then

$$\#\widehat{\pi}^{-1}(z) = \liminf_{l \to \infty} \lim_{\epsilon \to 0} \lim_{k \to \infty} \#^\epsilon \psi^{k-l} [\pi^{-1}(\rho_k'(z)) \cap \bigcap_j \psi^j(Y)] \qquad (4.7)$$

for all $z \in \widehat{X}$. To see this, observe first that a compactness argument shows that

$$\lim_{k \to \infty} \#^\epsilon \psi^{k-l} [\pi^{-1}(\rho_k'(z)) \cap \bigcap_j \psi^j(Y)] = \#^\epsilon \rho_l(\widehat{\pi}^{-1}(z)). \qquad (4.8)$$

Assume then that $\#\widehat{\pi}^{-1}(z) \geq N$ for some $N \in \mathbb{N}$, and let $y_1, y_2, \cdots, y_N$ be different elements of $\widehat{\pi}^{-1}(z)$. There is then a $K \in \mathbb{N}$ so large that the elements $\rho_i(y_1), \rho_i(y_2), \cdots, \rho_i(y_N)$ are different for all $i \geq K$. For such an i, the elements $\rho_i(y_1), \rho_i(y_2), \cdots, \rho_i(y_N)$ are δ_i-separated for some $\delta_i > 0$ and since $\rho_i(y_j) = \psi^{d-i}(\rho_d(y_j))$ for all $d > i$, we see that $\#^{\delta_i} \psi^{d-i}[\pi^{-1}(\rho_d'(z)) \cap \bigcap_j \psi^j(Y)] \geq N$ for all $d > i$. Hence $\lim_{k \to \infty} \#^{\delta_i} \psi^{k-i}[\pi^{-1}(\rho_k'(z)) \cap \bigcap_j \psi^j(Y))] \geq N$ and consequently

$$\lim_{\epsilon \to 0} \lim_{k \to \infty} \#^\epsilon \psi^{k-i} [\pi^{-1}(\rho_k'(z)) \cap \bigcap_j \psi^j(Y))] \geq N$$

for all $i \geq K$. It follows that the righthand side of (4.7) dominates the lefthand side. To prove the reversed inequality, consider an $\epsilon > 0$ and some $l \in \mathbb{N}$. If $\lim_{k \to \infty} \#^\epsilon \psi^{k-l}[\pi^{-1}(\rho_k'(z)) \cap \bigcap_j \psi^j(Y)] \geq N$ for some $N \in \mathbb{N}$, there is $K \in \mathbb{N}$ so large that $\#^\epsilon \psi^{k-l}[\pi^{-1}(\rho_k'(z)) \cap \bigcap_j \psi^j(Y)] \geq N$ for all $k \geq K$. We can therefore find, for any $k \geq K$, elements $y_1^k, y_2^k, \cdots, y_N^k \in \prod_{j=0}^\infty Y$ such that $\psi(\rho_j(y_i^k)) = \rho_{j-1}(y_i^k)$, $\pi(\rho_j(y_i^k)) = \rho_j'(z)$ for all i and all $j \leq k$, and such that the set

$$\{\rho_l(y_1^k), \rho_l(y_2^k), \cdots, \rho_l(y_N^k)\}$$

is ϵ-separated for all k. For some sequence $\{k_j\}$ in $\mathbb{N}$ the limits $\lim_{j \to \infty} y_i^{k_j} = y_i$ will all exist in $\prod_{j=0}^\infty Y$ and by construction they will lie not only in $\widehat{Y} = \varprojlim(Y, \psi)$, but actually in $\widehat{\pi}^{-1}(z)$. Furthermore, by construction the set $\{\rho_l(y_1), \rho_l(y_2), \cdots, \rho_l(y_N)\}$ will be ϵ-separated, so it follows that $\#\widehat{\pi}^{-1}(z) \geq N$. This proves (4.7).

There is a bijective correspondance between the $\widehat{\varphi}$-invariant Borel probability measures on $\widehat{X}$ and the φ-invariant Borel probability measures on X such that a φ-invariant Borel probability measure ν on X corresponds to the $\widehat{\varphi}$-invariant Borel probability measure μ on $\widehat{X}$ with the property that $\mu \circ \rho_k'^{-1} = \nu$ for all $k \in \mathbb{N}$. For such a μ Lebesgue's theorem on monotone convergence and Fatou's lemma combined with (4.7) gives us that

$$\int_{\widehat{X}} \log \#\widehat{\pi}^{-1}(z) \, d\mu(z)$$
$$\leq \liminf_{l \to \infty} \lim_{\epsilon \to 0} \int_{\widehat{X}} \log \left[\lim_{k \to \infty} \#^\epsilon \psi^{k-l} (\pi^{-1}(\rho_k'(z)) \cap \bigcap_j \psi^j(Y)) \right] d\mu(z). \qquad (4.9)$$

By compactness of Y the functions $z \mapsto \#^\epsilon \psi^{k-l}[\pi^{-1}(\rho_k'(z)) \cap \bigcap_j \psi^j(Y)]$ are uniformly bounded, so by Lebesgue's theorem on dominated convergence we have that

$$\int_{\widehat{X}} \log\Big[\lim_{k\to\infty} \#^\epsilon \psi^{k-l} \pi^{-1}(\rho'_k(z)) \cap \bigcap_j \psi^j(Y)\Big]\, d\mu(z)$$

$$= \lim_{k\to\infty} \int_{\widehat{X}} \log\Big[\#^\epsilon \psi^{k-l}\big(\pi^{-1}(\rho'_k(z)) \cap \bigcap_j \psi^j(Y)\big)\Big]\, d\mu(z). \tag{4.10}$$

Since

$$\int_{\widehat{X}} \log[\#^\epsilon \psi^{k-l}\big(\pi^{-1}(\rho'_k(z)) \cap \bigcap_j \psi^j(Y)\big)]\, d\mu(z)$$

$$= \int_X \log[\#^\epsilon \psi^{k-l}\big(\pi^{-1}(x) \cap \bigcap_j \psi^j(Y)\big)]\, d\nu(x),$$

where ν is the φ-invariant Borel probability measure on X corresponding to μ, we find that

$$\int_{\widehat{X}} \log\Big[\lim_{k\to\infty} \#^\epsilon \psi^{k-l}\big(\pi^{-1}(\rho'_k(z)) \cap \bigcap_j \psi^j(Y)\big)\Big]\, d\mu(z)$$

$$= \lim_{k\to\infty} \int_X \log\Big[\#^\epsilon \psi^{k-l}\big(\pi^{-1}(x) \cap \bigcap_j \psi^j(Y)\big)\Big]\, d\nu(x) \tag{4.11}$$

$$= \lim_{k\to\infty} \int_X \log\Big[\#^\epsilon \psi^{k}\big(\pi^{-1}(x) \cap \bigcap_j \psi^j(Y)\big)\Big]\, d\nu(x).$$

Since there is a uniform bound on $\#^\epsilon \psi^k\big(\pi^{-1}(x) \cap \bigcap_j \psi^j(Y)\big)$ and ν is a finite measure, we can use Fatou's lemma to conclude that

$$\lim_{k\to\infty} \int_X \log[\#^\epsilon \psi^k\big(\pi^{-1}(x) \cap \bigcap_j \psi^j(Y)\big)]\, d\nu(x)$$

$$\leq \int_X \log[\limsup_k \#^\epsilon \psi^k\big(\pi^{-1}(x) \cap \bigcap_j \psi^j(Y)\big)]\, d\nu(x). \tag{4.12}$$

However,

$$\int_X \log[\limsup_k \#^\epsilon \psi^k\big(\pi^{-1}(x) \cap \bigcap_j \psi^j(Y)\big)]\, d\nu(x)$$

$$= \int_{\widehat{X}} \log[\limsup_k \#^\epsilon \psi^{k-1}\big(\pi^{-1}(\rho'_1(z)) \cap \bigcap_j \psi^j(Y)\big)]\, d\mu(z)$$

$$= \int_{\widehat{X}} \log[\lim_k \#^\epsilon \psi^{k-1}\big(\pi^{-1}(\rho'_1(z)) \cap \bigcap_j \psi^j(Y)\big)]\, d\mu(z) \tag{4.13}$$

$$\leq \int_{\widehat{X}} \log \#\widehat{\pi}^{-1}(z)\, d\mu(z) \quad \text{(by (4.8))}.$$

It follows from (4.10)-(4.13) that

$$\int_{\widehat{X}} \log\Big[\lim_{k\to\infty} \#^\epsilon \psi^{k-l}\big(\pi^{-1}(\rho'_k(z)) \cap \bigcap_j \psi^j(Y)\big)\Big]\, d\mu(z)$$

$$\leq \int_X \log\Big[\limsup_k \#^\epsilon \psi^k\big(\pi^{-1}(x) \cap \bigcap_j \psi^j(Y)\big)\Big]\, d\nu(x)$$

$$\leq \int_{\widehat{X}} \log \#\widehat{\pi}^{-1}(z)\, d\mu(z),$$

for all $\epsilon > 0$ and all $l \in \mathbb{N}$. Combining with (4.9) and using Lebesgue's theorem on monotone convergence, we find that

$$\int_{\widehat{X}} \log \#\widehat{\pi}^{-1}(z)\, d\mu(z) = \int_X \log A_\pi(x)\, d\nu(x).$$

Since μ is $\widehat{\varphi}$-ergodic if and only if ν is φ-ergodic the theorem follows now from Theorem 4.6 and Theorem 2.1. $\qquad\square$

Corollary 4.11. *Let $\pi : (Y,\psi) \to (X,\varphi)$ be a factor map such that $D_r(\pi) < \infty$. There is then a $k \in \mathbb{N}$ and a φ-ergodic Borel probability measure μ on X such that $A_\pi(x) = k$ for μ-almost all x, and $D_r(\pi) = \log k$.*

Proof. This follows from Theorem 4.10 in essentially the same way as Theorem 3.8 follows from Theorem 2.1, using that $\#^\epsilon \psi^k(\pi^{-1}(\varphi(x))) \geq \#^\epsilon \psi^{k+1}(\pi^{-1}(x))$. $\qquad\square$

5. EQUIVALENCE RELATIONS BASED ON THE DEFECT

The subadditivity of the defect, Theorem 3.9, forms the basis for at least two equivalence relations among invertible dynamical systems which it seems worthwhile to investigate. They are both inspired by the work of Adler and Marcus in [AM]. The point of departure is the following lemma which is analogous to Proposition (2.14) of [AM].

Lemma 5.1. *Given invertible dynamical systems and factor maps,*

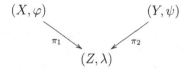

there is an invertible dynamical system (W, κ) *and a commuting diagram,*

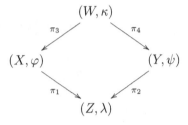

of factor maps such that $D(\pi_3) = D(\pi_2)$ *and* $D(\pi_4) = D(\pi_1)$.

Proof. Set $W = \{(x, y) \in X \times Y : \pi_1(x) = \pi_2(y)\}$, $\kappa = \varphi \times \psi$, $\pi_3(x, y) = x$, $\pi_4(x, y) = y$. Then $\#\pi_3^{-1}(x) = \#\pi_2^{-1}(\pi_1(x))$, so for any φ-invariant Borel probability measure μ we find that

$$\int_X \log \#\pi_3^{-1}(x) \, d\mu(x) = \int_Z \log \#\pi_2^{-1}(z) \, d\mu \circ \pi_1^{-1}(z) .$$

Since any λ-invariant Borel probability measure on Z has the form $\mu \circ \pi_1^{-1}$ for some φ-invariant Borel probability measure μ on X, we conclude from Theorem 2.1 that $D(\pi_3) = D(\pi_2)$. The equality $D(\pi_4) = D(\pi_1)$ follows in the same way. □

Definition 5.2. Two invertible dynamical systems, (X, φ), (Y, ψ), are *finitely equivalent* (resp. *strongly equivalent*) when there is an invertible dynamical system (Z, κ) and factor maps $\pi_1 : (Z, \kappa) \to (X, \varphi)$, $\pi_2 : (Z, \kappa) \to (Y, \psi)$ such that $D(\pi_1) + D(\pi_2) < \infty$ (resp. $D(\pi_1) = D(\pi_2) = 0$).

It follows from Lemma 5.1 and Theorem 3.9 that 'finite equivalence' and 'strong equivalence' are both equivalence relations for invertible dynamical systems.

Let us immediately extend the definition to cover general (non-invertible) dynamical systems.

Definition 5.3. Two dynamical systems, (X, φ), (Y, ψ), are *finitely equivalent* (resp. *strongly equivalent*) when there is a dynamical system (Z, κ) and factor maps $\pi_1 : (Z, \kappa) \to (X, \varphi)$, $\pi_2 : (Z, \kappa) \to (Y, \psi)$ such that $D_r(\pi_1) + D_r(\pi_2) < \infty$ (resp. $D_r(\pi_1) = D_r(\pi_2) = 0$).

Finite equivalence and strong equivalence are equivalence relations thanks to the sub-additivity of the reduced defect, Corollary 4.7, and the variational principle for the reduced defect, Theorem 4.10. The argument is basically the same as in the proof of Lemma 5.1.

Lemma 5.4. *Two invertible dynamical systems are finitely equivalent (resp. strongly equivalent) in the sense of Definition 5.2 if and only if they are finitely equivalent (resp. strongly equivalent) in the sense of Definition 5.3.*

Proof. When

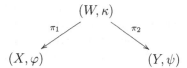

is a diagram of factor maps such that $D_r(\pi_1) + D_r(\pi_2) < \infty$ (resp. $D_r(\pi_1) = D_r(\pi_2) = 0$), and (X, φ) and (Y, ψ) are both invertible, by passing to natural invertible extensions we get also a diagram

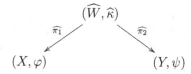

where $D(\widehat{\pi}_i) = D_r(\pi_i)$, $i = 1, 2$, by Theorem 4.6. Hence (X, φ) and (Y, ψ) are finitely equivalent (resp. strongly equivalent) in the sense of Definition 5.2. Since the other implication follows from 2) of Lemma 4.2, the proof is complete.

$\square$

Note that finitely equivalent dynamical systems must have the same topological entropy by Theorem 3.6 and/or Corollary 4.8.

Theorem 5.5. *Let (Σ_1, σ) and (Σ_2, σ) be irreducible twosided sofic shifts. Then the following conditions are equivalent:*

1) $h(\Sigma_1) = h(\Sigma_2)$.
2) (Σ_1, σ) and (Σ_2, σ) have a common finite-to-one extension which is a twosided irreducible subshift of finite type.
3) (Σ_1, σ) and (Σ_2, σ) are finitely equivalent.

Proof. 1) $\Leftrightarrow$ 2) follows from Theorem 3.6 and the entropy-classification of irreducible sofic subshifts, cf. Theorem 8.3.8 of [LM]. 2) $\Rightarrow$ 3) follows from 1) of Proposition 3.1. 3) $\Rightarrow$ 1) follows from Theorem 3.6. $\square$

Corollary 5.6. *Let (X, φ) and (Y, ψ) be boundedly finite-to-one factors of irreducible subshifts of finite type. Then (X, φ) and (Y, ψ) are finitely equivalent if and only if $h(\varphi) = h(\psi)$.*

In particular, we see that hyperbolic toral automorphisms are finitely equivalent if and only if they have the same entropy. Compare [AM].

Proposition 5.7. *Let (X, φ) and (Y, ψ) be invertible dynamical systems. Then the following are equivalent :*

1) (X, φ) and (Y, ψ) are finitely equivalent.
2) (X, φ^n) and (Y, ψ^n) are finitely equivalent for all $n \in \mathbb{Z}$.

3) (X, φ^n) and (Y, ψ^n) are finitely equivalent for some $n \in \mathbb{Z}$.

Proof. 1) $\Rightarrow$ 2) follows from the argument which proved Lemma 3.7 of [Th2], using the variational principle. It suffices therefore to show that 3) $\Rightarrow$ 1). Furthermore, it suffices to consider the case $n > 1$. Let (Z, σ) be a dynamical system and $\pi_1 : (Z, \sigma) \rightarrow (X, \varphi^n)$, $\pi_2 : (Z, \sigma) \rightarrow (Y, \psi^n)$ factor maps such that $D(\pi_1) + D(\pi_2) < \infty$. Let $Z_0, Z_1, Z_2, \cdots, Z_{n-1}$ be disjoint copies of Z and define $\widetilde{\sigma} : \bigcup_{j=0}^{n-1} Z_j \rightarrow \bigcup_{j=0}^{n-1} Z_j$ such that $\widetilde{\sigma}|_{Z_j} : Z_j \rightarrow Z_{j+1}$ is the identity when $j < n - 1$ and $\widetilde{\sigma}|_{Z_{n-1}} : Z_{n-1} \rightarrow Z_0$ is σ. Define $\widetilde{\pi}_1 : \bigcup_{j=0}^{n-1} Z_j \rightarrow X$ such that $\widetilde{\pi}_1|_{Z_j} = \varphi^j \circ \pi_1$ for all $j = 0, 1, 2, \cdots, n - 1$. Then $\widetilde{\pi}_1 : (\bigcup_{j=0}^{n-1} Z_j, \widetilde{\sigma}) \rightarrow (X, \varphi)$ is a factor map and $\#\widetilde{\pi}_1^{-1}(x) = n\#\pi_1^{-1}(x)$ for all $x \in X$. Hence $D(\widetilde{\pi}_1) = \log n + D(\pi_1)$ by Theorem 2.1. Similarly, we define a factor map $\widetilde{\pi}_2 : (\bigcup_{j=0}^{n-1} Z_j, \widetilde{\sigma}) \rightarrow (Y, \psi)$ such that $D(\widetilde{\pi}_2) = \log n + D(\pi_2)$. $\square$

Corollary 5.8. *Let (X, φ) and (Y, ψ) be dynamical systems. Assume that X and Y are finite-dimensional spaces. Assume that φ and ψ are periodic, i.e. that $\varphi^k = \mathrm{id}_X$, $\psi^m = \mathrm{id}_Y$ for some $k, m \in \mathbb{N}$. Then (X, φ) and (Y, ψ) are finitely equivalent if and only if there is a compact metric space Z and continuous surjections $\pi_0 : Z \rightarrow X$ and $\pi_1 : Z \rightarrow Y$ such that $\sup_{x,y} \max\{\#\pi_0^{-1}(x), \#\pi_1^{-1}(y)\} < \infty$.*

Proof. This follows immediately from Proposition 5.7. $\square$

Remark 5.9. In many cases it is easy to see that there is a Z satisfying the requirement in Corollary 5.8. On the other hand, it is also easy to give examples which shows that it does not always exist, so it would be nice to have general criteria for the existence of such a space Z.

Lemma 5.10. *Let (Σ_1, σ) and (Σ_2, σ) be two-sided mixing subshifts of finite type. Assume that $h(\Sigma_1, \sigma) \leq h(\Sigma_2, \sigma)$. Then the disjoint union $(\Sigma_1 \sqcup \Sigma_2, \sigma \sqcup \sigma)$ is finitely equivalent to (Σ_2, σ).*

Proof. Assume first that $h(\Sigma_1, \sigma) = h(\Sigma_2, \sigma)$. Then (Σ_1, σ) and (Σ_2, σ) are finitely equivalent by Lemma 5.5. It follows then easily that $(\Sigma_1 \sqcup \Sigma_2, \sigma \sqcup \sigma)$ is finitely equivalent to $(\Sigma_2 \sqcup \Sigma_2, \sigma \sqcup \sigma)$, which in turn is finitely equivalent to (Σ_2, σ). Assume next that $h(\Sigma_1, \sigma) < h(\Sigma_2, \sigma)$. By a wellknown formula for the topological entropy of a mixing subshift of finite type, cf. [LM], there is then an $N \in \mathbb{N}$ so large that

$$\#\{x \in \Sigma_1 : x \text{ has minimal period } n\} < \#\{x \in \Sigma_2 : x \text{ has minimal period } n\}$$

for all $n \geq N$. Let P be a prime larger than N. Let $\underline{P} = \{1, 2, \cdots, P\}$, and let $q : \underline{P} \rightarrow \underline{P}$ be cyclic permutation. Then $(\Sigma_1 \times \underline{P}, \sigma \times q)$ and $(\Sigma_2 \times \underline{P}, \sigma \times q)$ are irreducible subshifts of finite type, with entropy $h(\Sigma_1, \sigma)$ and $h(\Sigma_2, \sigma)$, respectively. In addition

$$\#\{x \in \Sigma_1 \times \underline{P} : x \text{ has minimal period } n \text{ under } \sigma \times q\}$$
$$< \#\{x \in \Sigma_2 \times \underline{P} : x \text{ has minimal period } n \text{ under } \sigma \times q\}$$

for all n. By Krieger's embedding theorem, cf. Theorem 10.1.1 of [LM], there is then an embedding $(\Sigma_1 \times \underline{P}, \sigma \times q) \to (\Sigma_2 \times \underline{P}, \sigma \times q)$. Since $(\Sigma_i \times \underline{P}, \sigma \times q)$ is finitely equivalent to (Σ_i, σ), and

$$(\Sigma_1 \times \underline{P} \sqcup \Sigma_2 \times \underline{P}, \sigma \times q \sqcup \sigma \times q)$$

to $(\Sigma_1 \sqcup \Sigma_2, \sigma \sqcup \sigma)$, we can assume from outset that we are given an embedding $\kappa : (\Sigma_1, \sigma) \to (\Sigma_2, \sigma)$. Define $\pi : (\Sigma_1 \sqcup \Sigma_2, \sigma \sqcup \sigma) \to (\Sigma_2, \sigma)$ such that $\pi|_{\Sigma_1} = \kappa$ and $\pi|_{\Sigma_2}$ is the identity. Then $\#\pi^{-1}(x) \le 2$ for all $x \in \Sigma_2$ and hence $D(\pi) \le \log 2$. Thus (Σ_2, σ) is finitely equivalent to $(\Sigma_1 \sqcup \Sigma_2, \sigma \sqcup \sigma)$, as asserted.

$\square$

Proposition 5.11. *The non-wandering parts of two subshifts of finite type are finitely equivalent if and only if they have the same entropy.*

Proof. By Lemma 5.10 the non-wandering part of a subshift of finite type is finitely equivalent to any of its irreducible components with the maximal entropy. Hence Theorem 5.5 gives the result. $\square$

Remark 5.12. I am not sure how sensitive finite equivalence is towards the wandering points of a subshift of finite type. However, entropy is certainly not the only invariant for finite equivalence of general subshifts of finite type. To illustrate the situation in the simplest cases, consider the graphs

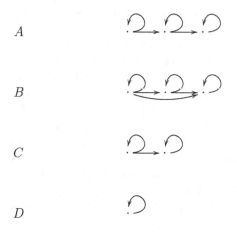

A

B

C

D

Of the four edge-shifts, all with zero entropy, given by these graphs, A and B are finitely equivalent, but no pair among B, C and D are finitely equivalent. So it appears as if finite equivalence is sensitive to how many irreducible components are connected, but ignores how they are connected. Presently I do not know how representative these very simple examples are.

Concerning strong equivalence it is clear that (X, φ) and (Y, ψ) can only be strongly equivalent when

$$\#\{x \in X : \varphi^n(x) = x\} = \#\{y \in Y : \psi^n(y) = y\}$$

for all $n \in \mathbb{N}$. In view of Krieger's Embedding Theorem and Boyle's Lower Entropy Factor Theorem, cf. Theorem 10.1.1 and Theorem 10.3.1 of [LM], respectively, it seems reasonable to ask if this condition is sufficient for irreducible subshifts of finite type. In this paper we concentrate the investigations on finite equivalence.

Lemma 5.13. *Let $\lambda \in \mathbb{T}^n$ and let $\underline{\lambda} : \mathbb{T}^n \to \mathbb{T}^n$ be the corresponding rotation, viz. $\underline{\lambda}(z) = \lambda z$. Then $(\mathbb{T}^n, \underline{\lambda})$ is finitely equivalent to $(\mathbb{T}^n, \underline{\lambda}^k)$ for all $k \in \mathbb{Z}$.*

Proof. Define $\pi_1 : \mathbb{T}^n \to \mathbb{T}^n$ by $\pi_1(z) = z^k$ and let π_2 be the identity map of $\mathbb{T}^n$. Then $\pi_1 : (\mathbb{T}^n, \underline{\lambda}) \to (\mathbb{T}^n, \underline{\lambda}^k)$ and $\pi_2 : (\mathbb{T}^n, \underline{\lambda}^k) \to (\mathbb{T}^n, \underline{\lambda}^k)$ are factor maps of defect $D(\pi_1) = \log n|k|$ and $D(\pi_2) = 0$, respectively. It follows from Lemma 5.1 that $(\mathbb{T}^n, \underline{\lambda})$ is finitely equivalent to $(\mathbb{T}^n, \underline{\lambda}^k)$. $\square$

Lemma 5.14. *Let $\lambda, \mu \in \mathbb{T}$, and let $\underline{\lambda} : \mathbb{T} \to \mathbb{T}$ and $\underline{\mu} : \mathbb{T} \to \mathbb{T}$ be the corresponding rotations of the circle. Then $\underline{\lambda}$ and $\underline{\mu}$ are equivalent if and only if there are numbers $k, m \in \mathbb{Z}$ such that $\lambda^m = \mu^k$, i.e. if and only if $\lambda = \mu$ modulo $\mathbb{Q}/\mathbb{Z}$.*

Proof. If such m and k exist it follows from Lemma 5.13 that $\underline{\lambda}$ and $\underline{\mu}$ are equivalent. Conversely, assume that $\underline{\lambda}$ and $\underline{\mu}$ are equivalent. If both λ and μ are rational (i.e. of finite order in the group $\mathbb{T}$), there is nothing to prove, so assume that λ is irrational. Let (X, ψ) be a common finite defect extension of $(\mathbb{T}, \underline{\lambda})$ and $(\mathbb{T}, \underline{\mu})$. Let $\pi : (X, \psi) \to (\mathbb{T}, \underline{\lambda})$ be a factor map of finite defect. It follows from Theorem 3.7 that there is a ψ-ergodic Borel probability measure ν on X such that $\nu \circ \pi^{-1}$ is Lebesgue measure on $\mathbb{T}$ and $\#\pi(x) = k$ for almost all $x \in \mathbb{T}$. Since $(\mathbb{T}, \underline{\mu})$ is a factor of (X, ψ), μ must be an eigenvalue for the unitary $T_\psi : L^2(X, \nu) \to L^2(X, \nu)$ induced by ψ. Let $f : X \to \mathbb{T}$ be the corresponding (continuous) eigenfunction. As is well-known, cf. Lemma 1 of [NP], we can identify the measure space (X, ν) with the space $\mathbb{T} \times \{1, 2, \cdots, k\}$ equipped with the product of Lebesgue measure on $\mathbb{T}$ with the homogeneous probability measure on $\{1, 2, \cdots, k\}$. In this picture $\psi(x, i) = (\lambda x, \sigma_x(i))$, where $\sigma : \mathbb{T} \to \Sigma_k$ is a Borel function taking values in the symmetric group. Hence

$$g(x) = \prod_{i=1}^{k} f(x, i)$$

is a Borel function $g : \mathbb{T} \to \mathbb{T}$ such that $g(\lambda x) = \mu^k g(x)$ for almost all $x \in \mathbb{T}$. Since the spectrum of the unitary on $L^2(\mathbb{T})$ induced by $\underline{\lambda}$ is $\{\lambda^z : z \in \mathbb{Z}\}$, we conclude that $\mu^k = \lambda^m$ for some $m \in \mathbb{Z}$. $\square$

Theorem 5.15. *Two orientation preserving homeomorphisms of the circle with irrational rotation numbers, α and β, are finitely equivalent if and only if there are integers, $n, m \in \mathbb{Z}$, such that $n\alpha - m\beta \in \mathbb{Z}$.*

Proof. By the Poincaré Classification Theorem, cf. Theorem 11.2.7 of [KH], any orientation preserving homeomorphism of the circle with irrational rotation number has the rigid rotation with the same rotation number (mod $\mathbb{Z}$) as a factor under a factor map π for which $\#\pi^{-1}(x) = 1$ for all x outside of a countable subset of the circle. Hence $D(\pi) = 0$ by Theorem 2.1 and we see that any orientation preserving homeomorphism of the circle with irrational rotation number is strongly equivalent to the rigid rotation with the same rotation number. Apply Lemma 5.14. $\qquad\square$

Remark 5.16. Theorem 5.15 is not true for orientation preserving homeomorphisms of the circle with rational rotation numbers. Indeed, it is not true that any homeomorphism of the interval $[0, 1]$ is finitely equivalent to the identity map. For example, it is not difficult to see that a homeomorphism of $[0, 1]$ for which the set of fixed points is infinite can not be finitely equivalent to one for which the set of fixed points is finite. So at least some characteristics of the fixed point set is preserved under finite equivalence, and presently I do not know exactly which. The problem is related to the problem mentioned in Remark 5.9. However, it is true that two orientation preserving homeomorphisms of the circle or two homeomorphisms of the interval are finitely equivalent when they have the same set of periodic points.

The method of proof in Lemma 5.14 can be extended to give a classification of minimal rotations of higher-dimensional tori as follows.

Theorem 5.17. *Let $\underline{\lambda}, \underline{\mu}$ be minimal rotations of the n-torus $\mathbb{T}^n$, $n \geq 1$. Then $(\mathbb{T}^n, \underline{\lambda})$ and $(\mathbb{T}^n, \underline{\mu})$ are finitely equivalent if and only if there is a continuous surjective endomorphism $B : \mathbb{T}^n \to \mathbb{T}^n$ and a natural number $k \in \mathbb{N}$ such that*

$$\underline{\lambda}^k \circ B = B \circ \underline{\mu}. \tag{5.1}$$

Proof. Assume first that B and k exist. Since B is constant-to-one (with a finite constant) we see immediately from (5.1) that B is factor map of finite defect showing that $(\mathbb{T}^n, \underline{\lambda}^k)$ and $(\mathbb{T}^n, \underline{\mu})$ are finitely equivalent. Hence $(\mathbb{T}^n, \underline{\lambda})$ and $(\mathbb{T}^n, \underline{\mu})$ are finitely equivalent by Lemma 5.13.

Assume then that $(\mathbb{T}^n, \underline{\lambda})$ and $(\mathbb{T}^n, \underline{\mu})$ are finitely equivalent, and write $\lambda = (\lambda_1, \lambda_2, \cdots, \lambda_n)$, $\mu = (\mu_1, \mu_2, \cdots, \mu_n)$, where $\lambda_i, \mu_i \in \mathbb{T}$ for all i. The argument from the proof of Lemma 5.13 shows that there is a $k \in \mathbb{N}$ such that λ_i^k is in the spectrum of $(\mathbb{T}^n, \underline{\mu})$ for all i. Since the spectrum of $(\mathbb{T}^n, \underline{\mu})$ is the set

$$\{\mu_1^{z_1} \mu_2^{z_2} \cdots \mu_n^{z_n} : z_1, z_2, \cdots, z_n \in \mathbb{Z}\},$$

we conclude that there is a $n \times n$-matrix $A = (A_{ij})$ with $\mathbb{Z}$-entries such that

$$\lambda_i^k = \mu_1^{A_{i1}} \mu_2^{A_{i2}} \cdots \mu_n^{A_{in}} \tag{5.2}$$

for all i. When B denotes the endomorphism of $\mathbb{T}^n$ given by A, (5.2) means that $\lambda^k = B(\mu)$, so the transitivity of $\underline{\lambda}$ (which implies the transitivity and hence the minimality of $\underline{\lambda}^k$) shows that B is surjective. $\qquad\square$

Theorem 5.17 can also be formulated as follows: Choose $\alpha, \beta \in \mathbb{R}^n$ such that $p(\alpha) = \lambda$, $p(\beta) = \mu$, where $p : \mathbb{R}^n \to \mathbb{T}^n$ is the canonical surjection. Then $(\mathbb{T}^n, \underline{\lambda})$ and $(\mathbb{T}^n, \underline{\mu})$ are finitely equivalent if and only if there is an invertible $n \times n$-matrix D over $\mathbb{Q}$ such that $D\beta - \alpha \in \mathbb{Q}^n$.

Before we turn to a few non-invertible dynamical systems, let us first observe that all surjective dynamical systems are finitely equivalent to their natural invertible extension. If namely (X, φ) is a dynamical system with φ surjective, the projection to the first coordinate, $p_0 : (\widehat{X}, \widehat{\varphi}) \to (X, \varphi)$, is a factor map which ensure a strong equivalence:

Proposition 5.18. *Let (X, φ) be a dynamical system with φ surjective. Then the reduced defect of the factor map $p_0 : (\widehat{X}, \widehat{\varphi}) \to (X, \varphi)$ is 0, and (X, φ) is strongly equivalent to $(\widehat{X}, \widehat{\varphi})$.*

Proof. Recall that $\widehat{X}$ is a closed subset of the infinite product $\prod_{i=0}^{\infty} X$, equipped with the metric

$$d_\infty((x_i), (y_i)) = \sum_{i=0}^{\infty} \frac{d(x_i, y_i)}{2^i}.$$

Then $p_0^{-1}(x) \subseteq \prod_{i=0}^{\infty} \varphi^{-i}(x)$ and $\widehat{\varphi}^k(p_0^{-1}(x)) \subseteq \prod_{i=0}^{\infty} \varphi^{k-i}(x)$ for all $x \in X, k \in \mathbb{N}$. Hence the d_∞-diameter of $\widehat{\varphi}^k(p_0^{-1}(x))$ tends to 0 as k tends to infinity, and $\limsup_k \#^\epsilon \widehat{\varphi}^k(p_0^{-1}(x)) = 0$ for all x and all $\epsilon > 0$. It follows that $A_{p_0} = 0$ and hence from Theorem 4.10 that $D_r(p_0) = 0$. $\qquad\square$

Remark 5.19. An alternative proof of Proposition 5.18 goes as follows : The natural extension $\widehat{p_0}$ of p_0 is a conjugacy and $D_r(p_0) = D(\widehat{p_0}) = 0$ by Theorem 4.6.

It follows from Proposition 5.18 that for any dynamical system (X, φ), the system $(\bigcap_j \varphi^j(X), \varphi)$ will be strongly equivalent to an invertible dynamical system. It seems to be generally agreed that the interesting dynamics of φ takes place in $\bigcap_j \varphi^j(X)$, so we may conclude that up to strong equivalence all interesting dynamics can be realized in invertible dynamical systems.

Theorem 5.20. *Two one-sided irreducible sofic subshifts are finitely equivalent if and only if they have the same entropy.*

Proof. The natural extension of a one-sided irreducible sofic subshift is a two-sided irreducible sofic subshift to which it is strongly equivalent by Proposition 5.18. Apply Theorem 5.5.   □

Corollary 5.21. *Let (X, φ) and (Y, ψ) be boundedly finite-to-one factors of irreducible one-sided subshifts of finite type. Then (X, φ) and (Y, ψ) are finitely equivalent if and only if $h(\psi) = h(\varphi)$.*

Proof. The necessity of equal entropy follows from Corollary 4.8. So assume that $h(\psi) = h(\varphi)$. The assumptions, combined with Theorem 2.1 and 2) of Lemma 4.2, show that both dynamical systems are finite reduced defect factors of irreducible one-sided subshifts of finite type. Apply Theorem 5.20.   □

Corollary 5.22. *Let (X, ψ) and (Y, φ) be expansive endomorphisms of compact finite-dimensional differentiable manifolds, X and Y, respectively. Then (X, ψ) and (Y, φ) are finitely equivalent if and only if $h(\psi) = h(\varphi)$.*

Proof. By Corollary 5.21 it suffices to show that both dynamical systems are finite-to-one factors of irreducible subshifts of finite type. First note that expansive endomorphisms are factors of full shifts by [S] and hence transitive. By Theorem 7.30 of [Ru] it suffices therefore to show that an expansive endomorphism is expanding in the sense of 7.26 of [Ru]. But this is not difficult, and was pointed out already in [CR].   □

The next class of dynamical systems we consider is the class of unimodal maps of the interval. Let $\psi : [0,1] \to [0,1]$ be a unimodal map with positive entropy and let $T : [0,1] \to [0,1]$ be the tent-map with the same entropy. Using their kneading theory, Milnor and Thurston constructed in [MT] a factor map $\lambda : ([0,1], \psi) \to ([0,1], T)$. The construction is reproduced in (e.g.) [KH], pp. 514-518. The reduced defect $D_r(\lambda)$ is either 0 or $+\infty$ depending on whether or not there exists a non-degenerate closed intervals $J \subseteq [0,1]$ and a natural number $m \in \mathbb{N}$ such that $\psi^m(J) = J$, $J, \psi(J), \psi^2(J), \cdots, \psi^{m-1}(J)$ are mutually disjoint and $h(\psi^m|_J) < h(\psi^m)$. To be precise we have :

(A) *If there is a non-degenerate closed interval $J \subseteq [0,1]$ and a natural number $m \in \mathbb{N}$ such that $\psi^m(J) = J$ and $h(\psi^m|_J) < h(\psi^m)$, then $D_r(\lambda) = \infty$.*

(B) *If there is no non-degenerate closed interval $J \subseteq [0,1]$ and natural number $m \in \mathbb{N}$ such that $\psi^m(J) = J$, $J, \psi(J), \psi^2(J), \cdots, \psi^{m-1}(J)$ are mutually disjoint and $h(\psi^m|_J) < h(\psi^m)$, then $D_r(\lambda) = 0$.*

To prove (A) assume first that such m and J exist. Then, in the notation of [MT],

$$l(\psi^k|J) \leq l(\psi^k|J \cup \psi(J) \cup \cdots \cup \psi^{m-1}(J))$$

$$\leq \sum_{n=0}^{m-1} l(\psi^k|\psi^n(J)) \leq \sum_{n=0}^{m-1} l(\psi^{k+n}|J),$$

from which it follows that

$$\limsup_k l(\psi^k|J)^{\frac{1}{k}} = \limsup_k l(\psi^k|J \cup \psi(J) \cup \cdots \cup \psi^{m-1}(J))^{\frac{1}{k}}. \tag{5.3}$$

Since $J \cup \psi(J) \cup \cdots \cup \psi^{m-1}(J)$ is ψ-invariant,

$$k \mapsto l(\psi^k|J \cup \psi(J) \cup \cdots \cup \psi^{m-1}(J))$$

is submultiplicative, and hence

$$\limsup_k l(\psi^k|J \cup \psi(J) \cup \cdots \cup \psi^{m-1}(J))^{\frac{1}{k}}$$
$$= \lim_k l(\psi^k|J \cup \psi(J) \cup \cdots \cup \psi^{m-1}(J))^{\frac{1}{k}}. \tag{5.4}$$

The same argument with ψ^m in the place of ψ shows that also

$$\limsup_k l(\psi^{mk}|J \cup \psi(J) \cup \cdots \cup \psi^{m-1}(J))^{\frac{1}{k}}$$
$$= \lim_k l(\psi^{mk}|J \cup \psi(J) \cup \cdots \cup \psi^{m-1}(J))^{\frac{1}{k}}. \tag{5.5}$$

By combining (5.3), (5.4) and (5.5) we find that

$$\limsup_k l(\psi^k|J)^{\frac{1}{k}} = \lim_k l(\psi^k|J \cup \psi(J) \cup \cdots \cup \psi^{m-1}(J))^{\frac{1}{k}}$$
$$= \lim_k l(\psi^{mk}|J \cup \psi(J) \cup \cdots \cup \psi^{m-1}(J))^{\frac{1}{mk}}$$
$$= (\lim_k l(\psi^{mk}|J \cup \psi(J) \cup \cdots \cup \psi^{m-1}(J))^{\frac{1}{k}})^{\frac{1}{m}}$$
$$= (\lim_k l(\psi^{mk}|J)^{\frac{1}{k}})^{\frac{1}{m}} = e^{\frac{h(\psi^m|J)}{m}} < e^{\frac{h(\psi^m)}{m}} = e^{h(\psi)} = \lim_k l(\psi^k)^{\frac{1}{k}}.$$

It follows that the radius of convergence of the power series $\sum_{n=0}^{\infty} l(\psi^n|J)t^n$ is strictly larger than that of $\sum_{n=0}^{\infty} l(\psi^n)t^n$, so by the definition of λ we must have that $\lambda(J)$ is a point. This point, x, is m-periodic under T and

$$A_\lambda(x) = \lim_{\epsilon \to 0}\left[\limsup_k \#^\epsilon \psi^k\left(\lambda^{-1}(x) \cap \bigcap_j \psi^j(Y)\right)\right] \geq \lim_{\epsilon \to 0}\left[\limsup_k \#^\epsilon \psi^k(J)\right] = \infty.$$

Hence $D_r(\lambda) = \infty$ by the variational principle for the reduced defect.

To prove (B) assume that there is no m and J with the properties stated in (B). We first argue that

if $x \in [0,1], m \in \mathbb{N}$, and $I \subseteq \lambda^{-1}(x)$ is a closed non-empty interval such that $\psi^m(I) = I$, then I contains only one point. $\tag{5.6}$

To prove (5.6) let I be an interval with the property specified in (5.6), and assume to reach a contradiction that I is non-degenerate. If none of the intervals $I, \psi(I), \psi^2(I), \cdots, \psi^{m-1}(I)$ contain the turning point, $\psi^m : I \to I$ is a homeomorphism and hence $h(\psi^m|_I) = 0 < h(\psi^m)$, and this is impossible by assumption. So one of the intervals contains the turning point. Since λ sends the turning point of ψ to the turning point of T we deduce that the turning point of T is periodic. The (argument of) Corollary (2.5.6) of [St] shows that in this case we also have that $h(\psi^m|_I) < h(\psi^m)$, which again is impossible by assumption. Hence (5.6) holds. To prove that $D_r(\lambda) = 0$ it suffices to show that $\lim_{n\to\infty} \operatorname{diam} \psi^n(\lambda^{-1}(x)) = 0$ for all $x \in [0,1]$ by Theorem 4.10. Since λ is nondecreasing, $\lambda^{-1}(x)$ is an interval, and the conclusion we seek is automatic if the sets $\psi^j(\lambda^{-1}(x)), j \in \mathbb{N}$, are all mutually disjoint, so we may assume that there are $k < l$ in $\mathbb{N}$ such that $\psi^k(\lambda^{-1}(x)) \cap \psi^l(\lambda^{-1}(x)) \neq \emptyset$. Set $J = \lambda^{-1}(T^k(x))$ and note that $\psi^k(\lambda^{-1}(x)) \subseteq J$. Since λ is constant on J and $\psi^l(\lambda^{-1}(x))$, and these sets intersect, we conclude that $T^{l-k}(T^k(x)) = T^k(x)$, so that $\psi^{l-k}(J) \subseteq J$. If follows that $I = \bigcap_{n\in\mathbb{N}} \psi^{(l-k)n}(J)$ is a closed non-empty interval in $J = \lambda^{-1}(T^k(x))$ such that $\psi^{l-k}(I) = I$. It follows therefore from (5.6) that I is a point. Since $J \supseteq \psi^{l-k}(J) \supseteq \psi^{2(l-k)}(J) \supseteq \psi^{3(l-k)}(J) \supseteq \cdots$ we deduce that $\lim_{n\to\infty} \operatorname{diam} \psi^{n(l-k)}(J) = 0$. Since the functions $\psi, \psi^2, \cdots, \psi^{l-k}$ are uniformly continuous it follows that $\lim_{n\to\infty} \operatorname{diam} \psi^n(J) = 0$. Hence $\lim_{n\to\infty} \operatorname{diam} \psi^n(h^{-1}(x)) = \lim_{n\to\infty} \operatorname{diam} \psi^{n-k}(J) = 0$, and we are done.

If we specialize a little we can say more. So assume now that $\psi : [0,1] \to [0,1]$ is a C^2 unimodal map which is not renormalizable and has a non-flat critical point. (For the definition of these notions see [NS].) Assume also that the set of periodic points is finite for each period. We call a non-degenerate closed interval $I \subseteq [0,1]$ *periodic* (of period m) when $\psi^m(I) = I$ and $I, \psi(I), \psi^2(I), \cdots, \psi^{m-1}(I)$ are mutually disjoint. I is a *periodic homterval* when it is a periodic interval of period m such that $\psi^m|_I$ is injective. For each $m \in \mathbb{N}$, let $I_1^m, I_2^m, \cdots, I_{L_m}^m$ be the maximal periodic homtervals of period m so that $L_m = 0$ for all large enough m by Theorem B on page 268 of [MS]. It follows that the total number of (maximal) periodic homtervals is finite since we assume that ψ has only finitely many periodic points of each period. Let $\tilde{I}$ denote the copy of $[0,1]$ obtained by collapsing each maximal periodic homterval to its left endpoint. There is then a continuous map $\tilde{\psi} : \tilde{I} \to \tilde{I}$ and factor maps $\tilde{\lambda} : (\tilde{I}, \tilde{\psi}) \to ([0,1], T)$ and $C : ([0,1], \psi) \to (\tilde{I}, \tilde{\psi})$ such that

$$\lambda = \tilde{\lambda} \circ C.$$

Lemma 5.23. $D_r(\tilde{\lambda}) = 0$.

Proof. It suffices to show that $\lim_{k\to\infty} \operatorname{diam} \tilde{\psi}^k(\tilde{\lambda}^{-1}(t)) = 0$ for all $t \in [0,1]$. Since $\tilde{\psi}^k(\tilde{\lambda}^{-1}(t)) = C(\psi^k(\lambda^{-1}(t)))$ it suffices to show that $\lim_{k\to\infty} \operatorname{diam} C(\psi^k(\lambda^{-1}(t)))$ 0. This is trivial when $\limsup_k \operatorname{diam}(\psi^k(\lambda^{-1}(t))) = 0$ so we may assume that $\limsup_k \operatorname{diam}(\psi^k(\lambda^{-1}(t))) > 0$. Then the intervals $\{\psi^i(\lambda^{-1}(t)) : i \in \mathbb{N}\}$ can not be mutually disjoint, so t must be pre-periodic, i.e. $T^k(t)$ must be periodic for some $k \in \mathbb{N}$. Let $t_1, t_2, \cdots, t_m$ be the orbit of $T^k(t)$ and set $I_i = \lambda^{-1}(t_i)$. As in the proof of (B) above it suffices to show that $\bigcap_k C(\psi^{km}(I_1))$ is a point. Set $I = \bigcap_k \psi^{km}(I_1)$ and note that $\psi^m(I) = I$ and $I, \psi(I), \psi^2(I), \cdots, \psi^{m-1}(I)$ are disjoint. If the turning point was in $\psi^j(I)$ for some j we would have that the turning point of T is periodic. By (2.5.6) of [St] this is impossible since ψ is non-renormalizable. It follows that I is a periodic homterval and hence that $C(I)$ is a point by construction of C. Since $C(I) = \bigcap_k C(\psi^{km}(I_1))$ we are done. $\qquad\square$

Lemma 5.24. $\psi|_{\Omega(\psi)}$ *is finitely equivalent to* $T|_{\Omega(T)}$.

Proof. By Lemma 6 of [JR], $\lambda(\Omega(\psi)) = \Omega(T)$, so $\tilde{\lambda}(C(\Omega(\psi))) = \Omega(T)$. It follows from Lemma 5.23 that $D_r(\tilde{\lambda}|_{C(\Omega(\psi))}) = 0$, so it suffices to show that $D_r(C|_{\Omega(\psi)}) < \infty$. But this follows from the construction of C because

$$\#C^{-1}(x) \cap \Omega(\psi) \le \sum_{j=1}^{m_0} \#\{t \in [0,1] : \psi^j(t) = t\},$$

where m_0 is so large that there are no periodic homtervals for ψ of period $> m_0$. $\qquad\square$

Lemma 5.25. *The reduced zeta-function of* ψ *(as defined in* [MT]*) is the same as the zeta-function of* T.

Proof. By definition the reduced zeta function only counts the monotone equivalence classes of periodic points of ψ. This is therefore exactly the zeta-function of $\tilde{\psi}$. It follows from Lemma 5.23 that $\tilde{\lambda}$ is a bijection on periodic points and hence we deduce that $\tilde{\psi}$ and T have the same zeta function. $\qquad\square$

Theorem 5.26. *Let* $\psi, \varphi : [0,1] \to [0,1]$ *be* C^2 *unimodal maps both of which are non-renormalizable and have a nonflat critical point. Assume also that they have only a finite set of periodic points for each period and non-zero entropy. Let* $\Omega(\varphi)$ *and* $\Omega(\psi)$ *denote the non-wandering parts of* φ *and* ψ, *respectively. Then the following are equivalent:*

1) $\varphi|_{\Omega(\varphi)}$ *and* $\psi|_{\Omega(\psi)}$ *are finitely equivalent.*
2) φ *are* ψ *have the same entropy.*
3) φ *are* ψ *have the same kneading determinant.*

Proof. 1) $\Rightarrow$ 2) follows from Corollary 4.8 and 3) $\Rightarrow$ 2) follows from [MT]. 2) $\Rightarrow$ 3) follows from Corollary 10.7 of [MT] combined with Lemma 5.25 and 2) $\Rightarrow$ 1) follows from Lemma 5.24. $\qquad\square$

Remark 5.27. Following Williams, [Wi], Franks and Richeson calls two dynamical systems, (Y, ψ) and (X, φ), *shift equivalent* (of lag m), when there are maps $r : (Y, \psi) \to (X, \varphi)$ and $s : (X, \varphi) \to (Y, \psi)$, not neccesarily surjective, but such that $r \circ s = \psi^m$ and $s \circ r = \varphi^m$ for some $m \in \mathbb{N}$,

[FR]. For invertible dynamical systems shift equivalence is the same as conjugacy, but not in general. Shift equivalence of (Y, ψ) and (X, φ) implies that $(\bigcap_j \psi^j(Y), \psi)$ and $(\bigcap_j \varphi^j(X), \varphi)$ are strongly equivalent. Indeed, if (Y, ψ) and (X, φ) are shift equivalent, say of lag m, via r and s as above, it follows that $r : (\bigcap_j \psi^j(Y), \psi) \to (\bigcap_j \varphi^j(X), \varphi)$ and $s : (\bigcap_j \varphi^j(X), \varphi) \to (\bigcap_j \psi^j(Y), \psi)$ are factor maps (i.e. surjective) and hence that the natural extensions $\widehat{r} : (\widehat{Y}, \widehat{\psi}) \to (\widehat{X}, \widehat{\varphi})$ and $\widehat{s} : (\widehat{X}, \widehat{\varphi}) \to (\widehat{Y}, \widehat{\psi})$ define a shift equivalence (of lag m) between $(\widehat{X}, \widehat{\varphi})$ and $(\widehat{Y}, \widehat{\psi})$. By using 6) of Proposition 3.1 this implies that $D(\widehat{s}) = 0$ and hence, by Theorem 4.6, that $D_r(s|_{\bigcap_j \varphi^j(X)}) = 0$.

In the preceding we have used the defect to investigate what seems to be a natural notion of finite equivalence for general dynamical systems. There is, however, also another more obvious application of the defect to dynamical systems which we would like to mention in closing: A dynamical system (X, φ) can be considered as a factor map from itself to itself; $\varphi : (X, \varphi) \to (X, \varphi)$, at least if φ is surjective. (If φ is not surjective, consider $\varphi : (\bigcap_j \varphi^j(X), \varphi) \to (\bigcap_j \varphi^j(X), \varphi)$ instead.) While the reduced defect is zero in this setting, the defect itself becomes a conjugacy invariant for dynamical systems, which carries information, not on the complexity of the dynamical system, but about 'how non-invertible' the system is. As an invariant it is in some cases more sensitive than the topological entropy. This is illustrated in the last example below.

Example 5.28. Let $n \in \mathbb{N}$, $n \geq 3$. For each natural number k, $k \geq \frac{n-2}{2}\sqrt{1 + n^{-2}}$, there is a (unique) piecewise linear map $\varphi_{n,k} : [0, 1] \to [0, 1]$ with slope n on all intervals of linearity, with $2k + 1$ turning points, such that $\varphi_{n,k}(t) = nt$, $t \in [0, \frac{1}{n}]$, and $\varphi_{n,k}(t) = -nt + n$, $t \in [1 - \frac{1}{n}, 1]$.

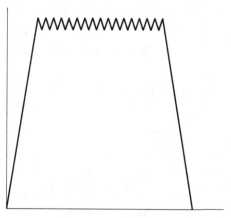

The graph of $\varphi_{n,k}$.

Set $c(n,k) = \frac{n-2}{2k}\sqrt{1+n^{-2}}$ which is $1 - \min_{t \in [\frac{1}{n}, 1 - \frac{1}{n}]} \varphi_{n,k}(t)$. When $\#\varphi_{n,k}^{-1}(t) > 2$, $\#\varphi_{n,k}^{-1}(\varphi_{n,k}^{j}(t)) = 2$ for all $j \geq 1$ such that $n^j c(n,k) < 1 - c(n,k)$. So if $k > \frac{(n^j+1)(n-2)}{2}\sqrt{1+n^{-2}}$, we find that $D(\varphi_{n,k}) \leq \frac{1}{j+1}\log(2k+2) + \frac{j}{j+1}\log 2$. On the other hand, when $n^{j+1}c(n,k) \geq 1$, we have a $j+1$-periodic point x_0 such that $\#\varphi_{n,k}^{-1}(x_0) = 2k+2$ and hence $D(\varphi_{n,k}) \geq \frac{1}{j+1}\log(2k+2) + \frac{j}{j+1}\log 2$ by 2) of Proposition 3.1. So for arbitrary $j \in \mathbb{N}$ we find that

$$D(\varphi_{n,k}) = \frac{1}{j+1}\log(2k+2) + \frac{j}{j+1}\log 2$$

when $\frac{(n^j+1)(n-2)}{2}\sqrt{1+n^{-2}} < k \leq \frac{n^{j+1}(n-2)}{2}\sqrt{1+n^{-2}}$.

REFERENCES

[AM]   R. Adler and B. Marcus, *Topological entropy and equivalence of dynamical systems*, Mem. Amer. Math. Soc. **219** (1979).

[BFF]   M. Boyle, D. Fiebig and U. Fiebig, *Residual entropy, conditional entropy and subshift covers*, Preprint (2000).

[B1]   R. Bowen, *Topological Entropy and Axiom A*, Proceedings of Symposia in Pure Mathematics **14** (1970), 23-41.

[B2]   ———, *Entropy for group endomorphisms and homogeneous spaces*, Trans. Amer. Math. Soc. **153** (1971), 401-414.

[CR]   E.M. Coven and W.L. Reddy, *Positively expansive maps of compact manifolds*, Global Theory of Dynamical Systems, LNM 819, Springer Verlag, Berlin, Heidelberg, New York, 1980.

[DT1]   S. Dorofeev and K. Thomsen, *Factors and subfactors arising from inductive limits of interval algebras*, Ergod. Th. & Dynam. Sys. **19** (1999), 363-381.

[DT2]   ———, *Factors from ergodic theory and group-subgroup subfactors*, J. Ramanujan Math. Soc. **12** (1997), 239-262.

[FR]   J. Franks and D. Richeson, *Shift equivalence and the Conley index*, Trans. Amer. Math. Soc. **352** (2000), 3305-3322.

[J]   V. Jones, *Index for subfactors*, Invent. Math. **72** (1983), 1-25.

[JR]   L. Jonker and D. Rand, *Bifurcations in One Dimension. I. The Nonwandering Set*, Invent. Math. **62** (1981), 347-365.

[KH]   A. Katok and B. Hasselblatt, *Introduction to the Modern Theory of Dynamical Systems*, Encyclopedia of Math. and its Appl. **54**, Cambridge Univ. Press (1995).

[LM]   D. Lind and B. Marcus, *An Introduction to Symbolic Dynamics and Coding*, Cambridge University Press (1995).

[MS]   W. de Melo and S. van Strien, *One-dimensional Dynamics*. Ergebnisse Series **25**, Springer Verlag, 1993.

[MT]   J. Milnor and W. Thurston, *On iterated maps of the interval*, LNM 1342, pp. 465-563, Springer Verlag, Berlin, 1988.

[NP]   D. Newton and W. Parry, *On a factor automorphism of a normal dynamical system*, Ann. Math. Statist. **37** (1966), 1528-1533.

[NS]   T. Nowicki and D. Sands, *Non-uniform hyperbolicity and universal bounds for S-unimodal maps*, Invent. Math. **132** (1998), 633-680.

[P]   W. Parry, *A finitary classification of topological Markov chains and sofic systems*, Bull. London Math. Soc. **9** (1977), 86-92.

[Ru]   D. Ruelle, *Thermodynamic Formalism*, Encyclopedia of Mathematics and its Applications 5, Addison-Wesley, Reading, Ma. 1978.

[S]   M. Shub, *Endomorphisms of compact differentiable manifolds*, Amer. J. Math. **91** (1969), 175-199.

[St]   S. van Strien, *Smooth dynamics on the interval*. In 'New Directions in Dynamical Systems', Cambridge Univ. Press (1987), pp. 57-119.

[Th1]   K. Thomsen, *The defect of factor maps*, Ergod. Th. & Dynam. Sys. **17** (1997), 1233-1256.

[Th2]   ———, *The variational principle for the defect of factor maps*, Israel J. Math. **110** (1999), 359-369.

[W]   P. Walters, *An Introduction to Ergodic Theory*, Springer Verlag, New York, Heidelberg, Berlin, 1982.

[Wi]   R. F. Williams, *Classification of one dimensional attractors*, Proceedings of Symposia in Pure Mathematics **14** (1970), 341-361.

DEPARTMENT OF MATHEMATICAL SCIENCES, UNIVERSITY OF AARHUS, NY MUNKEGADE BUILDING 530, 8000 ARHUS, DENMARK
E-mail address: matkt@imf.au.dk

ACTIONS OF AMENABLE GROUPS

BENJAMIN WEISS

ABSTRACT. This survey begins with a detailed treatment of the basic entropy theory for amenable groups that admit approximately invariant sets that tile the group. The recent pointwise theorems of Elon Lindenstrauss are given new proofs and there is also a brief discussion of the latest results obtained in the theory of actions with completely positive entropy.

CONTENTS

1. INTRODUCTION

Classical ergodic theory deals with the properties of measure preserving actions of $\mathbb{Z}$ and $\mathbb{R}$. That is to say either a single measure preserving transformation T of a measure space $(X, \mathcal{B}, \mu)$ or a one parameter flow T_t of such transformations. One of the first theorems in the subject is von-Neumann's ergodic theorem, which deals with the "time averages"

$$\frac{1}{n} \sum_{j=0}^{n-1} f(T^j x), \quad \frac{1}{n} \int_0^n f(T_t x) dt$$

of a square integrable function f and asserts the convergence of these averages to the projection of f onto the space of invariant functions.

Although von Neumann's original proof was based on the spectral representation theorem, later proofs by F. Riesz and others were based only on the fact that the sets $\{0, \ldots, n-1\}, [0, n]$ are approximately invariant. The class of groups that possess such sets are the **amenable groups** and in the last few decades many parts of the classical ergodic theory were extended to this more general setting. It is the purpose of this survey to give some idea of what has been accomplished in this direction. In the first section we shall introduce our main actors the amenable groups, and their actions and prove the mean ergodic theorem. Then in the next sections we shall give a detailed exposition of the basic Rokhlin lemma and the main results in the entropy theory of amenable groups. For the sake of simplicity we shall restrict attention to those groups that admit approximately invariant sets that tile the group. As was shown in [W] this includes a very large class of groups. For the general amenable group one needs to develop the theory of quasitiles as in [OW-1987] in order to obtain many of the results that we shall describe below. This adds additional complications which are not suitable for this type of survey.

Following that I shall discuss pointwise results especially the new work of E. Lindenstrauss [L] in which pointwise ergodic theorems are proven for all amenable groups. He used a probabilistic construction to establish Vitaly type covering lemmas and I will describe a deterministic construction that I found in collaboration with Don Ornstein. In the last section I shall survey some recent results concerning actions with complete positive entropy. In conclusion I wish to thank Elon Lindenstrauss who wrote up sections 2-4 as notes from a course that I gave several years ago at the Hebrew University, and Don Ornstein for permission on include our work described in sections 5-8.

2. AMENABLE GROUPS AND THE MEAN ERGODIC THEOREM

Throughout this survey G will be a countable amenable group. We will take as our definition of an amenable group the existence of a sequence of finite sets $\{F_n\}_1^\infty$, with the property that for all $g \in G$

$$\lim_{n \to \infty} |gF_n \triangle F_n|/|F_n| = 0.$$

This condition was shown to be equivalent to the earlier definition via the existence of invariant means on $\ell_\infty(G)$ by Erling Følner and the sets $\{F_n\}$ are sometimes called Følner sets for this reason. Clearly for $\mathbb{Z}$ we may take $F_n = \{0, 1, \ldots, n-1\}$, or for that matter $\{a_n, a_n + 1, \ldots, a_n + n - 1\}$ for any sequence of a_n. We shall use the convenient terminology that a set F in G is K, ϵ - invariant to mean that for all but an ϵ fraction of the elements f of F Kf is contained in F. For any finite set K and any positive ϵ all sufficiently large Følner sets will be K, ϵ - invariant.

It is easy to see that direct products of amenable groups are amenable, as are increasing unions of amenable groups and thus one sees that all abelian groups are amenable. If $H \triangleleft G$ and both H and G/H are amenable one can verify that G is also amenable, and this shows that all solvable groups are amenable. The archetypical example of a non amenable group is the free group on 2 generators. The fact that this group has no invariant mean is seen via "paradoxical decompositions" of it. These lie at the basis of the Hausdorff-Banach-Tarski paradoxical decompositions of the sphere and were the genesis of the initial interest in the phenomenon of amenability. for a more thorough discussion of these groups we refer to [G] and [P].

By an action of G on a probability space $(X, \mathcal{B}, \mu)$ we mean a homomorphism of G to the group $Aut(X)$ of measure preserving transformations of X. For economy of notation we shall usually denote this action, or mapping, from $G \times X$ to X by "gx", rather than by the $\mathbb{Z}$-notation which would call for $T^g x$ or $T_g x$. The action will be called **ergodic** if the only functions invariant under all g in G are the constants. We can now formulate the basic mean ergodic theorem in this setting:

Theorem 2.1 ("Mean Ergodic Theorem"). *If G is an amenable group acting on $(X, \mathcal{B}, \mu)$ and $\{F_n\}_1^\infty$, is any sequence of approximately invariant sets then for any $f \in L^2(X, \mathcal{B}, \mu)$*

$$\lim_{n \to \infty} \|\frac{1}{|F_n|} \sum_{g \in F_n} f(gx) - \hat{f}(x)\|_{L^2} = 0$$

where $\hat{f}$ is the projection of f onto the space of G-invariant functions $\mathcal{H}_{inv}$.

Proof. Since the result is clear for $f \in \mathcal{H}_{inv}$, it suffices to see what happens on its ortho complement $\mathcal{H}_1$ in L^2. If ϕ is a bounded function and $a \in G$, then clearly $\phi(ax) - \phi(x) \in \mathcal{H}_1$. It is also not hard to see that functions of this type span a subspace that is **dense** in $\mathcal{H}_1$. Indeed if

$$(\phi(ax) - \phi(x), \ f(x)) = 0$$

then since the action is measure preserving

$$\big(\phi(x), (f(a^{-1}x) - f(x))\big) = 0$$

and if that holds for all bounded ϕ then $f(a^{-1}x) = f(x)$ for μ - a.e. x, hence anything orthogonal to all such functions lies in $\mathcal{H}_{inv}$.

Finally, for a function of that form we have

$$|\frac{1}{|F_n|} \sum_{g \in F_n} (\phi(agx) - \phi(gx))| \leq \|\phi\|_\infty \cdot \frac{|aF_n \triangle F_n|}{|F_n|}$$

and by assumption this tends to zero as $n \to \infty$. Since these averages have norm at most one, this convergence to zero extends to the closed linear span of these functions, which as we have remarked is all of $\mathcal{H}_1$. Thus the proof of the theorem is completed.   □

The proof gives the result in a very uniform fashion. For a fixed function f, and $\varepsilon > 0$, one can specify a certain degree of approximate invariance needed for the average over F_n to be within ε of the projection $\hat{f}$. Even in situations where there is a pointwise convergence theorem no such uniformity holds. In this sense, the mean convergence result is stronger than the pointwise convergence due to G. D. Birkhoff (for $\mathbb{Z}$ and $\mathbb{R}$).

To conclude this brief introduction let me give some basic examples of an action of G. Let (A, P) be any probability space, where A is compact, for example $A = \{-1, +1\}$ with $P(-1) = p, P(+1) = 1-p$. Let $\Omega = A^G$ with the product topology and let μ be the product measure where in each component we take P. On Ω we can define two actions of G, the left shift L_g and the right shift, R_g defined by

$$(L_{g_0}\omega)(g) = \omega(g_0^{-1}g)$$
$$(R_{g_0}\omega)(g) = \omega(gg_0)$$

These actions both preserve the product measure μ, and are called the standard Bernoulli actions of G and correspond to the independent processes in the case of $\mathbb{Z}$. It is an easy exercise to verify that they are **ergodic**. Many more processes can be obtained by considering partitions of these spaces and the processes that they define. This means the following: Fix any partition of Ω into two sets ,B_{-1} and B_1 define a mapping from Ω to Ω by sending each point ω to its itinerary under the action of G. The itinerary of a point ω under the G action L_g with respect to a partition is that element of Ω whose g coordinate is zero or one according to set of the partition that the point $L_g\omega$ lies in. Clearly this mapping is equivariant with respect to the actions and the image of the product measure under this mapping is now a new invariant measure on Ω.

Another example begins with some unitary representations of G on a Hilbert space H. Fixing a vector $u \in H$ and defining

$$\psi(g) = (\rho(g)u, u)$$

we get a positive definite function on G. Now take Gaussian random variables indexed by g with covariance function given by $\psi(g)$. This defines a stationary Gaussian process and the action of G on the sample space of this process is clearly a measure preserving action of G.

Finally, in case G is residually finite, and $\hat{G}$ denotes it profinite completion, then G imbeds as a dense subgroup in the compact group $\hat{G}$ and taking Haar measure on $\hat{G}$ we get a natural action of G on the probability space $(\hat{G}, \text{Haar measure})$. This last action has discrete spectrum.

3. ROKHLIN TOWERS

Let G act on a probability measure space $(X, \mathcal{B}, \mu)$, preserving the measure μ. Recall that a map $T\colon X \to X$ is measure preserving if it is measurable (i.e. for every measurable set $B \in \mathcal{B}$, $T^{-1}B \in \mathcal{B}$) and furthermore for any such B,

$$\mu(T^{-1}B) = \mu(B).$$

We will need to assume an additional condition on the measure space we are working with:

Assumption: For any measure preserving map $T\colon X \to X$, if $\mu(B \bigtriangleup T^{-1}B) = 0$ for all $B \in \mathcal{B}$ then for μ almost every x, $Tx = x$.

We will make this assumption implicitly throughout the rest of this survey.

Remark: This assumption is satisfied for any reasonably nice measure space, for example if

$$(3.1) \qquad\qquad (X, \mathcal{B}, \mu) \cong ([0,1], \mathrm{Borel}, \nu),$$

for some Borel probability measure ν. All Borel measures on complete separable metric spaces are of this type. A measure space that satisfies (3.1) is called a **Lebesgue measure space**.

Recall also that we assume the group G is countable.

Definition 3.1. *We shall say that a (measure preserving) action of a group G on a measure space $(X, \mathcal{B}, \mu)$ is **free** if for every $g \neq e$ the set $\{x : gx = x\}$ has zero measure.*

We leave the proof of the following easy proposition to the reader

Proposition 3.2. *G acts freely on X iff for a.e. x the map $g \mapsto gx$ is one-to-one.*

We now state the Rokhlin Tower Lemma, and give a simple proof for a special case of this theorem for $G = \mathbb{Z}$. First let us recall what it means for a set to tile the group G. We will say that a set T **tiles** the group G if there is a set C, whose elements we will call **tiling centers** such that Tc is a partition of G as c ranges over all the elements of C. We shall use the same letter T both to denote a generic tile as well as the generic generator of a $\mathbb{Z}$ action. It should be clear from the context which is intended.

Theorem 3.3. *Let G act freely on $(X, \mathcal{B}, \mu)$, preserving the measure μ, with G an amenable group. Let $T \subset G$ be a finite set that tiles G. Then for all $\epsilon > 0$ there is a $B \in \mathcal{B}$ such that*

1. *the sets $\{tB : t \in T\}$ are disjoint, and*
2. *$\mu(\bigcup_{t \in T} tB) > 1 - \epsilon$.*

Sketch of proof for the case $G = \mathbb{Z}$ and $T = [0, \dots, N-1]$:

Assume there is a set $E \subset X$ such that $\mu(E) < \frac{\epsilon}{N}$ and such that $\mu(\bigcup_{i=0}^{\infty} T^i E) = 1$. If μ is ergodic one can take any set with small enough positive measure; for the general case this is still true but a little harder to show.

Define $E_0 = E$ and

$$E_k = T^k E \setminus \bigcup_{i=0}^{k-1} T^i E.$$

We take B to be

$$B = E_N \cup E_{2N} \cup E_{3N} \cup \dots .$$

Clearly $B \cap T^i B = \emptyset$ for $i \in [0, \ldots, N-1]$. Furthermore

$$\mu(\bigcup_{i=0}^{N-1} E_i \cup \bigcup_{i=0}^{N-1} T^i B) = \mu(\bigcup_{i=0}^{\infty} T^i E) = 1,$$

hence

$$\mu(\bigcup_{i=0}^{N-1} T^i B) \geq 1 - N\mu(E) > 1 - \epsilon.$$

$\square$

We now proceed to prove the general case, but before we do this, we would like to explain why we need to assume that T tiles G.

Definition 3.4. *Let G act on the space X, and let T be a finite subset of G. We shall say that a set $E \subset X$ is T-disjoint if for any $t \neq t' \in T$,*

$$tE \cap t'E = \emptyset.$$

Definition 3.5. *A subset $T \subset G$ will be called an R-set if for every free measure preserving action of G on a measure space $(X, \mathcal{B}, \mu)$, and every positive ϵ, there is a $B \in \mathcal{B}$ such that B is T-disjoint and*

$$\mu(TB) \equiv \mu(\bigcup_{t \in T} tB) \geq 1 - \epsilon.$$

Thus Theorem 3.3 can be restated as follows: every finite subset of an amenable group G that tiles G is an R-set. For any group G, if T is an R-set, it must tile G:

Proposition 3.6. *If G is any group, $T \subset G$ is an R-set, then T tiles G.*

Proof:

We first note that it is enough to show that any subset $K \subset G$ can be covered by disjoint translations of T. This is a standard reduction, that can be seen as follows: the subsets of G are in one-on-one correspondence with the points in $\{0,1\}^G$. The latter space is compact in the usual product topology. Let K_i be an increasing sequence of finite subsets of G with $\bigcup_{i=1}^{\infty} K_i = G$, and assume that each K_i can be tiled by T — i.e. there is a set E_i such that for any $g_i \in K_i$,

$$|T^{-1} g_i \cap E_i| = 1.$$

The points of $\{0,1\}^G$ that correspond to a subsequence of the E_i's (which we may as well assume to be the original sequence) converges to a point in $\{0,1\}^G$, that corresponds to a set $E \subset G$. This set gives us a tiling of G by $\{Tg : g \in E\}$ Indeed, every $g \in G$ is in K_i for all i large enough, hence $|T^{-1} g \cap E_i| = 1$ for all large i. Since $T^{-1} g$ is finite, eventually $T^{-1} g \cap E_i$ stabilizes and is equal to $T^{-1} g \cap E$.

Let K be a finite subset of G. To use the information that T is an R-set, we need a free T action. If G is finite, then the action of G on itself by left translations if free, and the uniform measure on G is invariant. For ϵ sufficiently small the defining property of an R set shows that R-sets tile.

For the more interesting case of infinite G, we take X to be $\{0,1\}^G$ with $\mathcal{B}$ the σ-algebra of Borel sets according to the product topology and μ the product measure with each of 0 and 1 having probability 0.5. As mentioned earlier, this is a measure preserving action which, for infinite G, is clearly free.

Taking $\epsilon \ll 1/|KT|$, we now use the fact that T is an R-set to find a subset $B_0 \subset X$ that is T-disjoint and such that $\mu(TB_0) > 1 - \epsilon$.

As $\mu(TB_0) > 1 - 1/|KT|$,

$$\bigcap_{g \in KT} g^{-1}(TB_0) \neq \emptyset.$$

In particular, there is an $x \in X$ such that $gx \in TB_0$ for every $g \in TK$, and this gives a cover of K by disjoint translations of T as follows:

$$K \subset \bigcup_{g \in TK: gx \in B_0} Tg.$$

$\square$

We now prove Theorem 3.3. We start with two lemmas:

Lemma 3.7. *Let G act freely on the measure space $(X, \mathcal{B}, \mu)$, and F be a finite subset of G. Let $B \in \mathcal{B}$ have positive measure. Then there is a $B_0 \subset B$ so that $\mu(B_0) > 0$ and B_0 is F-disjoint.*

Proof: We first prove the lemma when $F = \{e, f\}$, where $e \in G$ is the identity and $f \neq e$. We claim that under our assumptions, there is a set $C \subset B$ with positive measure such that $\mu(C \triangle fC) > 0$. Indeed, if no such C exist (and in particular $\mu(B \triangle fB) = 0$), then using our assumption on the measure space $(X, \mathcal{B}, \mu)$ we have that for almost every $x \in B$ $fx = x$ — a contradiction since we assumed G acts freely.

Take $B_0 = C \setminus fC$. Clearly

$$f(C \setminus fC) \cap (C \setminus fC) = \emptyset.$$

Furthermore, if $\mu(C \setminus fC) = 0$ then since the action is measure preserving $\mu(C \triangle fC) = 0$. Thus $\mu(B_0) > 0$.

The general case can be handled as follows. Let

$$F^{-1}F = \{f^{-1}g : f, g \in F\} = \{e, f_1, \dots, f_m\}$$

with $f_i \neq e$. Take $B_1 \subset B$ to be $\{e, f_1\}$ disjoint, $B_2 \subset B_1$ $\{e, f_2\}$ disjoint, etc., with all B_i ($i = 1, \dots, m$) having positive measure. For every $i \leq m$ and $j \leq i$ the set B_i is $\{e, f_j\}$ disjoint as a subset of B_j. Thus $B_m \cap f_k B_m = \emptyset$ for all $1 \leq k \leq m$. This shows that for every $f \neq g \in F$, the sets fB_m and gB_m are disjoint. Thus we can take $B_0 = B_m$. $\square$

Using inductive type reasoning, we can deduce from this the following:

Lemma 3.8. *Let G act freely on $(X, \mathcal{B}, \mu)$, and $F \subset G$ a finite set. Let B be a subset of X with positive measure. Then there exists a countable partition of B (up to measure zero) by sets which are F-disjoint.*

Proof: Consider

$$\mathcal{A} = \{A \in \mathcal{B} : A \subset B \text{ and } A \text{ is } F \text{ disjoint}\}.$$

According to the lemma,

$$\sup_{A \in \mathcal{A}} \mu(A) > 0$$

Take $A_1 \in \mathcal{A}$ such that $\mu(A_1) \geq \frac{1}{2} \sup \mu(A)$. Define inductively sets A_i ($i = 2$, ...) by

$$\mathcal{A}_{i-1} = \{A \in \mathcal{B} : A \subset B \setminus \bigcup_{j=1}^{i-1} A_j \text{ and } A \text{ is } F \text{ disjoint}\},$$

and take $A_i \in \mathcal{A}_{i-1}$ with

$$\mu(A_i) \geq \frac{1}{2} \sup_{A \in \mathcal{A}_{i-1}} \mu(A).$$

The sets A_i are disjoint, so $\sum_{i=1}^{\infty} \mu(A_i) \leq 1$ and $\mu(A_i) \to 0$. If $\mu(B \setminus \bigcup_{i=1}^{\infty} A_i) > 0$, there is an F-disjoint subset A of $B \setminus \bigcup_{i=1}^{\infty} A_i$ with positive measure. This set A is in all $\mathcal{A}_i$. Hence

$$0 < \mu(A) \leq \sup_{A \in \mathcal{A}_{i-1}} \mu(A) \leq 2\mu(A_i) \to 0 \qquad \text{as } i \to \infty$$

a contradiction. $\qquad\qquad\qquad\qquad\qquad\qquad\qquad\qquad\qquad\qquad\qquad\square$

Notice that if A_i are as above for $B = X$, then the FA_i's cover a.e. point exactly $|F|$-times — in particular, $\{FA_i\}_{i=1}^{\infty}$ evenly covers X.

Lemma 3.9. *Suppose that $\{U_i\}_{i=1}^{\infty}$ evenly cover X; then for any $\delta > 0$ there is a sub-collection $\{U_i\}_{i \in I}$ such that*

1. $\mu(\bigcup_{i \in I} U_i) \geq 1 - \delta$,
2. $\sum_{i \in I} \mu(U_i) \leq \delta^{-1}$.

Proof. Assume a.e. $x \in X$ is covered by exactly M of the U_i. Then for any U

$$\sum_{i=1}^{\infty} \mu(U_i)\mu(U) = M\mu(U) = \sum_{i=1}^{\infty} \mu(U_i \cap U)$$

hence there is an i such that $\mu(U)\mu(U_i) \geq \mu(U \cap U_i)$. We also sort the U_i's so that $\mu(U_1) \geq \mu(U_2) \geq \ldots$, omitting those U_i's with zero measure.

We define a sequence of increasing indexing sets I_k inductively as follows:

- Take $I_1 = \{1\}$ (and $A_0 = \emptyset$).
- Assume I_k is defined, and set $A_k = \bigcup_{i \in I_k} U_i$. If $\mu(A_k) \geq 1 - \delta$ we terminate the sequence and take $I = I_k$.

- Otherwise, since $\mu(A_k) < 1 - \delta$, there is an i such that

$$\mu(U_i \cap A_k) \le \mu(U_i)(1 - \delta).$$

Take i_{k+1} to be the smallest such k, and set $I_{k+1} = I_k \cup \{i_{k+1}\}$.

If the sequence I_k does not terminate at any finite stage, we set $I = \bigcup_{k=1}^{\infty} I_k$.
Take $V_k = U_{i_k} \setminus A_{k-1}$. The V_k's are disjoint, and

(3.2) $$\mu(V_k) = \mu(U_{i_k}) - \mu(U_{i_k} \cap A_{k-1}) \ge \delta\mu(U_{i_k}).$$

Thus

$$\sum_{i \in I} \mu(U_k) \le \delta^{-1}\mu(\bigcup_i V_i) \le \delta^{-1}$$

(in $\bigcup_i V_i$ i runs from 1 to the k where the sequence I_k was terminated,
or infinity if the sequence did not terminate.) Thus I satisfies (2.1) in the
statement of the lemma.

Assume that (1) does not hold. This is clearly impossible if the sequence
I_k was terminated at some finite stage. Set $A_\infty = \bigcup_{i \in I} U_i$, and we assume
that $\mu(A_\infty) < 1 - \delta$. There is a U_j so that

$$\mu(U_j \cap A_\infty) \le \mu(U_j)(1 - \delta)$$

hence for all k

$$\mu(U_j \cap A_k) \le \mu(U_j)(1 - \delta)$$

and so by the way we chose the i_k's $\mu(U_{i_k}) \ge \mu(U_j) > 0$ for all k. On the
other hand, by 3.2, $\mu(U_{i_k}) \to 0$ as $k \to \infty$ — a contradiction. $\square$

We are now ready to prove that if G is amenable, any tiling set is an R-set:

Proof of Theorem 3.3:

Suppose that G acts freely on $(X, \mathcal{B}, \mu)$, and that T tiles G. We are given
an $\epsilon > 0$, and want to find a T-disjoint set $B \subset X$ such that $\mu(TB) > 1 - \epsilon$.

Take $\delta = \epsilon/10$, and we take F to be a $(TT^{-1}, \delta^2/10)$-invariant subset of
G. Now we can assume that F is of the form TC, where C is T-disjoint,
that is, that F can be perfectly tiled by T. Indeed, find a $(TT^{-1} \cup T, \frac{\delta^2}{100|T|^2})$-
invariant set F'. Since T tiles G, there is a set C of minimal cardinality, so
that $F' \subset TC$ and

$$Tc \cap Tc' = \emptyset \qquad \text{for } c \ne c' \in C.$$

We take $F = TC$. Since

$$|TC \setminus F'| \le |TF' \setminus F'| \le \frac{\delta^2}{100|T|^2}|F|$$

we see that

$$|TT^{-1}F \triangle F| \le (|T|^2 + 1)|F \triangle F'| + |TT^{-1}F \triangle F'| < \frac{\delta^2}{10}.$$

By Lemma 3.8, we can find a countable partition $\{U_i\}$ of X by F-disjoint sets. As we have already mentioned, $\{FU_i\}_{i=1}^{\infty}$ evenly covers X, and so there is a set of indices $I = \{i_1, i_2, \ldots\}$ such that

$$\sum_{i \in I} \mu(FA_i) \le \delta^{-1}$$

$$\mu(\bigcup_{i \in I} FA_i) \ge 1 - \delta.$$

We now take K large enough so that $I = \{i_1, \ldots, i_K\}$ satisfies

$$\mu(\bigcup_{k=1}^{K} FA_{i_k}) > 1 - 2\delta.$$

We re-index the A_i's so that $i_k = k$ for all $k = 1, \ldots, K$.

For $k = 1, \ldots, K$ we define inductively T-disjoint sets B_k. We will show that $B = B_K$ will satisfy $\mu(TB) > 1 - \epsilon$, in addition to T-disjointness, thus completing the proof of the theorem. The sets $B_k \subset X$ are defined as follows:

1. $B_1 = CA_1$.
2. If B_k is defined, we define B_{k+1} by

$$(3.3) \qquad B_{k+1} = (B_k \setminus T^{-1}FA_{k+1}) \cup CA_{k+1}$$

Claim 1: all B_k are T-disjoint:

Since $F = TC$, the F-disjointness of A_i clearly implies that the CA_i are T-disjoint. In particular, for $i = 1$ we get that B_1 is T-disjoint.

Assume we know that B_k is F-disjoint. Take $t \ne t' \in T$, and assume

$$x \in tB_{k+1} \cap t'B_{k+1}.$$

As B_k is T-disjoint, without loss of generality $x \in t'CA_{k+1}$. Hence

$$t^{-1}x \subset B_{k+1} \cap t^{-1}t'CA_{k+1} \subset CA_{k+1},$$

contradicting the T-disjointness of CA_{k+1}.

Claim 2: $\mu(B_K) > 1 - \epsilon$:

Clearly, $FA_{k+1} \subset TB_{k+1} \subset TB_k \cup FA_{k+1}$. We wish to estimate $\mu(TB_k \cup FA_{k+1}) - \mu(TB_{k+1})$. If $x \in TB_k \setminus TB_{k+1}$ then $x \in TT^{-1}FA_{k+1} \setminus FA_{k+1}$. Hence

$$(3.4) \quad \mu(TB_k \cup FA_{k+1}) - \mu(TB_{k+1}) \le |TT^{-1}F \setminus F|\mu(A_{k+1}) \le \frac{\delta^2}{10}\mu(FA_{k+1}).$$

It is also clear that $\mu(TB_1) = \mu(FA_1)$. Hence using (3.4) for $k = 1, \ldots, K - 1$ we see that

$$\mu(TB_K) \ge \mu(\cup_{i=1}^{K}FA_i) - \frac{\delta^2}{10}\sum_{i=1}^{K}\mu(FA_i) \ge 1 - 2\delta - \frac{\delta^2}{10\delta} > 1 - \epsilon.$$

□

4. Entropy and the Shannon McMillan Theorem

We now wish to define a very important invariant for measure preserving amenable group actions — the entropy. We assume throughout this section that G acts *ergodically* on $(X, \mathcal{B}, \mu)$. As we have said in the introduction we will deal here with tiling Følner sets and not with general Følner sets.

We first define entropy for a *process* — which is the G-action on X equipped with a specific partition $\mathcal{P}$. Recall that a partition of X is simply a collection of disjoint sets whose union is X. If $\mathcal{P}$ is a finite partition $\{P_j\}$, and $F \subset G$ finite, we will denote by $\mathcal{P}^F$ the partition

$$\mathcal{P}^F = \{\cap_{f \in F} f^{-1} P_{j(f)}\}.$$

It is useful to think of $\mathcal{P}$ as a function that assigns to every $x \in X$ an element of $\{1, \ldots, |\mathcal{P}|\}$ — its $\mathcal{P}$-name. Then $\mathcal{P}^F$ is simply the map that assigns to every $x \in X$ the $\mathcal{P}$-names of all the points fx with $f \in F$.

In the case that $G = \mathbb{Z}$, with $F = \{0, 1, \ldots, n-1\}$, the partition $\mathcal{P}^F$ then corresponds to a vector $\xi_0, \xi_1, \ldots, \xi_{n-1}$ of n discrete random variables on the probability space $(X, \mathcal{B}, \mu)$ with $|\mathcal{P}|$ possible values for each. The random variables $\{\xi_i\}$ form a stationary stochastic process determined by $(X, \mathcal{B}, \mu, \mathcal{P}, \mathbb{Z})$.

After defining the entropy of a process we can obtain an invariant of the G action by taking the supremum over all finite partitions. This invariant is called the (measure theoretic) entropy of $(X, \mathcal{B}, \mu, G)$.

Our definition of entropy will depend on a particular choice of a tiling Følner sequence F_n. Latter we shall see that the choice of this sequence does not affect the numerical value of the entropy.

Definition 4.1. *Let $\mathcal{P}$ be a finite partition of X. Set, for $a \in (0, 1)$,*

$$b(F_n, a, \mathcal{P}) = \min\{|\mathcal{C}| : \ \mathcal{C} \subset \mathcal{P}^{F_n} \ \text{s.t.} \ \mu(\cup \mathcal{C}) \geq a\}.$$

Then the entropy $h(\mathcal{P}) = h(X, \mathcal{P}, G)$ is defined as

$$h(\mathcal{P}) = \lim_{\epsilon \to 0} \liminf_{n \to \infty} \frac{1}{|F_n|} \log b(F_n, 1 - \epsilon, \mathcal{P}),$$

(all our logarithms are base 2) and

$$h = h(X, \mathcal{B}, \mu, G) = \sup_{\mathcal{P} \ \text{finite partition of } X} h(\mathcal{P}).$$

Theorem 4.2 ("Shannon-McMillan" Theorem for Finite Partitions). *Let $\mathcal{P}$ be a fixed finite partition of X. For any $\epsilon > 0$, for any sufficiently invariant tiling set $F \subset G$ the atoms of $\mathcal{P}^F$ from the collection $\mathcal{C}$,*

$$(4.1) \qquad \mathcal{C} = \left\{ C \in \mathcal{P}^F : \ 2^{-(h(\mathcal{P}) + \epsilon)|F|} < \mu(C) < 2^{-(h(\mathcal{P}) - \epsilon)|F|} \right\},$$

cover at least $1 - \epsilon$ of X.

By F sufficiently invariant we mean that there are a finite set $K \subset G$ and $\delta > 0$ (that depend on ϵ), such that if F is (K, δ)-invariant, then the conclusion holds.

Before we proceed to prove this theorem, we would like to derive the following corollary:

Corollary 4.3. *For any tiling Følner sequence F'_n and any $a \in (0, 1)$,*

$$(4.2) \qquad h(\mathcal{P}) = \liminf_{n \to \infty} \frac{1}{|F'_n|} \log b(F'_n, a, \mathcal{P}) = \limsup_{n \to \infty} \frac{1}{|F'_n|} \log b(F'_n, a, \mathcal{P}).$$

Proof: Chose a very small $\epsilon \leq \min(a, 1 - a)/2$. For n large enough, F'_n will satisfy the requirements on F in Theorem 4.2. Let $\mathcal{C}$ be as in (4.1). Let $\mathcal{D} \subset \mathcal{P}^{F'_n}$ be a collection of $b(F'_n, a, \mathcal{P})$ sets that cover at least a of X. By the Shannon-McMillan Theorem, the sets in $\mathcal{D} \cap \mathcal{C}$ have total measure at least $a - \epsilon > a/2$ of X.

The measure of each set $C \in \mathcal{C}$ is at most

$$\mu(C) \leq 2^{-(h(\mathcal{P}) - \epsilon')|F'_n|}$$

and so

$$b(F'_n, a, \mathcal{P}) = |\mathcal{D}| \geq |\mathcal{D} \cap \mathcal{C}| \geq \frac{a/2}{\max_{c \in \mathcal{C}} \mu(C)} \geq a 2^{(h(\mathcal{P}) - \epsilon')|F'_n| - 1}$$

hence

$$\liminf_{n \to \infty} \frac{1}{|F'_n|} \log b(F'_n, a, \mathcal{P}) \geq h(\mathcal{P}) - \epsilon.$$

Conversely, since $\mathcal{C}$ covers $1 - \epsilon$ of X, there is a minimal subcollection $\mathcal{D}$ of $\mathcal{C}$ that covers a of X. Since for any $C \in \mathcal{C}$,

$$\mu(C) \geq 2^{-(h(\mathcal{P}) + \epsilon')|F'_n|},$$

we see that

$$b(F_n, 1 - \epsilon, \mathcal{P}) \leq |\mathcal{D}| \leq \frac{1}{\min_{c \in \mathcal{C}} \mu(C)} \leq 2^{(h(\mathcal{P}) + \epsilon')|F'_n|},$$

and

$$\limsup_{n \to \infty} \frac{1}{|F'_n|} \log b(F'_n, a, \mathcal{P}) \leq h(\mathcal{P}) + \epsilon.$$

As ϵ is arbitrarily small, (4.2) is established. $\qquad\square$

The following lemma, though rather simple, is actually quite deep, and will be used again and again.

Lemma 4.4. *Suppose that T tiles G (with the identity $e \in T$), and let $\epsilon > 0$. Then for any finite subset $F \subset G$, there is a collection $\mathcal{E}$ of centers of T-tilings of F so that for any $f \in F$,*

$$1 - \epsilon < \frac{|T|}{|\mathcal{E}|} \#\{E \in \mathcal{E} : f \in E\} < 1 + \epsilon.$$

Recall that E is a set of centers for a T tiling of F if $|T^{-1}f \cap E| = 1$ for every $f \in F$, and $E \subset T^{-1}F$.

Proof:

Choose $\tilde{F}$ to be much more invariant than F and T, and let $\tilde{E}$ be some fixed set of centers for a T-tiling of $\tilde{F}$. We take $\tilde{F}_0 \subset \tilde{F}$ to be all $g \in \tilde{F}$ such that $Fg \subset \tilde{F}$.

For any $g \in \tilde{F}_0$, we can induce a tiling on F from that of $\tilde{F}$ by taking as centers the set

$$E_g = (\tilde{E} \cap T^{-1}Fg)g^{-1}.$$

This is easily seen to be a tiling, since for any $f \in F$, $fg \in \tilde{F}$ and

$$|T^{-1}f \cap E_g| = |T^{-1}fg \cap \tilde{E}| = 1.$$

We take

$$\mathcal{E} = \{E_g : g \in \tilde{F}_0\}.$$

We need to show that $\mathcal{E}$ covers F nearly evenly. Let $f \in F$. Then

$$\#\{g \in \tilde{F}_0 : f \in E_g\} = \#\{g \in \tilde{F}_0 : f \in (\tilde{E}g^{-1} \cap T^{-1}F)\}|$$
$$= |f\tilde{F}_0 \cap \tilde{E}|.$$

We estimate this as follows:

$$|T| \times |\tilde{F}_0 \cap f^{-1}\tilde{E}| = |T(f\tilde{F}_0 \cap \tilde{E})|$$

since $\tilde{F} \subset T\tilde{E}$ and $e \in T$,

$$f\tilde{F}_0 \subset T(f\tilde{F}_0 \cap \tilde{E}) \subset TF$$

If $\tilde{F}$ is sufficiently invariant, the cardinality of all the above sets is within $\epsilon|\mathcal{E}|$ of $|\mathcal{E}| = |\tilde{F}_0|$. $\square$

We now continue with the proof of Theorem 4.2. For any $\epsilon > 0$ and F, we shall call those $P \in \mathcal{P}^F$ with

$$\mu(P) > 2^{-(h(\mathcal{P})-\epsilon)|F|}$$

ϵ-*fat*, and those $P \in \mathcal{P}^F$ with

$$\mu(P) < 2^{-(h(\mathcal{P})+\epsilon)|F|}$$

ϵ-*thin*. What we need to show that for F invariant enough, the ϵ-fat and ϵ-thin sets do not contain more than ϵ of X. That the ϵ-thin sets do not contain more than $\epsilon/2$ of X will be shown to be a corollary of the following lemma. If $F \subset G$ and T is a tiling set we define

$$C(F,T) = \#\{\text{sets of centers of tilings of } F \text{ by } T\} \leq \sum_{0 \leq i < |T^{-1}F|/|T|} \binom{|T^{-1}F|}{i}.$$

Lemma 4.5. *Let F' be a tiling set, $\mathcal{P}$ a finite partition of X and $\epsilon \in (0,1)$. Set $\beta := |F'|^{-1}$. Then for any $\epsilon' > 0$, any sufficiently invariant set F (depending on ϵ' and all the other parameters) satisfies*

$$(4.3) \qquad b(F, 1 - \epsilon', \mathcal{P}) \leq C(F, F') \, 2^{4\beta|F|} \, |\mathcal{P}|^{4\epsilon F} \, b(F', 1 - \epsilon, \mathcal{P})^{(1+\epsilon')\beta|F|}.$$

Proof. We can clearly assume $\epsilon' < \frac{\epsilon}{100}$. Let $\mathcal{C}$ be a collection of $b(F', 1 - \epsilon, \mathcal{P})$ sets from $\mathcal{P}^{F'}$ whose union has measure at least $1 - \epsilon$. Set $\Phi = X \setminus \cup \mathcal{C}$. By the Mean Ergodic Theorem, for any sufficiently invariant F, the set

$$X' := \{ x : \sum_{f \in F} 1_\Phi(fx) \leq 2\epsilon|F| \}$$

has $\mu(X') \geq 1 - \epsilon'$.

Now, we also require that F be invariant enough so that there is a collection $\mathcal{E}$ of centers of tilings of F by F' such that for any $f \in F$,

$$(1 - \epsilon')\frac{|\mathcal{E}|}{|T|} \leq \sum_{E \in \mathcal{E}} 1_E(f) \leq (1 + \epsilon')\frac{|\mathcal{E}|}{|T|}.$$

Our last requirement on F is that it is invariant enough so that

$$|F' F| \leq |F|(1 + \frac{\epsilon'}{10})$$

which also implies that every set E of centers for a tiling of F by F' satisfies

$$|F|/|F'| \leq |E| \leq (1 + \frac{\epsilon'}{10})|F|/|F'|.$$

We now estimate the number of $\mathcal{P}^F$ names needed to cover X'. For any $x \in X'$, for at most $2\epsilon|F|$ of the $f \in F$, $fx \in \Phi$. Since $\mathcal{E}$ cover F almost evenly, there is, for every x, an $E_x \in \mathcal{E}$ such that

$$\sum_{g \in E_x} 1_\Phi(gx) \leq 3\epsilon.$$

Since the $\mathcal{P}^{F' E_x}$-name of x is uniquely determined by the elements $\{P_{fg}\}_{f \in F', g \in E_x}$ defined by $P_{fg} \in \mathcal{P}$ and

$$(fg)x \in P_{fg},$$

the $\mathcal{P}^{F' E_x}$-name of x (hence also the $\mathcal{P}^F$-name of x) is determined by the $\mathcal{P}^{F'}$ names of $\{gx\}_{g \in E_x}$. Hence, the number of possible $\mathcal{P}^F$ names for $x \in X'$ is bounded by

$$\sum_{\substack{E \text{ is a set of centers} \\ \text{of a tiling of } F \text{ by } F'}} \# \left\{ \begin{array}{c} \text{choices of at most } 3\epsilon|E| \text{ elements } g \in E \\ \text{such that } gx \in \Phi \end{array} \right\}$$

$$\times |\mathcal{P}^{F'}|^{3\epsilon|E|} \times b(F', 1 - \epsilon, \mathcal{P})^{|E|}.$$

The lemma now follows from elementary combinatorics — there are $C(F, F')$ sets of centers of tilings of F by F', at most $2^{4\beta|F|}$ choices of $3\epsilon|E|$ elements from E, and

$$|\mathcal{P}^{F'}|^{3\epsilon|E|} \times b(F', 1 - \epsilon, \mathcal{P})^{|E|-\beta|F|} \leq |\mathcal{P}|^{4\epsilon|F|}.$$

$\square$

Remark: Notice that the constant in (4.3) can be estimated for sets F with $|F'^{-1}F| \leq 2|F|$, $\beta|F| \gg 1$, $\beta = |F'|^{-1} \ll 1$ by

$$C(F, F')2^{4\beta|F|} \leq 2^{\epsilon|F|}$$

Indeed, by Stirling's Formula,

$$n! = \left(\frac{n}{e}\right)^n \sqrt{2\pi/n}(1 + o(1)),$$

we see that for $k < n$

$$\begin{aligned}
log\binom{n}{k} &= n \log n - k \log k - (n - k)\log(n - k) + o(n) \\
&= -k\log(k/n) - (n - k)\log(1 - k/n) + o(n) \leq n(\phi(k/n) + o(1))
\end{aligned}$$

with $\phi(x) \to 0$ as $x \to 0$. Thus, if F' is sufficiently large, and F sufficiently larger than F',

$$C(F, F')2^{4\beta|F|} \leq 2^{4\beta|F|} \sum_{i \leq |F'^{-1}F|/|F'|} \binom{|F'^{-1}F|}{i} \leq C2|F|2^{|F|(2\phi(2\beta)+4\beta)} \leq 2^{\epsilon|F|}$$

Corollary 4.6. *Given $\epsilon > 0$, if F is invariant enough the ϵ-thin $\mathcal{P}^F$-names cover no more than $\epsilon/2$ of X.*

Proof. Let $\tilde{\epsilon} = \frac{\epsilon}{100(1+\log|\mathcal{P}|)}$. Take F_n to be an element of the Følner sequence of tiling sets we used to define the entropy (sufficiently large for the preceding remark to be true for $F' = F_n$, $\tilde{\epsilon}$, and any large enough F) and assume that

$$\log b(F_n, 1 - \tilde{\epsilon}, \mathcal{P}) \leq |F_n|(h(\mathcal{P}) + \tilde{\epsilon}).$$

Take F to be invariant enough so that Lemma 4.5 (and the remark following that lemma) holds for $F' = F_n$, $\tilde{\epsilon}$ and $\epsilon' = \tilde{\epsilon}$. We also assume $2^{-\epsilon|F|/2} \leq \tilde{\epsilon}$. Then

$$\begin{aligned}
b(F, 1 - \tilde{\epsilon}, \mathcal{P}) &\leq 2^{\tilde{\epsilon}|F|}b(F_n, 1 - \tilde{\epsilon}, \mathcal{P})^{(1+\tilde{\epsilon})|F|/|F_n|}|\mathcal{P}|^{4\tilde{\epsilon}|F|} \\
&\leq 2^{(h(\mathcal{P})+\tilde{\epsilon})(1+\tilde{\epsilon})|F|+\tilde{\epsilon}(1+\log(\mathcal{P}))|F|} \\
&\leq 2^{(h(\mathcal{P})+\epsilon/2)|F|}.
\end{aligned}$$

Let $\mathcal{C} \subset \mathcal{P}^F$ be a subcollection with $|\mathcal{C}| = b(F, 1 - \tilde{\epsilon}, \mathcal{P})$ and $\mu(\cup\mathcal{C}) \geq 1 - \tilde{\epsilon}$. The measure of the union of all the ϵ-thin sets of $\mathcal{C}$ can be at most

$$b(F, 1 - \tilde{\epsilon}, \mathcal{P})2^{-(h(\mathcal{P})+\epsilon)|F|} \leq 2^{-\epsilon|F|/2} \leq \tilde{\epsilon}.$$

The measure of all the ϵ-thin sets from $\mathcal{P}^F \setminus \mathcal{C}$ is at most

$$\mu(X \setminus \mathcal{C}) \leq \tilde{\epsilon}$$

hence the ϵ-thin sets from $\mathcal{P}^F$ cover at most $2\tilde{\epsilon} < \epsilon/2$ of X. □

To prove the other direction, we need a slightly more general version of Lemma 4.5:

Lemma 4.7. *Let F' be a tiling set, $\mathcal{P}$ a finite partition of X, $\beta := |F'|^{-1}$ and $0 < c < d < 1$. Then for any $\epsilon > 0$, any sufficiently invariant set F (again, depending on ϵ and all the other parameters) satisfies*

(4.4)
$$b(F, 1-\epsilon, \mathcal{P}) \leq C(F, F')\, 3^{4\beta|F|}\, |\mathcal{P}|^{4(1-d)F}\, b(F', c, \mathcal{P})^{c\beta|F|}\, b(F', d, \mathcal{P})^{\beta(d-c)|F|}.$$

Proof. Since the proof is very similar to that of Lemma 4.5, we omit some of the details. Let $\mathcal{C}$ be a collection of $b(F', c, \mathcal{P})$ sets from $\mathcal{P}^{F'}$ with total measure $\geq c$, and $\mathcal{D}$ a similar collection with $b(F', d, \mathcal{P})$ sets and total measure $\geq d$. Let $C = \cup\mathcal{C}$ and $D = \cup\mathcal{D}$. Using the Mean Ergodic Theorem, if F is sufficiently invariant then for all x in a set X' of measure at least $1 - \frac{\epsilon}{100}$,

$$\frac{1}{|F|} \sum_{f \in F} 1_C(fx) > c - \frac{\epsilon}{100}$$

$$\frac{1}{|F|} \sum_{f \in F} 1_D(fx) > d - \frac{\epsilon}{100}$$

We again need F to be invariant enough so that if E is a set of centers of a tiling of F by F' then $|E||F'| \leq (1 + \frac{\epsilon}{100})|F|$ We use Lemma 4.4 as in Lemma 4.5 to deduce from this that if F is invariant enough, we can find for every $x \in X'$ a set E_x of centers of a tiling of F by F' such that

$$\frac{1}{E_x} \sum_{f \in E} 1_C(fx) > c - \frac{\epsilon}{50}$$

$$\frac{1}{E_x} \sum_{f \in E} 1_D(fx) > d - \frac{\epsilon}{50}$$

To count how many $\mathcal{P}^F$ names cover X', it is again enough to sum over all sets E of centers of a tiling of F by F', the number of $|E|$-tuples of $\mathcal{P}^{F'}$-names that are needed to cover all those $x \in X'$ such that $E_x = E$ (the number of such E's is at most $C(F, F')$). If $E_x = E$, we let

$$E_{x,c} = \{f \in E : fx \in C\},$$
$$E_{x,d} = \{f \in E \setminus E_x^c : fx \in D\},$$
$$E_x' = E \setminus (E_x^c \cup E_x^d)$$

(there are at most $3^{|E|} \leq 3^{4|F|/|F'|}$ such partitions). For $f \in E_{x,c}$, we know that the $\mathcal{P}^{F'}$-name of fx is in $\mathcal{C}$, for $f \in E_{x,d}$ it is in $\mathcal{D}$, hence all those x with $E_{x,c} = E_c$, $E_{x,d} = E_d$ and $E_x' = E'$ have at most

$$|\mathcal{C}|^{|E^c|}|\mathcal{D}|^{|E^d|}|\mathcal{P}^{F'}|^{|E'|}$$

$\mathcal{P}^F$-names. Since $|\mathcal{C}| \le |\mathcal{D}| \le |\mathcal{P}^{F'}|$ and

$$|E_x| \le \beta|F|(1 + \frac{\epsilon}{50})$$

$$|E_{x,c}| \ge \beta|F|(c - \frac{\epsilon}{50})$$

$$|E_{x,c} \cup E_{x,d}| \ge \beta|F|(d - \frac{\epsilon}{50})$$

we have that

$$|\mathcal{C}|^{|E^c|}|\mathcal{D}|^{|E^d|}|\mathcal{P}^{F'}|^{|E'|} \le |\mathcal{C}|^{\beta c|F|}|\mathcal{D}|^{\beta(d-c)|F|}|\mathcal{P}^{F'}|^{4\beta(1-d)|F|},$$

hence X' can be covered by

$$C(F, F')3^{4\beta|F|}|\mathcal{C}|^{\beta c|F|}|\mathcal{D}|^{\beta(d-c)|F|}|\mathcal{P}|^{4(1-d)|F|}$$

$\mathcal{P}^F$-names.   □

Corollary 4.8. *Given $\epsilon > 0$, if F is tiling and invariant enough the ϵ-fat $\mathcal{P}^F$-names cover no more than $\epsilon/2$ of X.*

Proof. Let $\eta = \frac{\epsilon^2}{100 \log |\mathcal{P}|}$, $h = h(\mathcal{P})$. Assume there exists an exceedingly invariant tiling set F' such that the ϵ-fat $\mathcal{P}^F$-names cover more than $\epsilon/2$ of X — sufficiently invariant so that the η-thin $\mathcal{P}^{F'}$-names have collective measure smaller than η. We can clearly assume that $\beta^{-1} := |F'|$ is big enough so that for all invariant enough F,

$$C(F, F')\, 3^{4\beta|F|} < 2^{\eta|F|}.$$

Take $c = \epsilon/2$ and $d = 1 - \eta$. By our assumptions,

$$b(F', c, \mathcal{P}) \le 2^{(h-\epsilon)|F'|}$$

$$b(F', d, \mathcal{P}) \le 2^{(h+\eta)|F'|}$$

and so for any $\epsilon' > 0$, any invariant enough F satisfies

$$b(F, 1 - \epsilon', \mathcal{P}) \le C(F, F')\, 3^{4\beta|F|}\, |\mathcal{P}|^{4(1-d)|F|}\, b(F', c, \mathcal{P})^{c\beta|F|}\, b(F', d, \mathcal{P})^{(d-c)\beta|F|}$$
$$\le 2^{\eta|F|+c(h-\epsilon)|F|+(d-c)(h+\eta)|F|+4\log|\mathcal{P}|(1-d)|F|}$$
$$\le 2^{(h+(3+4\log|\mathcal{P}|)\eta-c\epsilon)|F|} \le 2^{(h-\epsilon^2/4)|F|},$$

i.e. if F_n is the tiling Følner sequence used to define the entropy,

$$h = \lim_{\epsilon' \to 0} \liminf_{n \to \infty} \frac{\log b(F, 1 - \epsilon', \mathcal{P})}{|F_n|} \le h - \frac{\epsilon^2}{4}$$

a contradiction.   □

This completes the proof of the "Shannon-McMillan" theorem for finite partitions in a monotilable amenable group.

5. COUNTABLE PARTITIONS AND ANOTHER APPROACH TO THE ENTROPY.

In this section we will first extend the Shannon-McMillan theorem to countable partitions and then we will describe another approach to the entropy of a process which is much closer to Shannon's original definition. For this we need to recall his basic definition of the entropy of partition. Notice that the development in the previous section made no explicit use of this fundamental quantity.

Definition 5.1. *Let Q be a finite or countable partition of X. Then*

$$H(Q) = -\sum_{q \in Q} \mu(Q) \log \mu(Q)$$

where it is understood that for $x = 0$, $x \log x = 0$.

In this section we prove a Shannon-McMillan theorem for countable partitions $\mathcal{P}$ with $H(\mathcal{P}) < \infty$. As we shall soon see, $H(\mathcal{P}) < \infty$ will imply that the entropy $h(\mathcal{P})$, which is defined exactly as in the case of a finite partition, is finite.

Theorem 5.2. *Let $\mathcal{P}$ be a countable partition with $H(\mathcal{P}) < \infty$. Then for any ϵ and invariant enough F,*

$$\frac{1}{|F|} \log b(F, 1 - \epsilon, \mathcal{P}) \leq (1 + \epsilon) H(\mathcal{P}),$$

and so, in particular,

$$h(\mathcal{P}) \leq H(\mathcal{P}).$$

We will need later the following stronger version of Theorem 5.2. Notice that if $G = \mathbb{Z}$ (i.e. the action is determined by one map $\theta: X \to X$), and $T = \{0, \ldots, N-1\}$ this stronger version is equivalent to Theorem 5.2 applied to θ^N. For $T = \{e\}$, Theorem 5.3 reduces to Theorem 5.2, since there is only one way to tile any set F by T.

Theorem 5.3. *Let T be a tiling subset of G, and $\mathcal{P}$ a countable partition such that $H(\mathcal{P}^T) < \infty$. Then for any ϵ and invariant enough F,*

$$\frac{1}{|F|} \log b(F, 1 - \epsilon, \mathcal{P}) \leq \frac{1 + \epsilon}{|T|} H(\mathcal{P}^T) + \frac{\log C(F, T)}{|F|}.$$

Proof: Define

$$\phi(x) = -\sum_{\mathbf{p} \in \mathcal{P}^T} 1_{\mathbf{p}}(x) \log \mu(\mathbf{p}) \in L_1(\mu).$$

The integral of ϕ is equal to $H(\mathcal{P}^T)$ and thus applying the Mean Ergodic Theorem, we see that if F is invariant enough there is a set X' (depending on F) of measure $\geq 1 - \epsilon/2$ such that for any $x \in X'$

$$\phi^F(x) := \sum_{f \in F} \phi(fx) \leq \left(1 + \frac{\epsilon}{100}\right) H(\mathcal{P}^T)|F|.$$

the function ϕ is non-negative; using Lemma 4.4, we see that if F is invariant enough there is a set of centers E_x of a tiling of F by T such that

$$\frac{1}{|F|}\phi^F(x) \geq \left(1 - \frac{\epsilon}{100}\right)\frac{1}{|E_x|}\phi^{E_x}(fx);$$

if $x \in X'$ we see that

$$\phi^{E_x}(fx) \leq \left(1 + \frac{\epsilon}{10}\right)H(\mathcal{P}^T)|E_x|.$$

Let E be a set of centers of a tiling of F by T. We give each set $\mathbf{P} = \bigcap_{f \in E} f^{-1}\mathbf{p}_f \in \mathcal{P}^{TE}$ ($\mathbf{p}_f \in \mathcal{P}^T$ for every f) a weight $w_E(\mathbf{P})$ defined by

$$w_E(\mathbf{P}) = \prod_{f \in E} \mu(\mathbf{p}_f) = 2^{-\phi^E(x)} \qquad \text{if } x \in \mathbf{P}.$$

Clearly

$$\sum_{\mathbf{P} \in \mathcal{P}^{TE}} w_E(\mathbf{P}) = 1,$$

hence there can be at most $2^{(1+\epsilon)H(\mathcal{P}^T)|E|}$ sets $\mathbf{P} \in \mathcal{P}^{TE}$ with $w_E(\mathbf{P}) \geq 2^{-(1+\epsilon)H(\mathcal{P})|E|}$. Denote the set of these $\mathbf{P}$ by $\mathcal{C}_E$. If $x \in X'$ we have ,

$$\phi^{E_x}(x) \leq (1 + \epsilon/10)H(\mathcal{P}^T)|E_x|,$$

and so $x \in \cup\mathcal{C}_E$. We can replace $|E_x|$ by $|F|/|T|$ and hence

$$b(F, 1 - \epsilon, \mathcal{P}) \leq \sum_E |\mathcal{C}_E| \leq C(F, T)2^{(1+\epsilon/100)H(\mathcal{P}^T)|F|/|T|}.$$

$\square$

Theorem 5.4 ("Shannon-McMillan" Theorem for Countable Partitions). *Let $\mathcal{P}$ be a fixed partition of X with $H(\mathcal{P}) < \infty$. For any $\epsilon > 0$, for any sufficiently invariant tiling set F, the atoms of $\mathcal{P}^F$ from the collection $\mathcal{C}$,*

$$\mathcal{C} = \left\{ C \in \mathcal{P}^F : 2^{-(h(\mathcal{P})+\epsilon)|F|} < \mu(C) < 2^{-(h(\mathcal{P})-\epsilon)|F|} \right\},$$

cover at least $1 - \epsilon$ of X.

Proof:

¿Beginning with the partition $\mathcal{P} = \{P_1, P_2, \dots\}$ we define

$$\mathcal{P}_{(n)} = \{P_1, P_2, \dots, P_{n-1}, \bigcup_{k=n}^{\infty} P_k\}$$

$$\mathcal{P}_{(n)}^* = \{\bigcup_{k=1}^{n-1} P_k, P_n, P_{n+1}, \dots\}.$$

The $\mathcal{P}_{(n)}$ are finite partitions, $H(\mathcal{P}_{(n)}^*) \to 0$ as $n \to \infty$, and $\mathcal{P} = \mathcal{P}_{(n)} \vee \mathcal{P}_{(n)}^*$. Another thing we would like to point out is that by definition, $h(\mathcal{P}) \geq h(\mathcal{P}_{(n)})$ for all n.

We take N to be large enough so that

$$H(\mathcal{P}^*_{(N)}) < \epsilon/10.$$

We will deduce Theorem 5.4 from the Shannon-McMillan theorem for the finite partition $\mathcal{P}_{(N)}$. Indeed, Let F be invariant enough so that the $\epsilon/10$-fat and $\epsilon/10$-thin $\mathcal{P}^F_{(n)}$-names cover at most $\epsilon/10$ of X.

Every ϵ-fat $\mathcal{P}^F$ name is a subset of an even fatter $\mathcal{P}^F_{(n)}$ name, so the ϵ-fat $\mathcal{P}^F$-names cover at most $\epsilon/10$ of X.

Since the $\epsilon/10$-thin $\mathcal{P}^F_{(n)}$-names cover at most $\epsilon/10$ of X, $1 - \epsilon/10$ of X can be covered by at most $2^{|F|(h(\mathcal{P}_{(N)})+\epsilon/10)}$ $\mathcal{P}^F_{(n)}$-names. From Theorem 5.2, if F is invariant enough,

$$b(F, 1 - \epsilon/10, \mathcal{P}^*_{(N)}) \le 2^{(1+\epsilon/10)H(\mathcal{P}^*_{(N)})|F|}.$$

Our last assumption on F is that it is large enough so that

$$2^{-\epsilon/2|F|} \le \epsilon/10.$$

Thus

$$b(F, 1 - \epsilon/5, \mathcal{P}) \le b(F, 1 - \epsilon/10, \mathcal{P}_{(N)}) \times b(F, 1 - \epsilon/10, \mathcal{P}^*_{(N)}) \le 2^{(h(\mathcal{P})+\epsilon/2)|F|}.$$

We take $\mathcal{P}' \subset \mathcal{P}^F$ to be any collection of $b(F, 1 - \epsilon/5, \mathcal{P})$ $\mathcal{P}^F$-names that cover at least $1 - \epsilon/5$ of X.

The ϵ-thin $\mathcal{P}^F$ names from the collection $\mathcal{P}'$ have total measure at most

$$b(F, 1 - \epsilon/5, \mathcal{P})2^{-|F|(h(\mathcal{P})+\epsilon)} \le 2^{-\epsilon/2|F|} \le \epsilon/10.$$

The ϵ-thin $\mathcal{P}^F$ names from $\mathcal{P}^F \setminus \mathcal{P}'$ cover at most $X \setminus \cup \mathcal{P}'$, a set of measure $\le \epsilon/5$. Thus the ϵ-thin $\mathcal{P}^F$ names cover at most $3\epsilon/10$ of X. $\qquad\square$

Classically, the entropy of a process is defined by using a specific Følner sequence F_n of tiling sets as follows:

Definition 5.5. *Let $\mathcal{P}$ be a (finite or countable) partition of X with $H(\mathcal{P}) < \infty$. Then the classical entropy of the process is*

$$\tilde{h}(\mathcal{P}) = \liminf_{n \to \infty} \frac{1}{|F_n|} H(\mathcal{P}^{F_n})$$

As we will see this definition is equivalent to the one we gave above and in particular is independent of the specific Følner sequence F_n.

We will first show that the sequence $\frac{1}{|F_n|} H(\mathcal{P}^{F_n})$ actually converges. We will need some of the classical properties of the entropy function H.

Proposition 5.6. *Let $\mathcal{Q}$ be a refinement of $\mathcal{P}$ (i.e. for every $P \in \mathcal{P}$ there is a $Q \in \mathcal{Q}$ with $P \subset Q$). then*

$$H(\mathcal{Q}) \ge H(\mathcal{P}).$$

Proof. This is immediate from the inequality

$$-x \log x - y \log y \geq -(x + y) \log(x + y)$$

for every x, $y > 0$ with $x + y \leq 1$. $\square$

Proposition 5.7. *Let* $\mathcal{P}$ *and* $\mathcal{Q}$ *be two partitions of* X *with* $H(\mathcal{P})$, $H(\mathcal{Q}) < \infty$. *Then*

$$H(\mathcal{P} \vee \mathcal{Q}) \leq H(\mathcal{P}) + H(\mathcal{Q})$$

Proof. First assume $\mathcal{P}$ and $\mathcal{Q}$ are finite. We shall use the fact that $-\sum_{i=1}^{N} x_i \log x_i$ is a concave function (we remind the reader that for $x = 0$ we take $x \log(x)$ to be 0). Thus

$$H(\mathcal{P} \vee \mathcal{Q}) = \sum_{P \in \mathcal{P}} \sum_{Q \in \mathcal{Q}} -\mu(P \cap Q) \log \mu(P \cap Q)$$

$$= \sum_{P \in \mathcal{P}} \left(\sum_{Q \in \mathcal{Q}} -\mu(P \cap Q) \log \mu(P \cap Q) + \mu(P) \log \mu(P) \right) - \sum_{P \in \mathcal{P}} \mu(P) \log \mu(P)$$

$$= H(\mathcal{P}) + \sum_{P \in \mathcal{P}} \mu(P) \sum_{Q \in \mathcal{Q}} -\frac{\mu(P \cap Q)}{\mu(P)} \log \frac{\mu(P \cap Q)}{\mu(P)}$$

$$\leq H(\mathcal{P}) + \sum_{Q \in \mathcal{Q}} -\left(\sum_{P \in \mathcal{P}} \mu(P) \frac{\mu(P \cap Q)}{\mu(P)} \right) \log \left(\sum_{P \in \mathcal{P}} \mu(P) \frac{\mu(P \cap Q)}{\mu(P)} \right)$$

$$= H(\mathcal{P}) + \sum_{Q \in \mathcal{Q}} -\mu(Q) \log \mu(Q) = H(\mathcal{P}) + H(\mathcal{Q}).$$

For the general case, set

$$\mathcal{P}_{(n)} = \{P_1, P_2, \ldots, P_{n-1}, \bigcup_{k \geq n} P_k\} \qquad n < |\mathcal{P}|$$

$$\mathcal{Q}_{(n)} = \{Q_1, Q_2, \ldots, Q_{n-1}, \bigcup_{k \geq n} Q_k\} \qquad n < |\mathcal{Q}|$$

We can now pass to the limit $n \to \infty$ as follows:

$$H(\mathcal{P} \vee \mathcal{Q}) = \sum_{P \in \mathcal{P}} \sum_{Q \in \mathcal{Q}} -\mu(P \cap Q) \log \mu(P \cap Q)$$

$$= \lim_{N \to \infty} \sum_{i=1}^{\min(|\mathcal{P}|, N-1)} \sum_{j=1}^{\min(|\mathcal{Q}|, N-1)} -\mu(P_i \cap Q_j) \log \mu(P_i \cap Q_j)$$

$$\leq \liminf_{N \to \infty} \sum_{P \in \mathcal{P}_{(N)}} \sum_{Q \in \mathcal{Q}_{(N)}} -\mu(P \cap Q) \log \mu(P \cap Q)$$

$$= \liminf_{N \to \infty} H(\mathcal{P}_{(N)} \vee \mathcal{Q}_{(N)})$$

$$\leq \liminf_{N \to \infty} H(\mathcal{P}_{(N)}) + H(\mathcal{Q}_{(N)}) \leq H(\mathcal{P}) + H(\mathcal{Q}).$$

      $\square$

An immediate corollary of the above two propositions is the following:

Corollary 5.8. *Let $\mathcal{P}$ be any partition with $H(\mathcal{P}) < \infty$. Then the function $a(F) = H(\mathcal{P}^F)$ from the finite subsets of G to $\mathbb{R}$ has the following properties:*

1. *$a(F) \geq 0$ for all F (a is non-negative).*
2. *$a(Fg) = a(F)$ for all $g \in G$ (a is translation invariant).*
3. *If $F' \subset F$ then $a(F') \leq a(F)$ (a is monotone).*
4. *For any disjoint F and F'*

$$a(F \cup F') \leq a(F) + a(F')$$

(a is subadditive).

Remark: It is not hard to show that a satisfies the following stronger inequality than simple subadditivity, namely that for every finite F and $F' \subset G$,

$$a(F \cup F') \leq a(F) + a(F') - a(F \cap F').$$

In his treatment of entropy for amenable group actions Moullin-Ollagnier [M] uses this property rather heavily. We will not use this fact and obtain thereby a stronger theorem which is needed in theory of mean dimension for amenable groups (see [LW]).

Theorem 5.9. *Let a be a function satisfying the properties listed in Corollary 5.8. Let F_n be a Følner sequence of tiling sets. Then $\lim_{n \to \infty} a(F_n)/|F_n|$ exists, and is independent of the choice of Følner sequence: in fact,*

$$\lim_{n \to \infty} \frac{1}{|F_n|} a(F_n) = \inf_{n \in \mathbb{N}} \frac{1}{|F_n|} a(F_n).$$

Proof. The independence from the choice of Følner sequence is a consequence of the existence of the limit since we can merge any two Følner sequences together to get a new one.

Thus we only need to show that for any $m \in \mathbb{N}$

$$\limsup_{n \to \infty} \frac{1}{|F_n|} a(F_n) \leq \frac{1}{|F_m|} a(F_m)$$

To see this let E_n be a set of centers for tiling of F_n by F_m, $E'_n \subset E_n$ be those $g \in E_n$ such that $F_m g \subset F_n$. The elements of $E_n \setminus E$ are in the F_m boundary of F_n, and thus if n is invariant enough,

$$|E_n \setminus E| \leq \frac{\epsilon}{|F_m|} |F_n|.$$

since $|E_n| \geq |F_n|/|F_m| \geq |E'_n|$ this means that

$$|E_n| \leq |E'_n| + \frac{\epsilon}{|F_m|} |F_n|. \leq (1 + \epsilon) \frac{|F_n|}{|F_m|}.$$

Thus,

$$\frac{1}{|F_n|}a(F_n) \le \frac{1}{|F_n|}a(F_m E_n) \le \frac{1}{|F_n|}\sum_{g\in E_n} a(F_m g) = \frac{|E_n|}{|F_n|}a(F_m)$$

$$\le (1+\epsilon)\frac{1}{|F_m|}a(F_m).$$

$\square$

Lemma 5.10. *For any partitions* $\mathcal{P}$ *and* $\mathcal{Q}$ *with* $H(\mathcal{P})$ *and* $H(\mathcal{Q})$ *finite,* $h(\mathcal{P}) \le h(\mathcal{P} \vee \mathcal{Q}) \le h(\mathcal{P}) + h(\mathcal{Q})$ *and* $\tilde{h}(\mathcal{P}) \le \tilde{h}(\mathcal{P} \vee \mathcal{Q}) \le \tilde{h}(\mathcal{P}) + \tilde{h}(\mathcal{Q})$.

Proof. For any finite F and $\epsilon > 0$, we clearly have the following inequalities:

$$b(F, 1-2\epsilon, \mathcal{P}) \le b(F, 1-2\epsilon, \mathcal{P} \vee \mathcal{Q}) \le b(F, 1-\epsilon, \mathcal{P})b(F, 1-\epsilon, \mathcal{Q}).$$

We take the logarithm of the above expression, divide by $|F|$, and pass to the limit first as F becomes an increasingly invariant tiling set, then as $\epsilon \to 0$, to get

$$h(\mathcal{P}) \le h(\mathcal{P} \vee \mathcal{Q}) \le h(\mathcal{P}) + h(\mathcal{Q}).$$

The corresponding inequality for $\tilde{h}$ is a consequence of the monotonicity and subadditivity of $H(\cdot)$. Indeed, using the identity $(\mathcal{P} \vee \mathcal{Q})^F = \mathcal{P}^F \vee \mathcal{Q}^F$, we have

$$H(\mathcal{P}^F) \le H(\mathcal{P}^F \vee \mathcal{Q}^F) \le H(\mathcal{P}^F) + H(\mathcal{Q}^F).$$

Again we divide by $|F|$ and take the limit as F is an increasingly invariant tiling set. $\square$

Lemma 5.11. *For any finite partition* $\mathcal{P}$ *and* $F \subset G$,

$$H(\mathcal{P}^F) \le \log b(F, a, \mathcal{P}) + (1-a)|F|\log|\mathcal{P}| + 1.$$

Proof. First we remark that the function

$$\tilde{H}(p_1, p_2, \ldots, p_N) := -\sum p_i \log p_i$$

has a unique maximum on the space $p_i \ge 0$, $\sum p_i = 1$ at the point

$$p_1 = p_2 = \cdots = p_N = 1/N.$$

Using the fact that the function H is concave, it is enough to check that the above point is a local maximum which is easily seen by differentiating $\tilde{H}$. In particular, we see that

(5.1) $$\max_{p_i \text{ as above}} \tilde{H}(p_1, p_2, \ldots, p_N) = \log N$$

Let $\mathcal{C}$ be a subcollection of $b(F, a, \mathcal{P})$ $\mathcal{P}^F$-names with total measure at least a, and $\mathcal{D}$ be the collection of all other $\mathcal{P}^F$-names. Let $\tilde{a} = \mu(\cup\mathcal{C})$. Then

$$H(\mathcal{P}^F) = \tilde{a}\sum_{P\in\mathcal{C}} -\frac{\mu(P)}{\tilde{a}}\log\frac{\mu(P)}{\tilde{a}} + (1-\tilde{a})\sum_{P\in\mathcal{D}} -\frac{\mu(P)}{1-\tilde{a}}\log\frac{\mu(P)}{1-\tilde{a}}$$
$$- \tilde{a}\log\tilde{a} - (1-\tilde{a})\log(1-\tilde{a}).$$

Since

$$\sum_{P \in \mathcal{C}} \frac{\mu(P)}{\tilde{a}} = \sum_{P \in \mathcal{D}} \frac{\mu(P)}{1 - \tilde{a}} = 1,$$

we can apply 5.1 three times and see that

$$H(\mathcal{P}^F) \le \log(|\mathcal{C}|) + \log(|\mathcal{D}|) + 1 \le \log b(F, a, \mathcal{P}) + (1 - a)|F| \log |\mathcal{P}| + 1$$

(we recall that all our logs are to base 2). $\qquad \square$

Theorem 5.12. *For any $\mathcal{P}$ with $H(\mathcal{P}) < \infty$,*

$$\tilde{h}(\mathcal{P}) = h(\mathcal{P}).$$

Remark: Recall our assumption throughout that $(X, \mathcal{B}, \mu, G)$ is ergodic. The above theorem is false for non-ergodic systems!

Proof. We first prove the theorem for finite $\mathcal{P}$.

Let F_0 be a very big tiling set such that

$$H(\mathcal{P}^{F_0}) \le |F_0|(\tilde{h}(\mathcal{P}) + \epsilon).$$

Using Theorem 5.3 if $F \subset G$ is invariant enough,

$$\frac{1}{|F|} \log b(F, 1 - \epsilon, \mathcal{P}) \le \frac{1 + \epsilon}{|F_0|} H(\mathcal{P}^{F_0}) + \frac{\log C(F, F_0)}{|F|}.$$

As we have seen, if F_0 is big and F invariant enough $C(F, F_0) \le 2^{\epsilon |F|}$. Passing to the limit we get that

$$h(\mathcal{P}) \le \tilde{h}(\mathcal{P}).$$

On the other hand, by Lemma 5.11,

$$\frac{1}{|F|} H(\mathcal{P}^F) \le \frac{1}{|F|} b(F, 1 - \epsilon, \mathcal{P}) + \epsilon \log(\mathcal{P}) + \frac{1}{|F|},$$

and so passing to the limit we see that

$$\tilde{h}(\mathcal{P}) \le h(\mathcal{P}).$$

For the general case, we again set

$$\mathcal{P}_{(n)} = \{P_1, P_2, \ldots, P_{n-1}, \bigcup_{k=n}^{\infty} P_k\}$$

$$\mathcal{P}_{(n)}^* = \{\bigcup_{k=1}^{n-1} P_k, P_n, P_{n+1}, \ldots\}.$$

The theorem follows from the following calculation

$$
\begin{array}{ccccccc}
 & & & & & & 0 \\
 & & & & & & \uparrow \\
h(\mathcal{P}_{(n)}) & \leq & h(\mathcal{P}) & \leq & h(\mathcal{P}_{(n)}) & + & h(\mathcal{P}^*_{(n)}) \\
\| & & & & \| & & \\
\tilde{h}(\mathcal{P}_{(n)}) & \leq & \tilde{h}(\mathcal{P}) & \leq & \tilde{h}(\mathcal{P}_{(n)}) & + & \tilde{h}(\mathcal{P}^*_{(n)}) \\
 & & & & & & \downarrow \\
 & & & & & & 0
\end{array}
$$

□

6. Approximately invariant tempered sets

Finitely generated nilpotent groups, such as $\mathbb{Z}^d$ have approximately invariant sequences that are increasing $F_{n-1} \subset F_n$ and have the additional property that

$$|F_n^{-1} F_n| \leq b|F_n|$$

for some constant b. Prior to [L] pointwise limit theorems were established only for such groups, see [OW-1983]. As was shown in [L] not all amenable groups can have such sequences. However, any approximately invariant sequence can be thinned out so that the following condition holds:

$$\left| \left(\bigcup_{j=1}^{n-1} F_j \right)^{-1} F_n \right| \leq b|F_n|$$

we will call such sequences **tempered**.

Suppose now that $X, \mathcal{B}, \mu$) is a probability space and that G acts on it in a measure preserving fashion. We usually write the action simply as $x \to gx$. Let $\mathcal{H}_{inv} \subset L^1(x, \mathcal{B}, \mu)$ be the G-invariant functions, and π the projection of L^1 onto this space: If $\mathcal{H}_{inv}$ is just the constants then $\pi(\phi) = \int \phi d\mu$. The basic ergodic theorem is the following:

Theorem 6.1. *For any approximately invariant tempered sequence $\{F_n\}$ and any $\phi \in L^1(X, \mathcal{B}, \mu)$ one has*

$$\lim_{n \to \infty} \frac{1}{|F_n|} \sum_{g \in F_n} \phi(gx) = (\pi \phi)(x) \qquad \mu - a.e.$$

where π is the projection onto $\mathcal{H}_{inv}$.

In the preceding sections we discussed in detail the entropy of the process defined by partitions $\mathcal{P}$, and the very basic Shannon-McMillan theorem. There is a pointwise refinement of this theorem that we will formulate now. Recalling that $\mathcal{P}^F = \bigvee_{g \in F} g^{-1} \mathcal{P}$ we will denote by $\mathcal{P}^F(x)$ the atom of $\mathcal{P}^F$ that contains x, i.e.

$$\mathcal{P}^F(x) = \{y \in X : \ gy \ \text{and} \ gx \ \text{lie in the same atom of} \ \mathcal{P} \ \text{for all} \ g \in F\}$$

The ergodic decomposition of μ is a measurable decomposition of X into invariant sets E_θ, such that if $\mu = \int \mu_\theta d\nu(\theta)$ is the disintegration of μ relative to this decomposition the system $(E_\theta, \mathcal{B}|E_\theta, \mu_\theta, G)$ is ergodic. The projection of ϕ onto the space of invariant functions, $\hat{\phi}$, is given by setting $\hat{\phi}$ on E_θ to be equal to the integral of ϕ with respect to μ_θ. Here is the general formulation of the Shannon-McMillan-Breiman theorem:

Theorem 6.2. *For any approximately invariant tempered sequence $\{F_n\}$ and any G-action $(X, \mathcal{B}, \mu, G)$ with measurable partition $\mathcal{P}$ one has*

$$\lim_{n\to\infty} \frac{-1}{|F_n|} \log \mu(\mathcal{P}^{F_n}(x)) = h(\mathcal{P}|E_\theta(x), \mu_\theta, G) \quad \mu - a.e.$$

where $E_\theta(x)$ denote the ergodic component to which x belongs.

   In the ergodic case of course the limit is constant and equals the entropy of $(\mathcal{P}, G)$ as above. In [L] , E. Lindenstrauss gave the first proof of these theorems (the latter only in the ergodic case). In the next three sections we will prove them by a somewhat different method that avoids the randomizations that are an essential feature of [L].

7. THE MAXIMAL INEQUALITY

   Throughout this section $\{F_n\}$ will be a tempered sequence in a group G that acts in a measure preserving fashion on $(X, \mathcal{B}, \mu)$. For an L^1 function ϕ one defines

$$(M\phi)(x) = \sup_n \frac{1}{|F_n|} \sum_{g\in F_n} |\phi(gx)|$$

The maximal inequality is the assertion that for some constant K, and all $\lambda > 0$

$(*)$ $$\mu\{x : M\phi(x) \geq \lambda\} \leq \frac{K}{\lambda}\|\phi\|_1$$

   Via the so-called transference principle, one reduces the proof of such an inequality to a combinatorial inequality relating to finite sets in the group G. This technique is sufficiently well known by now so that we will only sketch the reduction briefly.

   First one reduces the treatment to a finite version of M, namely

$$M_m\phi(x) = \sum_{n\leq m} \frac{1}{|F_n|} \sum_{g\in F_n} |\phi(gx)|$$

Next one takes a Følner set F that is much bigger than F_m, and considers, for a fixed λ, and x, the F-orbit of x, ϕ on this F-orbit and the set of $g \in F$ such that

$$M_m\phi(gx) \geq \lambda.$$

To this finite situation one applies the proposition to be proved below – and then averaging over all $x \in X$, and using the fact that the action is measure preserving will give the maximal inequality $(*)$.

We begin with an abstract combinatorial lemma.

Basic Lemma: In a finite set Ω we have finite sets $V_1, V_2, \ldots V_m$, all of the same size $|V_i| = v, v \geq 10$. A positive measure ϕ is defined on Ω and we assume:

(i) $\phi(V_i) \geq \lambda |V_i|, \quad 1 \leq i \leq m \qquad \lambda > 0$ fixed;

(ii) $\displaystyle\sum_{i=1}^{m} 1_{V_i}(\omega) \leq v \quad$ for all $w \in \Omega$;

then there is a subcollection $\{V_i : i \in I\}$ satisfying:

(a) $\phi\left(\displaystyle\bigcup_{i \in I} V_i\right) \geq \frac{1}{3}\lambda \cdot m$

(b) $|I| \cdot v \leq \frac{3}{\lambda}\phi\left(\displaystyle\bigcup_{i \in I} V_i\right)$

Proof. Beginning with $i(1) = 1$, inductively define $i(k+1)$ to be the least integer $\leq m$, greater than $i(k)$, such that

$$\phi\left(V_{i(k+1)} \setminus \bigcup_{1 \leq j \leq k} V_{i(j)}\right) \geq \frac{1}{2}\phi\left(V_{i(k_1)}\right)$$

if such an integer exists, otherwise stop and call $\{i(1), \ldots i(k)\} = I$. In order to verify (a), (b) consider two cases.

I. $k = |I| \geq \frac{m}{v}$. In this case clearly

$$\phi\left(\bigcup_1^k V_{i(j)}\right) \geq \frac{1}{2}\sum_1^k \phi(V_{i(j)}) \geq \frac{1}{2} \cdot k \cdot \lambda \cdot v \geq \frac{1}{2}\frac{m}{v} \cdot \lambda v > \frac{\lambda}{2}m$$

which gives (a). For (b) we stop after the second inequality and divide by $\lambda/2$.

II. $k < \frac{m}{v}$. In this case, many of the V_i have at least $\frac{1}{2}$ of their ϕ-mass inside $\bigcup_{i \in I} V_i$. Denote this latter set by U and $\bar{I} = \{i \leq m : i \notin I\}$. This means that

$$\int 1_{V_i}(\omega) 1_U(\omega) d\phi \geq \frac{1}{2}\int 1_{V_i}(\omega) d\phi, \qquad i \in \bar{I}.$$

sum over all $i \in \bar{I}$ and use (ii) to obtain

$$\frac{1}{2}\sum_{i \in \bar{I}} \phi(V_i) \leq v \cdot \phi(U)$$

using (i) and dividing by v we get

$$\phi\left(\bigcup_{i \in I} V_i\right) \geq \frac{1}{2}\lambda \cdot |\bar{I}| \geq \frac{1}{2}(1 - \frac{1}{v}) \cdot \lambda m > \frac{9}{20}\lambda m$$

which gives (a). This yields an upper bound on m of the form required in (b) and since $k < \frac{m}{v}$ this in fact gives us (b) in this case as well. ☐

Here is the main proposition needed to establish the maximal inequality. We have a sequence of finite sets $F_1, F_2, \ldots F_N$ in a group G that satisfy

$$\left| \left(\bigcup_{i=1}^{j} F_i \right)^{-1} F_{j+1} \right| \le b |F_{j+1}|, \qquad 1 \le j \le N - 1,$$

for some constant b. We are also given a large finite set Ω in G and a positive measure ϕ defined on Ω, and for a fixed $\lambda > 0$ we have disjoint sets $C_1, \ldots C_N$ in Ω that satisfy:

(I) $F_j C_j \subset \Omega, \quad 1 \le j \le N$
(II) $\phi(F_j c_j) \ge \lambda |F_i|, \qquad$ all $c_j \in C_j, \quad 1 \le j \le N.$

Proposition 7.1. *In the above situation*

$$\sum_{1}^{N} |C_i| \le \frac{6(b+1)}{\lambda} \phi(\Omega)$$

Proof. The basic lemma can be applied to the collection $\{F_N c : c \in C_N\}$ with the set C_N playing the role of the initial index set $\{1, 2, \ldots, m\}$. The hypothesis (i) follows from (II) while the crucial hypothesis (ii) is satisfied because all the sets $F_N c$ are translates of a single set in a group. Let us denote the resulting set of the basic lemma by D_N which is of course now a subset of C_N. Thus (a) and (b) can be written as follows:

(1) $\qquad |C_N| \le \frac{3}{\lambda} \phi(F_N D_N)$
(2) $\qquad |D_N| \cdot |F_N| \le \frac{2}{\lambda} \phi(F_N D_N).$

We take up next the collection $\{F_{N-1} c : c \in C_{N-1}\}$ and distinguish two cases:

I. For at least half of the elements c of C_{N-1}, $F_{N-1} c$ is disjoint from $F_N D_N$. In this case, we apply the basic lemma to those elements and construct a $D_{N-1} \subset C_{N-1}$ such that

(3) $\qquad F_{N-1} D_{N-1} \cap F_N D_N = \phi$
and
(4) $\qquad |C_{N-1}| \le \frac{6}{\lambda} \phi(F_{N-1} D_{N-1})$
(5) $\qquad |D_{N-1}| \cdot |F_{N-1}| \le \frac{2}{\lambda} \phi(F_{N-1} D_{N-1}).$

II. At least half of the elements c of C_{N-1} are such that $F_{N-1} c \cap F_N D_N \ne \phi$. In this case, those c's are contained in $(F_{N-1}^{-1} F_N) D_N$ and we simply record this fact and proceed to consider $\{F_{N-2} c : c \in C_{N-2}\}$.

After completing k steps of his procedure, for $N, N - 1, \ldots N - k + 1$, we have two kinds of indices $N - k + 1 \le j \le N$. Those for which we have constructed D_j's with $F_j D_j$ disjoint from the previous ones (in the descending order) and those that corresponded to case II.

We now take up $\{F_{N-k}c : c \in C_{N-k}\}$ and again distinguish two cases. In the first, when more than half of c's in C_{N-k} satisfy: $F_{N-k}c$ is disjoint from the previous $F_j D_j$'s, we construct a set $D_{N-k}c : C_{N-k}$ such that

(6) $F_{N-k}D_{N-k} \cap F_j D_j = \phi$, for all $j > N_k$ for which D_j is defined.

(7) $|C_{N-k}| \leq \frac{6}{\lambda}\phi(F_{N-k}D_{N-k})$

(8) $|D_{N-k}| \cdot |F_{N-k}| \leq \frac{2}{\lambda}\phi(F_{N-k}D_{N-k})$.

In the second, we record the fact that at least half of the C_{N-k}'s are contained in some set of the form $F_{N-k}^{-1}F_j d_j$, $N - k < j$, $d_j \in D_j$.

This process ends when we reach $F_1 C_1$. We now recall the definition of b, and estimate the size of all the C_j at levels where case II occurred using (2), (5), (8) etc. by

$$\frac{6b}{\lambda}\sum \phi(F_j D_j) \leq \frac{6b}{\lambda}\phi(\Omega)$$

while for the other levels, (1), (4), (7) etc. yield an upper bound of $\frac{6}{\lambda}\phi(\Omega)$. Combining these estimates gives the proposition. $\square$

As we have already explained, with this proposition, the standard argument via the transference principle now enables us to establish the desired maximal inequality with the constant $K = 6(b + 1)$.

8. THE ERGODIC THEOREM

As we said we will now give a brief sketch of how the pointwise ergodic theorem along a tempered sequence of almost invariant sets can be deduced from the maximal inequality.

Theorem 8.1. *If* $\{F_n\}$ *is a tempered sequence of almost invariant sets in* G, *and* (X, B, μ, G) *is a measure preserving action of* G *then for any* L^1-*function* ϕ ,

$$\lim_{n \to \infty} \frac{1}{|F_n|}\sum_{g \in F_n} \phi(gx) = \hat{\phi}(x) \qquad \mu - a.e.$$

where $\hat{\phi}$ *is the projection of* ϕ *on the* G-*invariant functions.*

Proof. It is standard to reduce to the case when the action is ergodic and $\hat{\phi}$ is simply the μ integral of ϕ. In that case, any function ϕ in L^1 with 0-integral may be approximated in L^1-norm by linear combinations of functions of the type $\psi(ax) - \psi(x)$ with $a \in G$ a fixed element and ψ bounded.

To establish this point — consider the decomposition $\phi = \phi_+ - \phi_-$ with ϕ_+ and ϕ_- nonnegative and $\phi_+ \cdot \phi_- \equiv 0$. Clearly one can find some sets E_+, E_- of equal positive measure and a constant $\alpha > 0$ such that $\phi_+ \geq \alpha 1_{E_\alpha}, \phi_- \geq \alpha 1_{E_-}$. By the ergodicity one finds some $a \in G$ such that $a^{-1}E_+ \cap E_-$ has positive measure, and thus one finds a function of the type $\psi(ax) - \psi(x)$ which may be subtracted from ϕ and reduces its L^1-norm. Repeating this procedure one readily shows that the difference can be made arbitrarily small in L^1-norm.

For functions of the form $\psi(ax) - \psi(x)$, with ψ bounded the ergodic theorem follows immediately from the fact that $\{F_n\}$ is a Følner sequence. Here the tempered condition is not needed. The same holds of course for finite linear combinations of such functions and thus we know that the theorem is true for a set of functions that is dense in L^1.

For $\phi \in L^1$ with $\int \phi d\mu = 0$, and $\epsilon > 0$, we can find a function ψ for which the ergodic theorem is valid with the limit equal to 0 and such that

$$\|\phi - \psi\|_1 < \epsilon^2/K$$

. Apllying the maximal inequality we will get that there is a set E_ϵ such that

(i) $\qquad \mu(E_\epsilon) \geq 1 - \epsilon$

(ii) $\qquad \limsup\limits_{n\to\infty} \frac{1}{|F_n|}|\sum\limits_{g\in F_n} \phi(gx)| \leq \epsilon \qquad x \in E_\epsilon.$

Repeating this for a sequence of ϵ's tending to zero rapidly enough gives a proof of the theorem. $\qquad\qquad\qquad\qquad\qquad\qquad\qquad\qquad\qquad\qquad\qquad\square$

9. THE SHANNON-MCMILLAN-BREIMAN THEOREM

In this section we will show how the proof of the Shannon-McMillan-Breiman theorem for the ergodic case in [OW-1983] can be modified so as to prove the same result along any tempered sequence of approximately invariant sets. The main novelty is the following version of the disjointification lemma from that paper. We recall the key notion of ε-disjointness.

Definition 9.1. *Sets $B_1, \ldots B_L$ in a finite set Ω are said to be ε-disjoint if for each $1 \leq j < L$ we have:*

$$\left|B_{j+1} \cap \left(\bigcup_{i=1}^{j} B_i\right)\right| < \varepsilon \cdot |B_{j+1}|.$$

As before $F_1, \ldots F_N$ are finite subsets in a group G satisfying the temperedness condition with constant b, Ω is a finite subset of the group, and C_j are disjoint subsets of Ω such that $F_j C_j \subset \Omega$, $\qquad 1 \leq j \leq N$.

Lemma 9.2. *(ε-disjointification) : In the above situation, for any $\frac{1}{4} > \varepsilon > 0$, there are subsets $D_j \subset C_j$ such that for $1 \leq j \leq N$*

(i) *$\{F_j d : d \in D_j\}$ are ε-disjoint*

(ii) *$F_j D_j \cap F_{j'} D_{j'} = \phi \qquad$ for $j \neq j'$.*

(iii) *$|\bigcup\limits_{1}^{N} F_j D_j| \geq \left(\frac{\varepsilon}{4+b}\right) \sum\limits_{j=1}^{N} |C_j|$*

Proof. The proof follow the same course as that of the basic lemma. We begin with the sets $\{F_N c : c \in C_N\}$ and find a maximal subcollection that is ε-disjoint, say $\{F_N d : d \in D_N\}$. We distinguish two cases

I. $|D_N| \geq \frac{1}{2}|C_N|$. Here each $F_N d$ contributes at least $(1 - \varepsilon)|F_N|$ new points and thus

$$|F_N D_N| \geq (1 - \varepsilon)\frac{|F_N|}{2} \cdot |C_N|$$

II. $|D_N| < \frac{1}{2}|C_N|$. Now by the maximality there are at $\frac{1}{2}|C_N|$ sets of the form $F_N c$ that have at least an ε fraction of their points in $F_N D_N$. This gives a lower bound for $|F_N D_N|$ of (recall that the maximal overlap is $|F_N|$)

$$\frac{\varepsilon}{2}|C_N|$$

and in both cases we can conclude

$$|F_N D_N| \geq \frac{1}{2}\varepsilon|C_N|.$$

Next we take up $\{F_{N-1}c : c \in C_{N-1}\}$ and distinguish, once again two cases.
I. For at least half of the $c \in C_{N-1}$, $F_{N-1}c$ is disjoint from $F_N D_N$. In this case we carry out the procedure in the previous step and construct a $D_{N-1} \subset C_{N-1}$ such that $F_{N-1}D_{N-1}$ is disjoint from $F_N D_N$ and

$$|F_{N-1}D_{N-1}| \geq \frac{1}{4}\varepsilon|C_{N-1}|.$$

II. At least half of C_{N-1} is contained in $F_{N-1}^{-1}(F_N D_N)$. Here the main point is that since the $\{F_N d : d \in D_N\}$ are ε-disjoint we can conclude that

$$|F_{N-1}^{-1}(F_N D_N)| \leq \frac{b}{1 - \varepsilon}|F_N D_N| < 2b|F_N D_N|$$

Note that in this case D_{N-1} is empty. This process is continued all the way down to $\{F_1 c : c \in C_1\}$ as in the proof of the basic lemma. By the construction the sets $F_k D_k$ are disjoint and $\sum_1^N |C_k|$ is bounded from above by

$$\left(\frac{4}{\varepsilon} + 4b\right)\sum_1^N |F_k D_k| \leq \frac{4 + b}{\varepsilon}\sum_1^N |F_k D_k|$$

This gives the lower bound in (iii).   □

The proof in [OW-1983] of the SMB depends only on the disjointification lemma and the pointwise ergodic theorem. The lemma we have just proved can serve as a substitute, we need only remark that the over counting due to the ε-overlaps is controlled exactly by ε, while we are still free to use enough levels so that the fact that the constant in (iii) is very small doesn't matter at all.

This discussion serves to prove theorem 2 in the ergodic case. To treat the non ergodic case we will use what we have already done. The fact that

$$h_\theta(x) = h(\mathcal{P}|E_\theta(x), \mu_\theta, G)$$

is a lower bound for a.e. $x \in E_\theta(x)$ can be deduced rather easily as follows. Fix a θ and a $\delta > 0$, and for each n let $\mathcal{C}_n$ denote the collection of atoms A_n of $\mathcal{P}^{F_n}$ that satisfy

$$\mu(A_n) \geq 2^{\delta \cdot |F_n|} \mu_\theta(A_n)$$

Upon dividing by $2^{\delta|F_n|}$ and summing over all $A_n \in \mathcal{C}_n$ one sees that

$$\mu_\theta(\mathcal{C}_n) \leq 2^{-\delta|F_n|}$$

and then the Borel-Cantelli lemma implies that μ_θ-a.e. x lies in only finitely many of the sets $\cup \mathcal{C}_n$, and hence

$$\frac{-\log \mu(\mathcal{P}^{F_n}(x))}{|F_n|} \geq h_\theta(x) - \delta \qquad \mu_\theta - a.e. x \in E_\theta$$

by the SMB theorem applied to E_θ. Since δ was arbitrary this gives the lower bound that we claimed.

For the upper bound a more involved counting argument is necessary. Here is a rough sketch of the argument. Fix a level h_0 of the entropy function h_θ for which the ν-measure of the θ's for which $h_\theta < h_0$ is positive and a $\delta > 0$.

We shall describe a set of atoms $\mathcal{D}_n$ of $\mathcal{P}^{F_n}$ whose number satisfies

$$|\mathcal{D}_n| \leq 2^{h_0 + \delta}|F_n|$$

with the property that for most of those θ's μ_θ-a.e. x in E_θ eventually belongs to $\cup \mathcal{D}_n$. The Borell-Cantelli lemma can be used to show that for μ-a.e. x in that set

$$\frac{-\log \mu(\mathcal{P}^{F_n}(x))}{|F_n|} \leq h_0 + 2\delta.$$

Repeating this argument for a countable dense set of values of h_0, for a sequence of δ's tending to zero and for all k_0, will give the desired upper bound.

For a more detailed argument the main point is to explain the construction of the $\mathcal{D}_n$'s. As before, only the case when each F_k tiles the group will be treated. The general case requires the quasi-tiling machinery for which once again the reader is referred to [OW-1987]. Fix another small number η which will be specified at the end. By the Shannon-McMillan theorem for ergodic actions, for each θ and all sufficiently large k one can find a set $\mathcal{C}_\theta$ contained in $\mathcal{P}^{F_k}$ that satisfies:

(i) $|\mathcal{C}_\theta| \leq 2^{h_0 + \eta}|F_k|$
(ii) $\mu_\theta(\cup \mathcal{C}_\theta) > 1 - \eta$

We can choose a fixed k_0 so that set of θ's for which we can find a $\mathcal{C}_\theta$ satisfying the above almost exhausts, in ν measure, the set of θ's that we are considering now. The pointwise ergodic theorem that we have already established implies that for μ_θ a.e point x, and n sufficiently large $\mathcal{P}^{F_n}(x)$ is $(1 - 2\eta)$ covered by F_{k_0} names that come from the collection $\mathcal{C}_\theta$. It is elements of $\mathcal{P}^{F_n}$ like this that we will take to build up the required $\mathcal{D}_n$. In order to estimate the number of elements in $\mathcal{D}_n$ we need to get into the situation where the F_{k_0} names are seen disjointly. This is where the fact that we have a tiling of the group by F_{k_0} simplifies the picture.

There are in fact $|F_{k_0}|$ different tilings of F_n, and ignoring edge effects, which we may by the almost invariance, a simple averaging argument will give that for at least one of them we have a $1 - 3\eta$ covering of $\mathcal{P}^{F_n}(x)$ by F_{k_0} names from the collection $\mathcal{C}_\theta$. For the rest of the $\mathcal{P}^{F_n}(x)$ we will use the fact that $\mathcal{P}$ is a finite partition to estimate the number of different ways that those places can be filled in. Now the number of ways of choosing $3\eta|F_n|$ places from $|F_n|$ is given by a binomial coefficient which is easily estimated by Stirling's formula while since we are treating F_{k_0} as being fixed so are the the total number of possibilities for the collections $\mathcal{C}_\theta$. It follows that if we choose η so that

$$-\eta ln(\eta) - (1 - \eta)ln(1 - \eta) < \frac{1}{2}\delta$$

and we take for $\mathcal{D}_n$ all the names that we have described , then indeed

$$|\mathcal{D}_n| \leq 2^{h_0+\delta}|F_n|$$

This completes our proof of the SMB theorem for non-ergodic actions of discrete amenable groups.

10. COMPLETELY POSITIVE ENTROPY AND FREE ACTIONS

As soon as the entropy of G actions is defined two special kinds of action are singled out – the **zero entropy actions** and the actions with completely positive entropy (CPE) i.e. actions for which $h(\mathcal{P}, G) > 0$ for any nontrivial partition $\mathcal{P}$. The relation between these classes is best expressed using the notion of **disjointness** introduced by H. Furstenberg [F-1967]. We say that two actions $(X_i, \mathcal{B}_i, \mu_i, G)$, $i = 1, 2$ are **disjoint** if whenever they are both factors of a third action $(X, \mathcal{B}, \mu, G)$ via maps $\pi_i : X \to X_i$, the σ-algebras $\pi_i^{-1}(\mathcal{B}_i)$ are **independent**.

The two prime examples of this notion are the disjointness between any weakly mixing action and any action with pure point spectrum, and the disjointness between zero entropy actions and CPE actions. In both of these cases the fact that two such transformations cannot have any common factor is obvious. Clearly this is a first obstacle to disjointness, since actions with a common factor can be combined as the relatively independent product over this common factor so that they are both common factors of a larger system without being independent. however, it is known that in general this property is not enough to ensure disjointness. For $\mathbb{Z}$-actions the disjointness between zero entropy and CPE was originally established using ideas of Pinsker which were based on the past and future of stationary stochastic processes. In [GTW] there is a development of these results without using the past, which is valid for all amenable groups.

Some of the properties of CPE actions are very well understood. These include mixing properties and their spectral character. For $\mathbb{Z}$-these were established in the work of V. Rokhlin and Y. Sinai [RS] using the past in a very essential way. It is only recently that these results were extended to amenable groups by means of a rather striking reduction to the case of $\mathbb{Z}$. In the remainder of this section I shall give an overview of these developments.

In the background is the basic work of H. Dye in which he showed that any two non periodic, finite measure preserving ergodic actions of $\mathbb{Z}$, $(X_i, \mathcal{B}_i, \mu_i, T_i)$, $i = 1, 2$ are **orbit-equivalent**. This means that there is an invertible measure preserving mapping $\theta : X_1 \to X_2$ that takes T_1-orbits to T_2-orbits. More precisely, for μ_1-a.e. $x_1 \in X_1$, and any $n \in \mathbb{Z}$, there is a $k(n, x_1)$ such that

$$\theta(T_1^n x_1) = T_2^{k(n, x_1)}(\theta(x_1))$$

and for fixed x_1, the range of $k(\cdot, x_1)$ is all of $\mathbb{Z}$. Dye showed in fact that if G is any infinite abelian group then any free ergodic action of G is orbit equivalent to an ergodic $\mathbb{Z}$-action. D. Ornstein and I extended this result to all amenable groups (cf. [CFW] where a more general theorem is established). For many years it was thought that this kind of result cannot be used in the usual study of ergodic actions, since orbit equivalence unifies all actions – and in particular the entropy of an action is not preserved.

However, in the relative theory it turns out that various properties are preserved under an orbit equivalence of the base. To explain this let us recall what the relative theory is all about. Whenever we have an ergodic action $(X, \mathcal{G}, \mu, T_g)$ and a factor map $\pi : X \to Y$ to another action $(Y, \mathcal{C}, \nu, T_g)$ we can view X as an extension of the Y system. V. Rokhlin showed that one can view X as a product measure space $Y \times Z$, (Z, ρ), and the action $T_g : X \to X$ as a skew product

$$T_g(y, z) = (T_g y, S(g, y)x)$$

where $S : G \times Y \to M(Z, \rho)$ is a cocycle from the G-action on Y to the measure preserving transformations of Z. The cocycle equation

$$S(g_1 g_2, y) = S(g_1, T_{g_2} y) S(g_2, y)$$

expresses the fact we have a G-action on $Y \times Z$. In case $S(g, y)$ is independent of y then the cocycle defines a G-action on (Z, ρ) and the action on X is the product of this with (Y, T_g). The relative theory is concerned with the properties of this cocycle $S(g, y)$. The two main examples where the relative theory has been developed and used extensively should clarify this. When (Z, ρ) has the structure of a compact metric space and the maps $S(g, y)$ are isometries one says that the extension X over Y is a compact extension. The **relatively independent product** of X over Y is the action on $Y \times Z \times Z$ defined by $T_g(y, z_1, z_2) = (T_g y, S(g, y)z_1, S(g, y)z_2)$. The extension is said to be **relatively weakly mixing** if this relative product is ergodic. It was shown by H. Furstenberg [F-1977] that if an extension fails to be relatively weakly mixing it is because there is an intermediate extension that is relatively compact. These notions, developed by R. Zimmer [Z] and H. Furstenberg [F-1977] played an important role in the application of ergodic theoretic methods in combinatorics.

In the Bernoulli theory, the basic idea is that of an extension being relatively Bernoulli which means that by a proper choice of the cocycle defining the extension, the cocycle is constant, and the resulting action on Z is isomorphic to a Bernoulli system.

For the relative entropy theory we need to define the entropy of an extension. This is easily done if all systems have finite entropy, since then we simply define the relative entropy to be the difference between the entropies of X and the factor Y. However it is important to give a more direct definition in terms of the cocycle $S(g, y)$. This will enable us to define the relative entropy of a partition $\mathcal{P}$ of X over the factor Y. For a partition $\mathcal{P}$ of X, which is now taken to be identified as $Y \times Z$, we denote by $\mathcal{P}_y$ the partition of Z obtained by restricting $\mathcal{P}$ to $\{y\} \times Z$. If F is a finite set in the group G, we denote by $\mathcal{P}_y^F$ the following join

$$\mathcal{P}_y^G = \bigvee_{g \in F} S(g, y)^{-1} \mathcal{P}_{T_g y}$$

which we think of as being in the fiber $\pi^{-1}(y)$ over y. The relative entropy is the average growth rate of these joins as F increases to G through a sequence of approximately invariant sets.

Once we have a notion of relative entropy we can define relatively CPE as the statement that any partition $\mathcal{P}$ of X that does not come from a partition of Y, i.e. $\mathcal{P}_y$ is not trivial for a set of y of positive measure, has **positive** relative entropy. Suppose that we have an orbit equivalence between the G-action on Y and the action of another group, say $\mathbb{Z}$. Using the orbit equivalence we can think of the mapping τ that generates the $\mathbb{Z}$ action as being defined on Y with the same orbits as that of the G action. This defines a G-valued cocycle on $\mathbb{Z} \times Y$ since $\tau^n y = R(n, y) y$ where $R(n, y) \in G$. It is clear that we can use the same cocycle S that defined a G-extension on X to define a skew product over the $\mathbb{Z}$-action on Y, say $\hat{\tau}(y, z) = (\tau y, S(R(1, y), y) z)$. In [RW] it was shown that the relative entropy of these two extensions, one of them a G-action and the other a $\mathbb{Z}$-action are the same! In retrospect this should not be so surprising since the relative entropy expresses a property that depends only on the cocycle $S(g, y)$.

Indeed the earlier properties that I mentioned, relatively weakly mixing, relative compactness and relative Bernoulli are all preserved under such orbit equivalences of the base. This observation was first used in [RW] to show that for amenable groups, CPE actions are strongly mixing. In fact we established there a much stronger property that we called **uniform mixing** which is in fact equivalent to CPE.

Definition 10.1.   1. *For a finite set $D \subset G$, we say that a subset $E \subset G$ is D-separated if the sets $\{Dg : g \in E\}$ are disjoint.*

2. *An action $(X, \mathcal{B}, \mu, G)$ is said to be uniformly mixing if for every finite partition $\mathcal{P}$, and every $\varepsilon > 0$ there is a finite set $D \subset G$, such that for any finite D-separated set $E \subset G$, we have*

$$\frac{1}{|E|} H(\bigvee_{g \in E} T_g^{-1} \mathcal{P}) \geq H(\mathcal{P}) - \varepsilon.$$

Clearly this uniform mixing implies mixing of all orders, but it even implies CPE. The converse, is not at all evident and constitutes the main result of [RW]. The proof goes by fixing an auxiliary action (Y, G) and then taking a CPE action $(Z, \mathcal{A}, \rho, G)$ and forming the product action $X = Y \times Z$. One verifies that this is a relatively CPE extension. While this seems clear the proof is not at all straightforward. Next one applies the result alluded to above to find an orbit equivalence between the G action on Y and a $\mathbb{Z}$-action there. Then one forms the skew product of this $\mathbb{Z}$-action with the cocycle given by the original action on Z. It is then shown that this is still relatively CPE, and then classical results from the $\mathbb{Z}$-theory show that this extension is relatively uniformly mixing (the definition of this should be clear from the above). This passes through the orbit equivalence to show that the G-extension is relatively uniformly mixing and finally this implies that the original G-action is uniformly mixing.

Very recently, A. Dooley and V. Golodets succeeded in using the same proof pattern to extend the Rokhlin-Sinai description of the spectral nature of CPE systems to free actions of amenable groups. Countable multiplicity Lebesgue spectrum is replaced by a countable number of copies of the regular representation of G on $l^2(G)$. I will not describe these results in any detail but do wish to point out a fairly simple observation that should have been made long ago. We say that an action of a group G is **effective** if T_g is not the identity for $g \neq e$. This is a much milder restriction than freeness, and in fact the set of $g's$ for which T_g is the identity is a normal subgroup of G, say H, and any noneffective action can be thought of as an effective action of the group G/H.

An action may be noneffective and still be CPE if the group H is finite and the action of G/H is CPE. However if a CPE action is effective then it must be free! This observation means that the freeness in the hypotheses of [DG] may be replaced by effectiveness.

References

[CFW]   A. Connes ,J. Feldman and B. Weiss, *An amenable equivalence relation is generated by a single transformation* Ergodic Theory Dynamical Systems **1** (1981), 431–450 (1982).

[DG]   A. H. Dooley and V. Ya. Golodets *The spectrum of completely positive entropy actions of countable amenable groups* preprint, 2001.

[F-1967] H. Furstenberg, *Disjointness in ergodic theory, minimal sets, and a problem in Diophantine approximation* Math. Systems Theory **1** 1967 1–49.

[F-1977] H. Furstenberg, *Ergodic behavior of diagonal measures and a theorem of Szemerdi on arithmetic progressions* J. Analyse Math. **31** (1977), 204–256.

[G] F. Greenleaf, *Invariant means on topological groups and their applications* Van Nostrand Mathematical Studies, No. 16 New York-Toronto, 1969.

[GTW] E. Glasner, J-P. Thouvenot and B. Weiss, *Entropy theory without a past* Ergodic Theory Dynamical Systems **20** (2000), no. 5, 1355–1370.

[K] J. C. Kieffer, *An entropy equidistribution property for a measurable partition under the action of an amenable group* Bull. Amer. Math. Soc. **81** (1975), 464–466.

[L] E. Lindenstrauss, *Pointwise theorems for amenable groups* Invent. Math. **146** (2001), no. 2, 259–295.

[LW] E. Lindenstrauss and B. Weiss, *Mean topological dimension* Israel J. Math. **115** (2000), 1–24.

[M] J. Moulin-Ollagnier, *Ergodic theory and statistical mechanics.* Springer Lecture Notes in Mathematics, 1115 (1985).

[OW-1983] D. Ornstein and B. Weiss, *The Shannon-McMillan-Breiman theorem for a class of amenable groups* Israel J. of Math. **44**, (1983), 53-60.

[OW-1987] D. Ornstein and B. Weiss, *Entropy and isomorphism theorems for actions of amenable groups* Journal d'Analyse Math. **48**, (1987), 1–141.

[P] A. Paterson, *Amenability.* AMS Mathematical Surveys and Monographs, 29 (1988).

[RS] V. A. Rokhlin and Ya. G. Sinai, *The structure and properties of invariant measurable partitions* Dokl. Akad. Nauk SSSR **141** (1961) 1038–1041.

[RW] D. Rudolph and B. Weiss, *Entropy and mixing for amenable group actions* Ann. of Math. (2) **151** (2000), 1119–1150.

[W] B. Weiss, *Monotileable amenable groups* in Topology, Ergodic Theory, Real Algebraic Geometry , V. Turaev and A. Vershik ed. AMS Translations II-202 257-262.

[Z] R. Zimmer, *Ergodic actions with generalized discrete spectrum* Illinois J. Math. **20** (1976), 555–588.

INSTITUTE OF MATHEMATICS, THE HEBREW UNIVERSITY, JERUSALEM 91904, ISRAEL

E-mail address: `weiss@math.huji.ac.il`